IET ENERGY ENGINEERING SERIES 143

Medium Voltage DC System Architectures

Other volumes in this series:

Medium Voltage DC System Architectures

Edited by
Brandon Grainger and Rik W. De Doncker

The Institution of Engineering and Technology

Published by The Institution of Engineering and Technology, London, United Kingdom

The Institution of Engineering and Technology is registered as a Charity in England & Wales (no. 211014) and Scotland (no. SC038698).

The Institution of Engineering and Technology
Michael Faraday House
Six Hills Way, Stevenage
Herts SG1 2AY, United Kingdom

www.theiet.org

British Library Cataloguing in Publication Data
A catalogue record for this product is available from the British Library

ISBN 978-1-78561-844-4 (hardback)
ISBN 978-1-78561-845-1 (PDF)

Typeset in India by MPS Limited
Printed in the UK by CPI Group (UK) Ltd, Croydon

Contents

About the editors

Brandon Grainger is an assistant professor and Eaton faculty fellow at the University of Pittsburgh Swanson School of Engineering, USA. He is also associate director of the Energy GRID Institute and co-director of the Advanced Magnetics for Power and Energy Development (AMPED) consortium. His research interests include medium to high voltage power electronics, general power electronic converter design, wide bandgap semiconductor device utilization, solid-state transformers, and electric vehicle motor drives. He has worked for ABB, ANSYS Inc., Mitsubishi, and Siemens. He is a member of the IEEE Power Electronics Society and Industrial Electronics Society.

Rik W. De Doncker is a full professor at RWTH Aachen University, Germany. He is the director of the Institute for Power Electronics and Electrical Drives (ISEA). He also serves as director of the E.ON Energy Research Center of RWTH Aachen University and the BMBF German Federal Government Research Campus Flexible Electrical Networks. He is co-director of the RWTH Center for Wind Drives (CWD), the Center for Mobile Propulsion (CMP), and the Research Center Railways (RCR). Before joining RWTH he worked as Senior Scientist at General Electric CRD, Schenectady, NY and was CTO of Silicon Power Corporation, Malvern, PA. He is an IEEE fellow and member of several IEEE societies, and recipient of the 2013 IEEE Newell Power Electronics Technical Field Award, and the 2020 IEEE Medal in Power Engineering. His current research topics include electric vehicle propulsion systems, DC-to-DC converters, high power semiconductor devices, energy storage systems, and hybrid medium voltage switches and circuit breakers.

Preface

Growing concerns about air quality and climate change force us to rethink the way our societies are provided with energy. The rate at which fossil fuels are consumed has reached its limits, and actions are undertaken at a large scale to stop global warming, as it impacts the life of all of us. The urgency of this energy transition has moved to the top of the political agenda of many nations. In 2015, 195 nations agreed to ratify the United Nations Paris Agreement to limit global warming by two degrees compared to the preindustrial time. To accomplish this, targets have been set to reduce CO_2 emissions. While in other countries, the debate is still ongoing as to how to reach the climate change targets, with the Green Deal, the European Commission stipulates that, compared to 1990, by 2030 its CO_2 emissions should be reduced by 55%. By 2050 Europe wants to become a CO_2-neutral society.

Saving energy by improving efficiency will not be sufficient to reach such challenging targets. It forces the engineering community to rethink the entire value chain of all sectors and provide not only technical, but also economically and ecologically viable solutions. These measures also have to be acceptable by society as they may impact the way we live, travel, produce, and transport goods, recycle materials, and so on. With such constraints, electrical energy comes ever more to the foreground as the main carrier of energy. Indeed, once it is generated electrical energy provides a great many advantages. It can be converted efficiently to many other forms of energy, for example to mechanical work using electro-mechanical machines (motors), and vice versa using generators. Electrical technology has developed to the point that it can provide almost any other form of energy in a reliably, ecologically sound, and cost-effective way. LED lighting consumes far less energy than the incandescent light bulb. The cost of photovoltaic systems that convert light into electrical energy has dropped by orders of magnitude over the past decades. As homes and buildings become less energy-hungry, modern heating and cooling systems use heat-pumps with performance coefficients well above four. The entire transportation and mobility sectors are rapidly moving towards full electrification. Together with state-of-the-art powerful batteries, electric propulsion systems are becoming mainstream. 2020 marks the year that more electric passenger cars were sold in Europe than diesel cars.

Clearly, none of these applications would come to fruition if it wasn't for power electronics. Similar to photovoltaic (PV) solar cells, the cost of power electronic inverters over the past 25 years has decreased twenty-five-fold. Silicon-based technology has set in motion a chain reaction in the sense that PV cells and most power electronic devices are produced from sand (silicon dioxide) and energy. As the energy comes from PV and sand is abundantly available there is virtually no limit on the amount of PV and power semiconductors that can be produced at every decreasing cost. Such a cost regression cannot be found in other renewable power sources, such as hydropower or wind turbines. Wind turbines use large amounts of cement and steel. Their generators and transformers use steel and copper. All these materials are used in other sectors (electrical equipment uses about 30% of the global copper production). In addition, these materials are limited in exploration and production. At the current rate of copper consumption, the US Geological Survey predicts that copper produced with the current mining and refining technologies will be exhausted by 2065. It can be anticipated that ultimately the prices of these valuable materials will continue to increase over time. Hence, the statement made by power electronic specialists "more silicon and less copper and steel" is well-founded.

This is where DC technology can play an important role. Indeed, with power electronics, we are not bound any longer to classical 50/60 Hz AC transmission and distribution systems. Electromagnetic (passive) devices, such as transformers, generators, inductors, and capacitors that operate at low frequencies tend to become bulky. For a given frequency, their size scales with the apparent power. As their apparent power is proportional to frequency, higher frequency operation leads to smaller components for a given apparent power. This can be illustrated in switched-mode power supplies for consumer electronics and personal computers (PCs). Whereas first generation PCs used 50/60 Hz transformers followed by a rectifier, modern day power supplies operate at 140 kHz thereby increasing the power density from 10 W/kg up to 1 kW/kg. Wide bandgap semiconductors, currently in production, will enable even higher frequencies, further reducing the size of passive components and material usage. Recently, SiC DC converters with power densities up to 50 kW/kg have been reported.

In addition, as long as AC distribution is used, multiple power converter stages are needed. Indeed, all electrical potentials that can be found in nature present themselves as voltage sources, e.g. PV systems, batteries, electrolyzers, and so on. Even wind turbines are DC sources. Indeed, the machine side converter rectifies the variable frequency of the wind turbine generator to a regulated, constant DC voltage. Still today, most PV parks and wind farms use medium-voltage AC collector fields. This enforces the use of AC grid connected inverters and 50/60 Hz transformers. To provide grid compatibility, the grid side pulse-width modulated (PWM) converter has to produce current waveforms with low THD. Hence, filter components and high switching frequencies are needed. In addition, reactive power control, stable control, and fault ride through capability are additional functionalities that add to the complexity of the control algorithms. Hence, it comes as no surprise that grid-side converters in wind turbines are causing most interruptions. Compared to a DC collector field, studies have consistently shown improvements of energy output, reliability, and a reduction of costs.

These considerations have led many power electronic specialists and grid planners to explore more DC applications. The need for charging electrical vehicles in the urban environment poses challenges to the classical (radial) medium voltage AC grid. The problem is not the amount of energy that is required to charge all electric passenger cars, but rather the peak power that can overload transformers. To overcome these issues, intelligent charging control can help a long way, but ultimately the distribution grid has to overcome bottlenecks, either by increasing the number of transformers or by interconnecting substation transformers. The latter makes more sense as transformers in cities are on average loaded not much higher than 30%–50% of their installed capacity (for power quality reasons and the fact that loads vary over the day). Interconnections in distribution grids be done by back-to-back AC converters that may have built-in DC/DC converters for galvanic separation, that is, the solid-state transformer. This concept, is the so-called underlay grid and provides horizontal power flow in existing AC grids. In addition, the grid side inverters can provide power quality services which are needed when the AC grid gets supplied by volatile renewable sources.

These and other industrial applications of MVDC technology are covered in this book, of which the authors hope that it motivates readers to explore the benefits of DC technology. In addition, the following chapters provide a deep insight of state-of-the-art power electronic DC conversion systems that have been developed over the past decades and have reached a high level of maturity.

In Chapter 1, Dr. Rik De Doncker (RWTH Aachen University, Germany) opens the research novel by providing the motivation and arguments for the global pursuits in medium voltage DC technology. In Chapter 2, Dr. Robert Cuzner (University of Wisconsin-Milwaukee, USA) and Dr. JiangBiao He (University of Kentucky, USA) provide an exhaustive overview of power electronic converter impacts on medium-voltage DC (MVDC) system architectures that arise in various applications. These opportunities arise in grid applications, as will be discussed in the introduction to the book, but in other areas including shipboard and aircraft electrification. These DC technology areas have penetrated the culture of various electrical manufacturers but this is the first time the editors have seen the technological advances described in Chapter 2. The authors also provide insight to future DC applications in the renewable energy market, specifically for solar and wind power conversion. The chapter is very foundational and sets the tone of the book for the forthcoming chapters in this reference book.

In Chapter 3, Dr. Aditya Shekhar and Dr. Pavol Bauer (Delft University of Technology, Netherlands) provide a thorough and analytical approach for refurbishing existing AC infrastructure for DC operation. They also consider the case where AC and DC based links in the system architecture operate in parallel and consider the capacity and economic impacts of such a decision. When a system is faulted, the authors evaluate approaches for reconfiguring these AC–DC link architectures, which leads into the discussion of hybrid AC–DC distribution systems. The value this chapter brings to the engineering community is it's the first that the editors have seen that provides quantifiable advantages of using DC.

With the system architectures evaluated, the book begins to discuss the technological building blocks that will make these medium voltage DC architectures function, the power electronic systems and their control. In Chapter 4, Dr. Rik De Doncker and Dr. Jingxin Hu (RWTH Aachen University, Germany) thoroughly describe the modeling and control of single input, single output bidirectional, DC-to-DC converters referred to as a dual active bridge converter (DAB). The

DAB is a pioneering soft-switching topology, first developed in 1988 for NASA to explore robust and highly efficient DC power converters for the International Space Station. Nowadays, the single-phase and the three-phase DAB serves as the basis of many high power, solid state transformer designs being developed globally. Specifically, for three-phase DABs, advanced control techniques are organized in this chapter including instantaneous flux and current control, anti-saturation control for the high frequency transformers found in the DAB, fault current limiting control, and dead-time compensation. With the foundation set based upon two-level three-phase DAB designs, the authors provide a small treatment for the three-level, three-phase DAB to conclude their chapter.

With the foundation set with single-input converter designs, in Chapter 5, Dr. Gregory Kish (University of Alberta, Canada), presents a thorough discussion on multiport DC power converters for MVDC applications. Dr. Kish defines the medium voltage magnitudes as this varies from industrial sector and discusses the modular multilevel converter in detail, which found its first commercial application in the high voltage direct current (HVDC) market. After discussing two port converters briefly, he provides a thorough description of multiport DC converters and desired features. Finally, Dr. Kish ends the chapter with convincing case studies conducted in simulation for the interested reader.

In Chapter 6, Dr. Brandon Grainger and Dr. Zachary Smith (University of Pittsburgh, USA) provide an exhaustive treatment of controlling the DAB converter with modern, adaptive control procedures. For most power engineering applications, the default control technique is to use proportional-integral (PI) controllers because of their robust behavior. However, for constant power based loads that are naturally unstable, the system dynamics will not be regulated with traditional PI controllers well. In this chapter, the authors translate a procedure from modern reference control theory and apply it to stabilizing and controlling the DAB that interfaces a constant power load. Techniques for choosing the control parameters are provided with a basic simulation study illustrating the effectiveness of the modern reference control techniques compared to proportional-derivative (PD) controls typically used to attempt to stabilize dynamic systems. As the focus of this chapter is on dynamic behavior, and assuming that the reader of this book has reviewed Chapter 5 on multiport designs, the authors mathematically show the modes of operation of the multiport DAB with coupled inductors. The reader will gain appreciation of how such a simple power converter topology has a wide range of operating modes depending upon the phase relationship between each port.

With magnetics being a critical component of medium voltage design, power conversion technology, the editors found that a chapter on the subject would be necessary. In Chapter 7, Dr. R. Byron Beddingfield (North Carolina State University, USA) and Dr. Paul Ohodnicki (University of Pittsburgh, USA) discuss magnetic coupling in multi-winding components, discuss the realization of non-ideal magnetic components including medium frequency resistance, magnetic core loss, leakage flux, and parasitic capacitance. Material and magnetic component characterization is a necessity for most designers and, therefore, the chapter authors discuss this in significant detail. All sections are supported with magnetic measurement results for these medium frequency,

DC-to-DC applications. Finally, the authors discuss some specialized knowledge with regards to laminated magnetic materials and winding configurations with transformers in mind.

System stability is a topic that must be covered for any text that covers system architectures. In Chapter 8, Dr. Fred Wang, Dr. Yaosuo Xue and Dr. Le Kong cover the current, foundational techniques for analyzing power converter stability. The larger set of impedance-based stability techniques include small-signal impedance models, Middlebrook criterion, and the Nyquist criterion. Each technique comes with examples that are useful for the beginner in grasping their utilization. All engineers know that small signal models have their limitations, thus, Dr. Wang and his team also discuss commonly used large-signal model approaches including the Lyapunov direct method, Takagi-Sugeno fuzzy model method, and mixed potential function and stability criterion. Auxiliary topics such as methods for improving system stability with hardware and methods of measuring small-signal impedances are also discussed.

In Chapter 9, Dr. Xiaoqing Song, Dr. Pietro Cairoli (ABB Corporate Research, USA) and Dr. Marco Riva (ABB Medium Voltage Products—Technology Center, Italy) provide an extensive overview of the circuit breaker technologies suitable for MVDC system protection, including mechanical circuit breakers, solid state circuit breakers and hybrid circuit breakers. The fundamentals, benefits, and design considerations of the three MVDC circuit breaker technologies are introduced and discussed. A further classification of each type of circuit breaker is provided with the circuit topology and brief description of their operating principles. The advantages and disadvantages of the different topologies are also discussed and compared.

In Chapter 10, Dr. Lisa Qi (ABB Corporate Research, USA) and John Lindtjørn (ABB Marine and Ports, Norway) contribute a chapter on how DC technology is used on marine vessel electric systems. In this chapter, the authors discuss medium voltage and low voltage DC architectures, the benefits and advantages of using DC for ships such as optimized combustion engine performance, ease of power source integration, and the opportunity for high fault tolerant power systems. DC system stability and DC protection are engineering challenges addressed in Chapters 8 and 9, respectively, and now covered in the context of electric ship design in Chapter 10. The benefits of using energy storage in the DC system and ABB's commercially available vessel control system are covered as well as an overall summary of lessons learned from field operation. Thus Chapters 9 and 10 serve as chapters showing how manufacturers are currently making use of DC and, now, not just an academic exercise.

In Chapter 11, we conclude the book by allowing each author to provide their final comments on their respective area while providing several insights to their vision for the future of the technology tied to medium voltage DC system architectures.

Pittsburgh, Pennsylvania,
Brandon Grainger USA
University of Pittsburgh

Aachen, Germany
Rik De Doncker
RWTH Aachen University

Chapter 1

Medium voltage DC technologies—key enabler for a flexible, multi-terminal underlay distribution grid

Rik W. De Doncker[1]

No doubt, the liberalization of the energy market and awareness of climate change set in motion significant structural changes of the energy supply system. Over the past two decades, the electrical energy supply system changed in many developed countries from a top-down grid architecture with large-scale central power stations towards a decentralized system with many medium- and small-scale distributed power generators. Vast amounts of renewable, dispersed, but volatile power generator systems (mostly wind and PV) are being installed, next to combined heat and power (CHP) systems. The latter provides stability, energy storage and operate at high exergy* levels. Nowadays, using power electronic conversion systems storage and conversion of electrical energy in other useful energy forms can be performed with high efficiency, zero local emissions, and at low cost. Hence, electrification of all sorts of sectors, for example, indoor climatization, mobility, and transportation is accelerating [1–7].

On the other hand, compared to fossil fuels, electrical energy carriers have relatively low energy density. In addition, despite the massive savings of primary energy and high efficiencies of modern electrical conversion systems, electricity demand is steadily increasing. In light of the renewable and volatile energy sources, several measures are needed to provide a robust and secure supply of the electrical energy. Next to fully automated, digital demand side management systems, all sorts of energy storage (in form of heat, coal, gas and batteries) and more flexible grids, in particular distribution grids, are needed. Even the architecture of the distribution grid must evolve from the classical radial grid structure towards a multi-terminal grid to better interconnect prosumers. The potential efficiency gains, material, and cost savings that can be realized by using DC technology in solid-state substations

[1]E.ON Energy Research Center and BMBF Research CAMPUS Flexible Electrical Networks, RWTH Aachen University, Aachen, Germany
*Exergy – in thermodynamics, it is defined as the maximum amount of work (useful energy) that can be extracted from a system.

and distribution grids are so substantial that a radical change towards DC technology cannot be ignored.

1.1   Towards a CO_2 neutral energy supply

Market liberalization led to a massive decentralization of energy producers, as compared to the centralistic power production with large-scale central power stations that were in place in many nations. This market liberalization also paved the way to install photovoltaic systems and wind turbines. As these renewable sources became mass produced and cost effective, the political desire grew to develop a so-called CO_2-neutral society. In his treatise on the "War of Currents" [3], the author argues that, from an engineering point of view, constructing a CO_2 neutral, that is, a 100% sustainable electrical power generation system, is technically feasible:

> *"Taking the European power system as an example and extrapolating the increase of new installation capacities of wind, PV and CHP systems between 2010 and 2011 [5,6], a CO_2-neutral electrical power system could be built in approximately 15 years (by 2025). Note that, in order to realize a CO_2 neutral energy supply, it is anticipated that despite all energy saving measures the electrical power consumption will continue to increase up to 5000 TWh in the next decades. This is a direct consequence of the electrification of many sectors."*

A concept of a CO_2 neutral energy supply system, combining the electrical, heat and gas grids for Europe, is shown in [1–3], illustrating the high level of coupling between the energy grids and other sectors, such as building climatization and mobility. Of course, such a complex intertwined energy supply system requires a high level of automation. Together with power electronics, low-cost wireless communication and digital technologies this high level of automation became practical over the last decade.

Figure 1.1 illustrates the so-called *four grids* concept. Note that the specific properties of each energy grid complement each other, making them indispensable. Physics teaches us that electrical energy can be generated and converted efficiently, without local emissions, in many other forms of energy. Electro-mechanical converters, such as generators, motors and drives, are used, for example, in power stations, wind turbines and railway and electric vehicles. Electrical conversion systems, such as voltage transformers, power electronic converters, power supplies, etc., are all around us and reach efficiencies well above 95%. Also, solid-state electricity generation from light using photo-voltaic cells offers by far the lowest levelized cost of energy (LCOE[†]). Vice versa light generation in lighting fixtures (LEDs) provide the highest efficiencies at lowest costs. On the other hand, comparing the energy density of electrical storage systems (batteries and capacitors)

[†]LCOE – Levelized Cost of Energy computes the ratio of the lifetime cost, i.e. capital and operational costs, of an energy source or plant over its energy production. In Europe, it is often expressed in €/kWh.

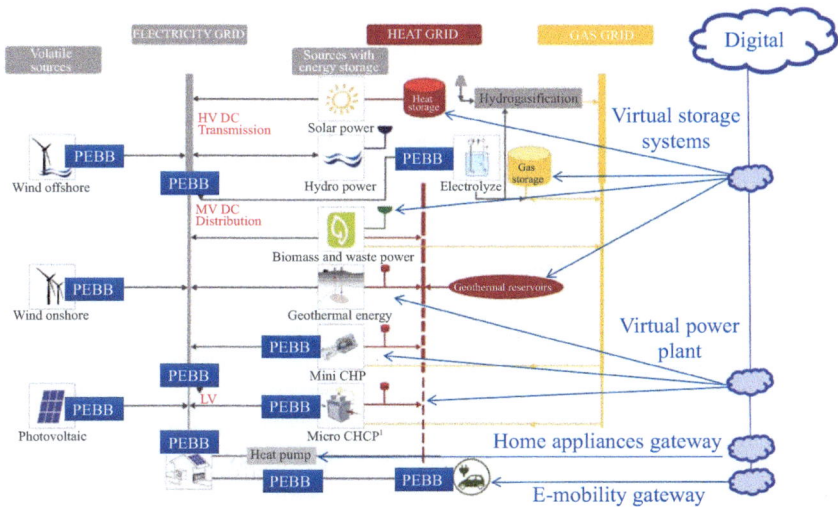

Figure 1.1   Single line diagram representing the four grids for a CO₂ neutral electrical energy production, showing (from left to right), the electrical transmission & distribution grid (grey), the dispersed heat grid (red), the gas distribution grid (yellow), and the digital automation & communication grid (blue)

with heat and chemical (gas, fuels) storages one should realize that the energy density of electrical storages (batteries) is lower, a fact that often leads to higher costs when large amounts of energy need to be stored. Clearly, when all primary energy comes from renewables, massive energy storage, transmission and distribution of electrical energy over continents will be needed. Obviously, both solutions, that is, storage and transmission & distribution are, from an economical viewpoint, in competition with each other. In many developed countries and in large cities, approximately 40% of all primary energy is used for heating and cooling. As a consequence, heating, fossil fuel and gas infrastructures are commonly available and historically accepted by society.

To provide large-scale energy storage engineers look for so-called dual use of different energy storages. Dual use means that the storage capacity is already fully or partially paid for by existing applications, for example, batteries of electric vehicles with bi-directional chargers [8,9]. To increase comfort, highly insulated homes and buildings install ventilation and heat pumps to provide not only heating but also cooling [10,11]. With high-temperature power electronics heat pump systems are available with CoP-numbers[‡] above 4, up to 10, making a heat pump undeniably ecologically a better solution than burning fossil fuels or gas. Most heat pumps systems have medium size up to very large heat storages (water tanks,

---

[‡]CoP – Coefficient of performance is the ratio of heat or cooling production versus electrical energy consumption of a heat pump.

underground or geothermal heat exchangers) that can be "charged" similar as a battery, when the electricity production from renewables reaches high peaks. On the other hand, fuel and (liquified) gas tanks can store massive amounts of energy over much longer periods. Therefore, they are often considered as strategic energy storages to prevent strong price fluctuations in the global market. As many countries are abandoning nuclear fusion base load power plants, the development and demonstration of large-scale (>100 MW) electrolyzers is currently on-going to close the power-to-gas (hydrogen) loop. That way, using gas engines, gas turbines or fuel cells a stable, base load energy supply can be realized using CHP power plants. Note that the type and size of these CHP units will vary depending on the local availability of a heat or gas grid, social policies and user preferences.

From the perspective of the electrical grid and taking this CO_2 neutral scenario with sector coupling into account, and considering the fact that the cost of electricity production, in particular PV, is less than the cost to transmit and distribute it, the move to decentralized energy production seems unstoppable. However, some technical factors work against a fully decentralized energy production. For example, even when (tandem) PV cells would reach their maximum theoretical efficiency (more than 60%) any time soon, all the roof tops and facades in cities do not provide sufficient areas for PV installations to cover all local electrical needs. Clearly, optical and acoustic noise issues make it hard to imagine that wind turbines and large heat pump systems are installed in (historic) downtowns. Furthermore, all engineered products follow scaling laws that state that larger units are more efficient than smaller units, reducing their LCOE. Even when the reliability of small and large units would be the same (larger units are usually better engineered and more reliable), the maintenance cost of many more small units could become a show stopper. In conclusion, the degree of dispersion of renewable power generators is an economical optimization of many technical cost factors, which are influenced by many local factors. The latter include meteorological, geological, legacy and social aspects, even governmental policies that are rather random and can change over time. From an engineering viewpoint, three major cost categories stand out, namely cost of power generation, energy storage costs and transmission, and distribution costs.

Against this background, and considering the fact that the market price of electricity always ends up balanced among these three categories, one can postulate that in a liberalized market the "one-third rule" applies.[§] This means that about one-third of the installed electrical energy production and power base is installed at the transmission level, one-third at the medium-voltage distribution level[‖] and one-third at the low-voltage networks in buildings and homes [2]. By 2018, the distribution of the installed capacities in Germany reached almost exactly this one-third rule.

[§]The "one-third rule" proposed by the author is a reasonable estimation of how the installed capacities of electrical power generation and storage systems in any CO_2 neutral supply system distribute over the voltage levels.

[‖]According to IEC standards 100 kV_{ac} and above is defined as high-voltage, which applies usually to transmission systems. Medium-voltage distribution starts at 1 kV_{ac} up to 100 kV_{ac}. Typical low-voltage distribution systems have rms phase voltages of 115V_{ac}, 230 V_{ac}, and 490 V_{ac}.

1.2 Transition towards flexible medium voltage distribution grids

One can ask the question why electrical distribution grid structures have to change when electrical power production changes from high-power central power stations towards more decentralized small-scale power plants that may include a large fraction of volatile, renewable power generators. To provide an answer one has to consider two elements that play a critical role. First, considering the one-third rule, explained above, one can easily see that the scenario shown in Figure 1.1 leads to situations in which the medium-voltage distribution grid has to handle energy flows from the top (transmission system), from the bottom (low-voltage side), as well as horizontally. Second, the classical AC distribution grid is designed as a radial grid, which is illustrated in Figure 1.2(a). Indeed, using the technology of the early 1900s, the simplest way to provide a reliable supply of electrical energy and to protect the AC grid against short circuits was to use circuit breakers or fuses and limit fault currents by the short circuit impedance of transformers. A higher transformer impedance meant a lower short circuit current, but a larger voltage-drop when operating at nominal power. Power quality guidelines typically stipulate a maximum voltage deviation from the nominal voltage below 5%. As the electrical energy used to flow only from the high-voltage transmission system towards the distribution grid, coordination of protection gear and power quality was less complex with a radial grid.

Clearly, when power flow reversal occurs, for example, due to PV energy feed in from the low-voltage side or horizontally from wind turbines at the medium-voltage level, the protection system may not be able to cope with the increased short circuit capacity of all power supplies. In addition, most transformers are set such that at no-load their secondary (output) voltage is at +5% of nominal. At full load, as classical AC loads tend to be inductive-resistive, the voltage may drop at the end of a line to −5%. It's not difficult to realize that when many consumers install PV systems, that is, become prosumers, the voltage at the transformer can exceed easily the +5% limit above the nominal voltage.

Aside of the transformer impedance, the voltage drop caused by the reactance in low-voltage AC cables, even when they are transposed to reduce the reactance, is not insignificant. Often to limit this voltage drop the cable currents are typically limited to about 30% of their maximum continuous thermal current rating. Hence, in most distribution networks, AC cables are gravely underutilized. In other words, the ampacity (current rating) of the cables can be up to three times the peak power that is delivered to the end-users, which is a major cost factor. Hence, power quality regulations, that is, voltage limits, tend to limit gravely the maximum peak installed renewable (PV) power generation in urban AC distribution cable networks.

Next to PV installations, electrification of the heating and cooling sector with heat pumps and the mobility sector poses additional power quality challenges on the classical low-voltage AC grid. The electrification of the heating and cooling sector requires considerably more energy as compared to the mobility sector, while the latter requires more peak power. Well-insulated private homes have modern heat pump systems with a performance coefficient up to five and draw a maximum

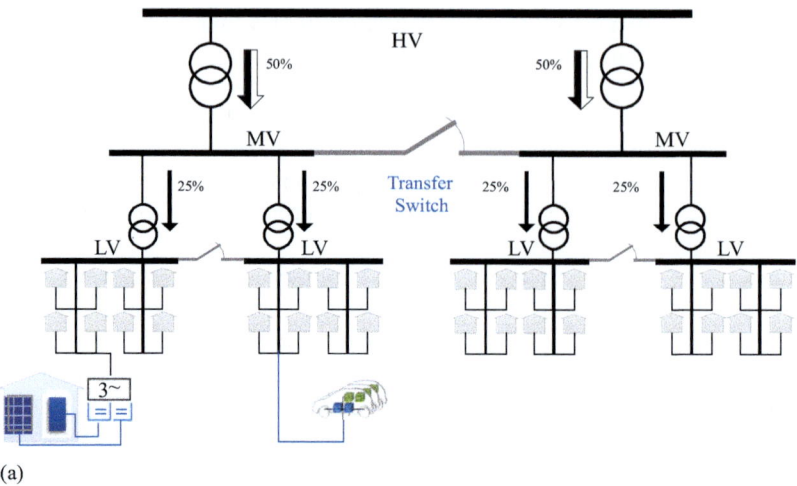

(a)

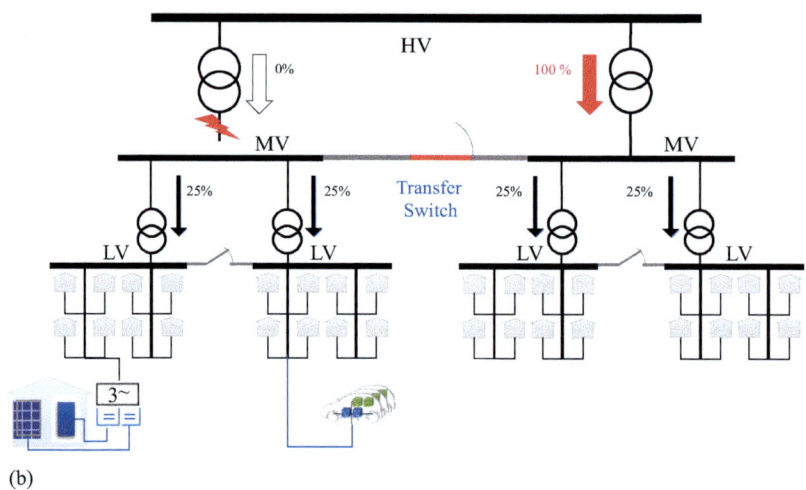

(b)

Figure 1.2   *(a) Classical AC distribution grids have a radial topology. To provide redundancy transfer switches are used, creating the well-known open ring bus structure. Energy from renewables or prosumers at a given voltage level can only be exchanged to other consumers by reversing the energy flow and feeding into the next higher voltage level. (b) When a fault occurs in an open ring bus network, power can be restored by closing the transfer switch (mechanical switches perfume a "break before make action" and require several cycles causing an interruption, solid state transfer switches can transfer power at subcycle speed). This high level of reliable energy supply requires transformers and cables to be substantially derated*

power typically of 1.6–3.6 kW. Hence the amount of electrical energy needed for a single-family home may double as compared to a gas or oil heating system. However, this increase of electricity use may not affect the distribution grid much as the average loading of the grid is typically below 50%. Indeed, most heat pump systems have large hot water storages so that the heat pump can operate at off peak hours. Hence, in several countries the regulator implemented dynamic tariffs to prevent electrical grid overloading and encourages home owners to install roof top PV systems that directly feed the heat pump.

The electrification of the mobility sector in an urban environment is more critical for the distribution grid. Even though the peak installed capacity of end-users can be as high as 27.7–43.5 kVA (i.e., 40–63 A in typical four-wire 400 V_{ac} line-to-line voltage network), typical planning guidelines for AC cables and transformers foresaw an average peak-power consumption of about 3.7 kVA (16 A at 230 V) per end-user. To prevent fire hazards single-phase 230 V AC chargers for electric vehicles draw maximum 10 A (2.3 kW) from standard home wall plugs. A three-phase wall box charger can deliver 11–22 kW to the vehicle. In [12] an exemplary case study of an urban low-voltage network with 64 homes connected to a 250 kVA transformer was performed. The study illustrates that at certain times of day, even when only six EVs (with battery capacities ranging from 40 to 90 kW) are simultaneously being charged at 11 kW, the transformer would be overloaded over longer periods of time. The only solution DSOs have today is to install larger transformers when space permits or provide intelligent energy demand-side services. Smart control may avoid transformer and cable thermal overload conditions, but the problem of power quality is harder to fix. Typically, around noon time when PV panels are providing electricity, most electric vehicles are parked at the workplace of the car owners. Unless all homes with PV systems have a heat pump with a large hot water storage, during a sunny summer day most of the PV energy will be fed into the grid. Hence, the AC voltage of the low-voltage line may increase well above the maximum allowable voltage limit. In the evening, when PV panels deliver little or no energy, the electrical consumption traditionally peaks when people arrive at home. An under-voltage situation is hard to prevent when EV car owners simultaneously plug in their charger as soon as they arrive at home in the evening.

In summary, dispersed generation of electrical energy, combined with an increased number of prosumers, makes the coordination of protection systems of classical radial AC grids a difficult task. Power quality requirements pose limits on the amount of renewable energy sources and electric charging capacity that can be installed in the AC (radial) distribution grid. Solid-state tap changing transformers [13] and VAR compensators [14,15] may alleviate the problem but cause additional losses and costs. Alternatively, the feed-in power from PV and wind turbines must be curtailed at certain moments to maintain safe operating limits. As PV and wind turbines are feeding their power into the AC grid via PWM inverters, curtailing and VAR compensation are technically speaking options, but the implementation of these dispatching functions can become rather complex and may lead to stability problems. Moreover, prosumers are not willing to pay for power quality, which is the task of the DSO.

To provide a secure and reliable supply of electrical energy, radial AC grids have been built with a high level of redundancy. As can be seen in Figure 1.2(a) transfer switches are installed between substations. These transfer switches are normally open. Hence, they form the so-called open ring bus structure. In case of a fault in one of the feeders, for example, when the MV transformer top left is disconnected as illustrated in Figure 1.2(b), the transfer switch can be closed to provide power from the healthy transformer on the right. Clearly, this transformer now has to provide twice the power that it normally needs to handle (assuming the transformers and the low-voltage loads on both sides have similar ratings). Hence, under normal operating conditions distribution grid transformers are typically not used more than 50% of their nameplate apparent power. In reality the average power transmitted through a distribution transformer is well below that level, because most loads vary during the day and transformers are installed based on an averaged peak load per consumer (as explained above).

Data from DSOs in Europe and Japan show that even in big cities, distribution transformers deliver on average an amount of energy that corresponds to an annual basis of about 30% of their nominal rating. This explains why distribution transformers, designed for 25 years of operation at full power, have in practice longer lifetimes, often more than 40 years.

Hence, to provide a high reliability, classical AC distribution grids are wildly overdesigned. Even without taking these redundancy factors into account and ignoring reactive power, it was estimated in [2,3] that to transmit and distribute 1 GW of active power, about 25 000 ton of copper and Si-steel is used in transformers. According to the US Geological Survey [16], at the current global consumption of copper ore, the known mines show reserves for about 47 years. No doubt, prices per kVA of transformers will increase when copper becomes scarce. So, not only do AC grids have a limit to what amount of renewable, dispersed energy they can distribute, their costs to maintain power quality, reliable operation and stability can only increase in the coming decades. In addition to climate goals, there is an important global social aspect to this as almost 10% of the world population still has no access to electricity [17]. It should be the task of electrical engineering to find affordable solutions to provide electricity, and a good standard of living to all people on this planet. The following section shows how power electronics can make the medium-voltage DC distribution grid not only more flexible and more efficient, but contrary to AC grids, reduce costs.

1.3   Cellular MVDC distribution grids for a CO_2 neutral energy supply

To overcome the problems of classical AC distribution grids, DC technology based on modern power electronics can provide technical solutions that are, compared to AC solutions, less expensive and are socially more acceptable. The latter stems from the fact that DC technology lends itself better to underground cable networks. VDE Task Force studies show that DC cables can be installed in existing

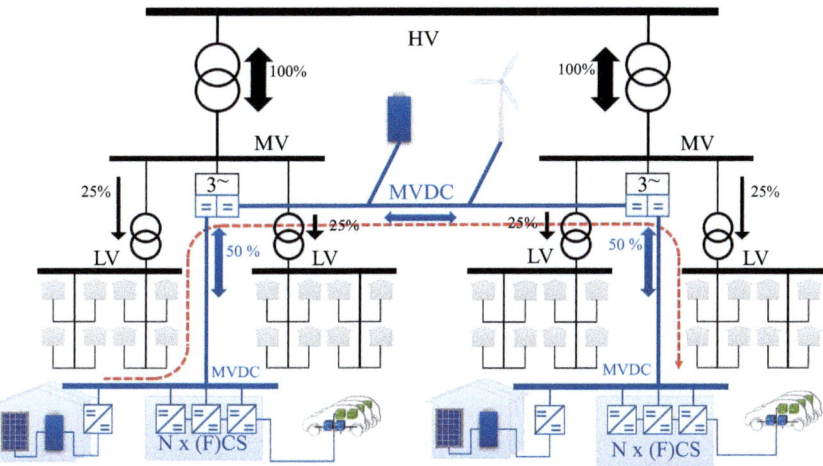

Figure 1.3   Interconnecting prosumers via an MV and/or LV DC link allows a continuous rerouting of energy between different zones of the distribution grid. Moreover, large PV systems, wind farms, battery storage systems, factories and electrolyzers can be connected to the medium-voltage DC link. The AC grid side inverters provide VAR compensation and active damping to the AC grid

infrastructure [18], do not require transposition and, for the same cross section of conductor, can transmit 1.5 up to two times the power of a three-phase cable. In addition, as the frequency in DC-to-DC converters, that is, the so-called solid-state DC transformers, can be chosen much higher than the AC grid frequency the transformer size can be substantially reduced [19–22]. Furthermore, DC networks can be interconnected as the DC converters can dynamically control the power flow. Hence, linking substations together via a DC link, as exemplary shown in Figure 1.3, enables a continuous flow of energy between the substations and allows a flexible control of this power flow. Such a hybrid AC-DC grid provides a high level of reliability and improved power quality when multiple substations are interconnected.

The AC grid side inverters can provide VAR compensation (power quality) and active damping (stability) to the AC grid. As explained in the previous section, all distribution transformers are not experiencing peak power at the same time as loads shift during the day. Hence, the power reserve one substation transformer has can be shifted by the DC underlay grid to the consumers that are fed by another substation. In addition, the bi-directional DC-to-DC converters can regulate a constant voltage at the secondary side even when the input voltage of the primary side varies over a wide range. Hence, power quality (voltage control) at the end-user is largely decoupled from the grid planning. This flexible power flow control is particularly interesting for DSOs, as it allows redistribution of energy within their

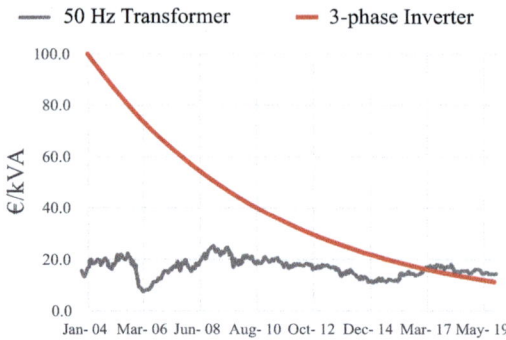

Figure 1.4   Estimated price (€/MVA) evolution of 50 Hz transformers (bottom curve) versus medium-voltage VSI PWM inverters. Price transformers (30% Cu, 70% Fe) based on evolution of copper and cold rolled steel materials [23], multiplied with factor (typically 1.4) to account for Cu-wire and lamination production and a factor (typically 3) for enclosure, production, transportation and installation of the transformer. Price of converters based on sales prices of medium-voltage drive converters for wind turbines, including control, communication hardware, enclosures, transportation and installation

own networks at all voltage levels with high efficiency. Consequently, the tariffs to use high-voltage transmission can often be avoided.

Taking into account that the cost of power electronic inverters, which is the basic building block of many power electronics circuits, has declined over the past 25 years by a factor of 25, the price per kVA of a 50 Hz transformer is currently higher than that of a power electronic inverter, as illustrated in Figure 1.4. In [2,3] it was concluded that;

> *"A full DC grid system can save at least 10 000 ton/GW of transformer materials, leading to (transformer) cost savings (price) of ca. 87 M€/GW."*

Hence, it is anticipated that the DC-link concept for the hybrid gird, presented in Figure 1.3, keeps on expanding. The DC grid builds a so-called *"DC underlay grid,*¶*"* gradually eliminating the AC grid. The trend towards more DC technology is on-going, not only in medium-voltage distribution grids, but also in ultra-high voltage HVDC transmission grids, as well as in low-voltage networks at touch-safe voltage levels, for example, power by ethernet for LED lighting and USB wall plugs. Another argument that favors more DC distribution is the fact that batteries and solar cells are fundamentally DC sources. Even wind turbines have an intermediate DC-link to decouple the variable speed (frequency) of the turbine from the

---

¶The term "underlay grid" was touted in reference to the ultra-high voltage UHVDC overlay grid. As the underlay grid can avoid power bottlenecks in the transmission grid, it becomes questionable if an UHVDC is needed due to its high development costs and diminishing social acceptance.

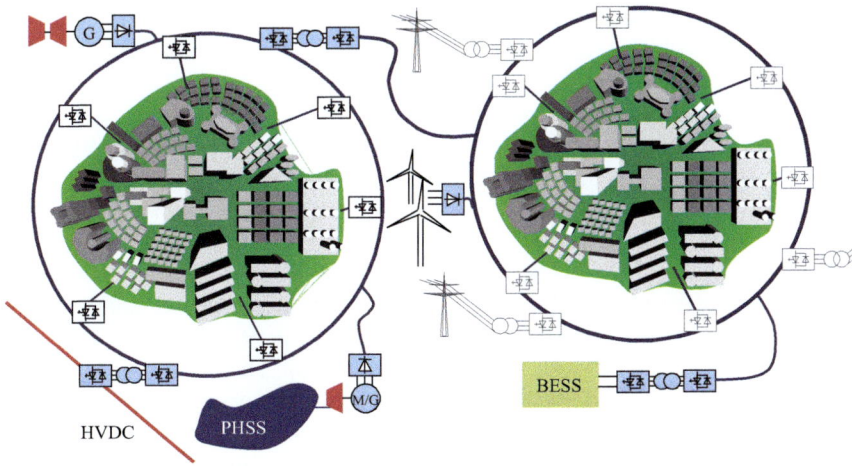

Figure 1.5   *The future energy grids will comprise multi-terminal HVDC networks that feed power, via electronic transformers, in a medium-voltage DC ring bus cable network. The DC-to-DC converters provide an interface to DC power sources, such as PV, electrolyzers and battery energy storage systems (BESS). Wind turbines, combined heat and power (CHP) units and pumped hydro storage systems (PHSS) are linked to the MVDC bus via robust and low-cost converters*

fixed grid frequency. Hence, with DC distribution grids, multiple low-frequency AC-to-DC conversion steps can be avoided.

One can envision that ultimately, a flexible, digitally controlled, cost effective, and more efficient transmission and distribution grid will evolve towards a full DC grid. Similar to the AC transmission grid, such DC distribution grids can be built cellular, that is, interconnected instead of radial (see Figure 1.5). Indeed, the DC converters control locally the power quality and electronically they limit short circuits [24,25].

Note that galvanic isolated power electronic DC converters are easily connected in series or in parallel so that power levels are easily scalable, which offers a high degree of redundancy. Using mass produced standard components, that is, so-called Power Electronic Building Blocks (PEBBs), the cost of DC converters will drop in the coming decades. Price reduction due to standardization and mass production can already be observed in low-voltage LED and PC power supplies. Hence, as power electronic technology always trickles upwards towards higher voltage and power levels, it can be anticipated that the cost of next generation medium-voltage DC converters will reduce further as voltage and power levels become standardized. It is not unrealistic to expect prices of isolated bi-directional DC converters to become as low as 20 €/kW in the coming decade. The creation of standardization working groups and committees (DKE, DIN, CIGRE, IEEE, VDE, etc.) illustrates this growing interest in DC technology for medium-voltage DC grids [26–28].

One can truly state that power electronic DC-to-DC converters were Edison's missing link to realize a multi-terminal DC grid that is more flexible, cheaper and efficient than the existing AC grid technology. These power electronic DC-to-DC conversion systems are sometimes called electronic DC transformers. Among the many types of DC converters, the so-called (three-phase) Dual Active Bridge Converter (DABC), which was proposed in 1988 as an alternative power supply systems for NASA space station [29,30], stands out for its simplicity and robustness. The basic concept of the DABC is explained in subsequent chapters of this book.

Due to its galvanic isolation, the DC-to-DC converter can be used as a power electronic building block (PEBB) to link medium-voltage systems to high-voltage systems or low-voltage DC systems (see Figure 1.6). The modularity offers high potential to define standard PEBBs that can be mass-produced. Hence, a "plug-and-play" concept can be readily realized with DABCs for MVDC grids. Furthermore, as the current can be fully controlled by the DC-to-DC converters (due to its buck-boost capability), intelligent DC substations can be realized to control power flow in a multi-terminal DC distribution grid. Studies have shown that such inter-connected multi-terminal system can provide very high part-load efficiency [31], can avoid overload conditions and limit short circuit currents. Innovative concepts of linking DC systems with existing AC systems (hybrid grids) allows higher uti-lization of existing AC infrastructure, while improving reliability. Indeed, multi-terminal DC systems allow power flow control between multiple AC substations

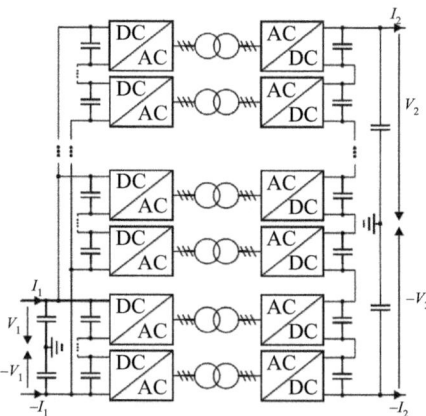

Figure 1.6   Galvanically isolated DC converters can be stacked in series or connected in parallel. A parallel input and series output DC transformer that connects a low-voltage primary side (left) to a medium-voltage secondary (right) is exemplarily shown. Series production of such standardized power electronic building blocks, reduces costs, improves reliability and avoids high re-engineering costs for each voltage and power level

(see previous section), which allows increased utilization of the infrastructure of existing AC distribution grid, while providing higher levels of redundancy.

No doubt, the transition from consumers to prosumers, and likewise from AC to DC wiring in single homes and apartments will take a long time. This may create the impression that the transition to DC grids or hybrid grids (mix use of AC and DC) in the low-voltage distribution grid may take a very long time. Understandably, DC technology is often met with a lot of skepticism, even in the technical community. Some would say that this transition towards DC will never happen. On the other hand, the fact that DC technology lends itself better to use underground cables it is expected that it will be socially more acceptable. In addition, despite the long life-span of transformers and cables, DSOs operating in large cities claim that 80% of the distribution infrastructure is renewed within 15 years. This relatively fast turn-around is caused primarily by the changing load centers in cities. Under such conditions and due to the fact that MVDC systems provide great flexibility and controllability up to the end-user, the transition from low-voltage AC to MVDC in urban grids could come a lot faster than has been realized up to now.

Clearly, the transition to DC technology will accelerate when the economical and ecological benefits of DC technology can also be found in the low-voltage wiring in homes and buildings. New energy-efficient homes will have electrically driven heat pumps and ventilation systems. Newbuilt homes possibly have PV installations with battery storages, while future homeowners may have at least one electric vehicle (EV) and want fast(er) charging. An interesting paradox can be noted. Indeed, these modern homes and buildings will be CO_2 neutral, but will require more electricity to run heat pumps or to charge EVs. The amount of electrical energy produced by self-production (PV) or taken from the grid will depend on the economical optimizations end-users will perform. Nevertheless, it is the author's belief that both systems (self-consumption and energy from the grid) will exist side by side at approximately equal capacity levels, as electricity prices will settle between prosumers and grid operators.

Already today many appliances, such as vacuum cleaners, washing machines, dishwashers and refrigerators are based on brushless DC motor drive technology. Internally, these appliances use DC power. As they are connected to AC grids, they need power electronic rectifiers with unity power factor corrector (UPFC) capability. UPFCs are required to comply with AC grid noise and harmonic current guidelines and standards. These AC-to-DC interfaces tend to be complex and remain relatively expensive. It can be estimated that about 50%–60% of the overall power electronic converter cost in appliances is caused by the grid side UPFC. Hence, major cost savings can be realized in DC distribution systems by avoiding the complex inverter systems that provide unity power factor and low-harmonic current content.

Future smart homes will use DC-to-DC bi-directional converter modules to connect to the DC distribution grid, which would further reduce material costs of the distribution grid substantially. All large micro-generators, storages and loads can be connected to this DC building- or domestic-grid, such as PV modules,

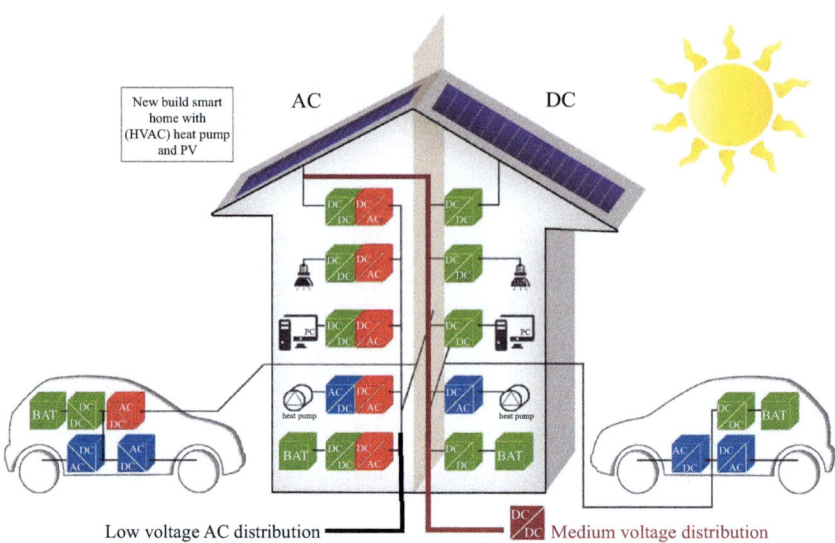

*Figure 1.7   Future "smart homes" with PV and heat pumps (right side of picture)
will use DC bus systems, eliminating the need for multiple UPFCs (red
boxes) and can provide fast charging to EVs. Note that the building
can be connected to both the AC and DC distribution grids via existing
PV power electronic inverters (top left)*

batteries, heat pumps, washing machines and fast DC chargers for EVs. Studies
show that using EV batteries as bi-directional storage devices in homes could
actually extend the operating lifetime of the battery [8]. Hence, using bi-directional
chargers [9] would be beneficial. If the vehicles use low-voltage batteries with DC-
DC converters (as illustrated in Figure 1.7), then a direct connection to the DC grid
is possible without any additional power electric infrastructure. This proposed DC
grid for the all-electric home (eHome) provides to "prosumers" interfaces with a
high level of automation by distributed intelligence [32], demand side management
at lower cost and higher flexibility [33].

1.4   Conclusions

In summary, the main cost advantages of DC multi-terminal networks with a large
number of decentralized power generators are:

- Higher efficiency in energy conversion systems, due to the fact that less conversion steps are needed and DC-to-DC converters can have less losses than
50 Hz transformers.
- Higher power capacity in DC cables.
- Lower infrastructure costs in the distribution grid and less material use (less
copper and Si-steel) in energy conversion systems (electronic transformers).

- Lower CO_2 footprint of the entire value chain (substantially less components and material consumption).
- Higher reliability of the decentralized power generators that require otherwise AC grid side PWM VSIs. Less maintenance cost.
- Potential of further cost reductions, especially with new materials (SiC, magnetic materials). [34]

The following chapters of this book describe the technical concepts of MVDC grids in detail, highlighting the advantages of the new era of power electronic fed electrical grids. Furthermore, the reader may be convinced that power electronics has come a long way and is ready for implementation in electrical MVDC grids, as power electronics is omnipresent in homes, buildings, factories, mobility, appliances, industrial drives, PV, wind farms, etc. Using zero Hertz in the electrical grid augments its capacity, using high-frequencies in DC transformers reduces material consumption. The technology exists today to provide all people on this planet with a near CO_2 neutral electrical energy supply.

Acknowledgement

The author wishes to thank the E.ON ERC gGmbH Foundation for supporting early on the development of the 5 MW DC-to-DC DABC demonstrator. Since 2014, work on DC medium voltage grids and DC-to-DC high power converters continues at the BMBF *Forschungscampus Flexible Electrical Networks (FEN)*. Furthermore, the author is indebted to all PhD assistants at RWTH Aachen University, in particular ISEA, E.ON ERC|PGS & E.ONERC|ACS and IAEW, who contributed to the development of DC-to-DC converter technology, the design and control of medium-voltage DC grid technology.

References

[1]   R.W. De Doncker, "Towards a Sustainable Energy Supply – The New Landscape of Energy Technologies," *Panasonic Technical Journal*, Vol. 57, No. 4, Jan. 2012.

[2]   R.W. De Doncker, "Power Electronic Technologies for Flexible DC Distribution Grids," *2014 International Power Electronics Conference (IPEC-Hiroshima 2014 – ECCE ASIA)*, Hiroshima, 2014, pp. 736–743.

[3]   FESTSCHRIFT E.ON ERC – 10 Years Research and Development on Urban Energy Systems, July 2017 Chapter III, pp. 74-93, https://www.eonerc.rwth-aachen.de/cms/E-ON-ERC/Das-Center/Aktivitaeten-und-Publikationen/~pajj/Festschrift-des-E-ON-ERC/lidx/1/

[4]   European Green Deal, https://ec.europa.eu/info/strategy/priorities-2019-2024/european-green-deal_en

[5]   European Wind Energy Association, "Wind in Power – 2010 European Statistics," Feb. 2011, http://www.ewea.org and http://www.eurobserv-er.org/windenergie-barometer-2020/

[6]   European Photovoltaic Industry Association, "Market Outlook 2010," Feb. 2011, on-line: www.epia.org and http://www.eurobserv-er.org/photovoltaic-barometer-2020/

[7]   EUROSTAT, Energy, Transport and Environment Indicators, "Pocketbooks 2010, ISSN 1725-4566, regular updates on-line: http://epp.eurostat.ec. europa.eu/

[8]   André Hackbarth, Benedikt Lunz, Reinhard Madlener, Dirk Uwe Sauer, Rik W. De Doncker, "Plug-in Hybrid Electric Vehicles for CO2-Free Mobility and Active Storage Systems for the Grid, Part I," E.ON ERC Series Series ISSN 1868-7415 , Dec. 2010, Volume 2, Issue 3.

[9]   M. Rosekeit, R. De Doncker, "Galvanically Isolated Bidirectional Charger for Electric Vehicles – Analysis and Implementation," *APEC 2011, Conference Proceedings*, 2011.

[10] Kristian Huchtemann, Dirk Müller, "Evaluation of Field Study Data on Domestic Heat Pump Systems," E.ON ERC Series ISSN 1868-7415, April 2011, Volume 3, Issue 2.

[11] W. Nijs, P. Ruiz, I. Gonzales, "Heat Roadmap Europe, A Low-carbon Heating and Cooling Strategy to 2050," Report 2017, see www.heatroadmap.eu

[12] M. Stieneker, R.W. De Doncker, "Medium-voltage DC distribution grids in urban areas," *2016 IEEE 7th International Symposium on Power Electronics for Distributed Generation Systems (PEDG)*, Vancouver, BC, 2016, pp. 1–7.

[13] J. Schwarzenberg, R.W. De Doncker "15 kV Medium Voltage Static Transfer Switch," *IEEE IA Conference, 1995, Thirtieth Annual Meeting*, conference records, Vol. 3, pp. 2515–2520, Oct. 1995

[14] C. Schauder, E. Stacey, M. Lund, L. Gyugyi, L. Kovalsky, A. Keri, A. Mehraban, A. Edris, "AEP UPFC Project: Installation, Commissioning and Operation of the ±160 MVA STATCOM (Phase I)," *IEEE Transactions on Power Delivery*, Oct. 1998, Vol. 13. No. 4, pp. 1530–1535.

[15] N. Hingorani, "Introducing Custom Power," *IEEE Spectrum*, June 1995, Vol. 32, No. 6, pp. 41–48.

[16] US Geological Survey, https://www.usgs.gov/centers/nmic/copper-statistics-and-information

[17] World Bank Data, https://data.worldbank.org/indicator/EG.ELC.ACCS.ZS

[18] VDE-Studie Gleichspannung in der elektrischen Energieverteilung, https://shop.vde.com/de/etg?orderby=15

[19] C. Meyer, "Key Components for Future Offshore DC Grids," PhD dissertation, Institute for Power Electronics and Electrical Drives (ISEA), RWTH Aachen University, Germany, 2007.

[20] S.P. Engel, N. Soltau, H. Stagge, R.W. De Doncker, "Improved Instantaneous Current Control for High-Power Three-Phase Dual-Active Bridge DC-DC Converters," *IEEE Transactions on Power Electronics*, 2013.

[21] R. Lenke, "A Contribution to the Design of Isolated DC-DC Converters for Utility Applications," PhD thesis, RWTH Aachen University, E.ON ERC| PGS, 2012, ISBN 978-3-942789-05-9.

[22] R. Lenke, S. Rhode, F. Mura, R. De Doncker, "Characterization of Amorphous Iron Distribution Transformer Core for use in High-Power Medium-frequency Applications," in *IEEE Energy Conversion Congress and Expo*, pp. 1060–1066, Sept. 2009.

[23] Historical data of copper and steel prices, e.g. www.infomin.com or www.metalprices.com or www.lme.com

[24] J. Hu, S. Cui, R.W. De Doncker, "DC Fault Ride-Through of a Three-Phase Dual-Active Bridge Converter for DC Grids," in *2018 International Power Electronics Conference (IPEC-Niigata 2018–ECCE Asia)*, Niigata, 2018, pp. 2250–2256, doi: 10.23919/IPEC.2018.8507672.

[25] J. Hu, *Modulation and dynamic control of intelligent dual-active-bridge converter based substations for flexible dc grids*, Aachen: Dissertation, E. ON Energy Research Center, RWTH Aachen University, 2019, ISBN 978-3-942789-68-4.

[26] "Medium Voltage Direct Current (MVDC) Grid Feasibility Study," WG C6.31. See CIGRE webpages: www.cigre.org

[27] "German Standardization Roadmap – Low Voltage DC," Version 1, Feb. 2016. This DKE Roadmap for developing DC guidelines can be downloaded from the DKE webpages: www.dke.de/Gleichstrom-Roadmap

[28] ETSI standards and guidelines can be found at www.etsi.org/standards.

[29] R.W. De Doncker, D.M. Divan, M.H. Kheraluwala, "Power conversion apparatus for DC/DC conversion using dual active bridges," U.S. patent US5027264 A, Sep. 29, 1989/June 1991.

[30] R.W. De Doncker, D. Divan, M. Kheraluwala, "A Three-Phase Softswitched High-Power-Density dc/dc Converter for High-Power Applications," *IEEE Transactions of IA*, Vol. 27, No. 1, Jan/Feb 1991, pp. 63–73.

[31] Florian Mura, Rik W. De Doncker, "Design Aspects of a Medium-Voltage Direct Current (MVDC) Grid for a University Campus," 8th International Conference on Power Electronics (ICPE 2011 – ECCE Asia), Jeju, Korea, May 2011.

[32] A. Monti, F. Ponci, "Power Grids of the Future: Why Smart Means Complex," Complexity in Engineering, 2010. COMPENG'10, pp. 7, 11, 22–24 Feb. 2010.

[33] A. Riccobono, M. Ferdowsi, J. Hu, *et al.* "Next Generation Automation Architecture for DC Smart Homes," *2016 IEEE International Energy Conference (ENERGYCON)*, Leuven, pp. 1–6, 2016.

[34] G. Wang, X. Huang, J. Wang, T. Zhao, S. Bhattacharya, A.Q. Huang, "Comparisons of 6.5kV 25A Si IGBT and 10-kV SiC MOSFET in Solid-State Transformer application," IEEE Energy Conversion Congress and Exposition (ECCE), pp. 100,104, Sept. 2010.

Chapter 2

Power electronic converters impacts on MVDC system architectures*

Robert M. Cuzner[1] and JiangBiao He[2]

Present-day Renewable Energy Source (RES) installations, such as offshore wind power, wind and solar farms, combine the enabling Power Electronic Converter (PEC) technology (i.e., AC–DC, DC–AC, DC–DC converters) with conventional power distribution components, such as transformers and switchgears. Microgrids are introduced at points of usage in utility based Medium Voltage AC (MVAC) distribution grids as a means of increasing RES insertion locally, and thereby reducing carbon emissions, reducing energy costs and increasing grid resiliency, leaving the existing grid infrastructure intact. In these systems, a significant portion of the PEC hardware and controls, with associated cost, complexity and losses, are dedicated to the interface between energy sources, i.e., wind generator, solar photovoltaic (PV) array, Battery Energy Storage (BES), and the MVAC distribution system.

(Figure 2.1) shows typical Distributed Energy Resource (DER) interfaces to a three-phase MVAC system. The fundamental components of the DER to MVAC interface are the PEC(s), low-frequency step-up transformer and a current breaking static switch for connection or disconnection to or from the MVAC grid. In most implementations, because of the need to phase synchronize with the grid, the static switch combines an electromechanical circuit breaker (EMCB), for fault protection, combined with Silicon Controlled Rectifier (SCR)-based transfer switches. The SCR-based transfer switches control any inrush associated with small phase differences between the AC voltages synthesized by the PEC and the AC grid voltages during connection.

In addition to the provision of voltage step-up and connection or disconnection functionality by the transformer and EMCB, respectively, both of these components provide the critical functionality of galvanic isolation, also indicated in

*The material of this chapter is based, in part, by work supported in part by the National Science Foundation, Grant No. 1650470, the Office of Naval Research, Grant N00014-20-1-2667. Any opinions, findings, and conclusions or recommendations expressed in this material are those of the author(s) and do not necessarily reflect the views of these institutions.
[1]Department of Electrical Engineering and Computer Science, University of Wisconsin-Milwaukee, Milwaukee, WI, USA
[2]Department of Electrical and Computer Engineering, University of Kentucky, Lexington, KY, USA

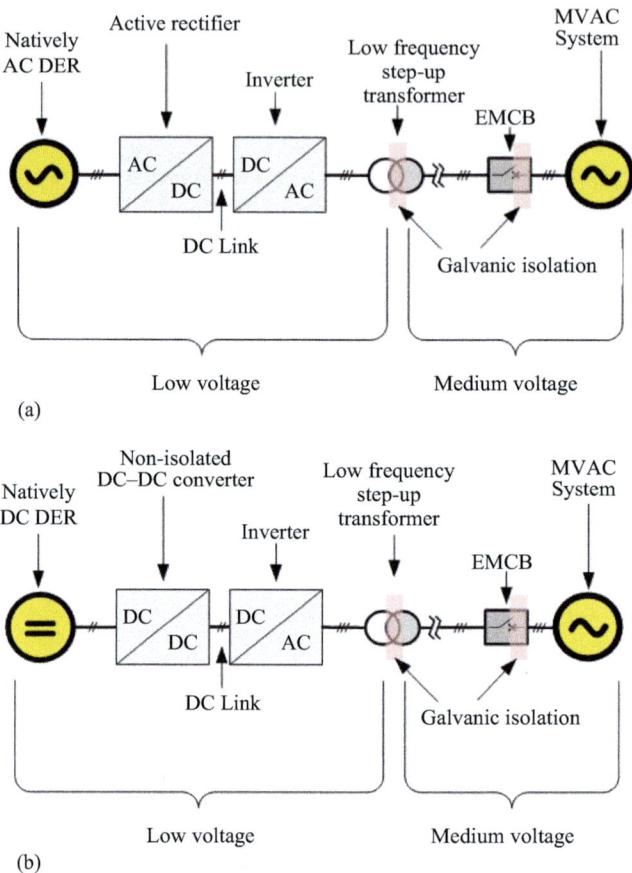

Figure 2.1   Distributed energy resource interfaces to MVAC system: (a) natively AC DER interface and (b) natively DC DER interface

(Figure 2.1). Galvanic isolation refers to the capability to prevent current flow by any means—either intentional or non-intentional—between parts of the system. The transformer provides galvanic isolation between electrical grounding paths on the primary and secondary sides of the transformer. This feature is essential to both human and equipment safety of the system.

The EMCB provides galvanic isolation between the two sides of the switch whenever it is in the opened state and has the additional capability of breaking the current path while it is opening. This limits the current associated with the sudden inception of a short-circuit Line-to-Line (LL), Line-to-Ground (LG), Line-to-Line-to-Ground (LLG) and Multiple Line-to-Ground (MLG) faults.

There are increasingly apparent costs and risks associated with the growth in insertion levels of PEC-interfacing, non-dispatchable Distributed Energy Resources (DERs) and their interactions with electrical grids [1]. On the other hand, since nearly all PEC implementations are built on the backbone of a voltage-stiff capacitor bank (or DC link) it would make sense to re-arrange the flow of power through PECs in order to take advantage of inherent energy buffer provided by the DC link and eliminate the complexity associated with the MVAC interface. A wide range of applications are emerging which utilize a shared MVDC bus for energy collection from PEC-interfacing energy sources and/or for distribution to PEC-interfacing energy users [2–9]. Equivalent DER interfaces to a MVDC system are shown in (Figure 2.2). Since the DC link is moved to the common point of interface for all converters, a minimum number of power conversion stages are required. Distribution components are removed, except for the addition of a No Load Switch (NLSw) for connection or disconnection to or from the MVDC system.

The apparent simplicity of such systems may be misleading, however, because the PECs themselves must take on additional functionalities of the distribution

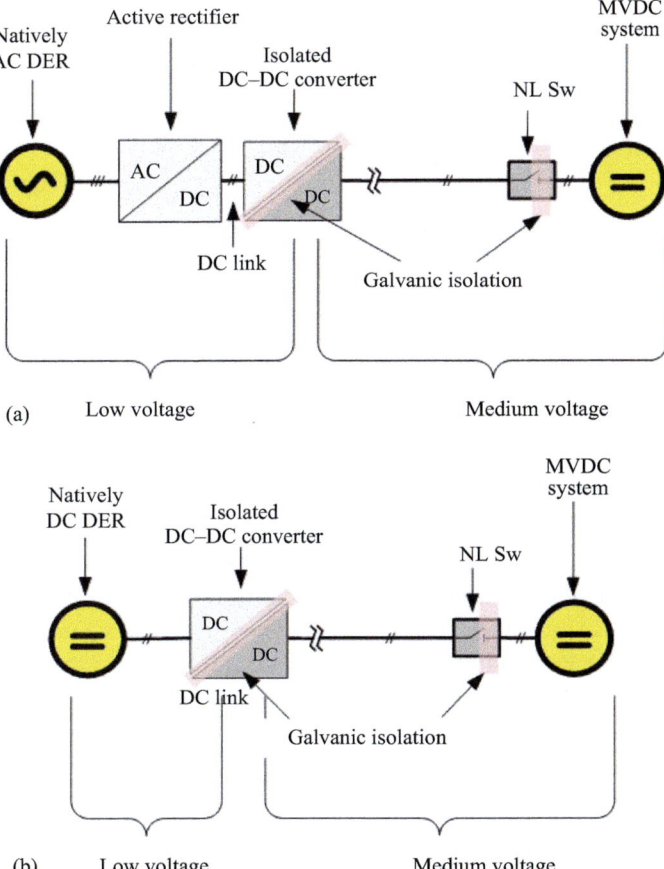

Figure 2.2   Distributed energy resource interfaces to MVDC system: (a) natively AC DER interface and (b) natively DC DER interface

components, whereas the PECs are only responsible for power conversion (synthesis of power and energy from one form to another) and power/energy flow control. For example, in the MVDC system, the isolated DC–DC converter provides functionalities including power/energy flow control, voltage step-up and galvanic isolation between grounding of the DER and the MVDC system. The isolated DC–DC converter may also include MVDC-side fault current limiting, to enable opening of the NLSw. If the latter capability cannot be provided by the PEC, then an intervening power electronics-based Current Limiter (CL) must be added. The CL must have the capability to quickly drive the currents to zero in both the supply and return lines between the NLSw and the MVDC system. Note also that the one-line descriptions of (Figure 2.2) denote that NLSw's are installed in both lines of the DC system in order to ensure complete galvanic isolation.

The additional complexities associated with the allocation of power distribution component functionalities to the PECs will inform decisions associated with the PEC implementation, such as power electronic circuit topology, passive filter interfaces, current de-ratings, controls, and so on. Furthermore, the PEC circuit topologies will also inform the architectural implementation of the MVDC system.

This chapter is dedicated to the unique challenges and opportunities associated with the design of MVDC-interfacing PECs and their influences on MVDC architectural implementations.

2.1 Changing power distribution system paradigms

MVDC systems offer benefits when compared to the conventional MVAC systems well beyond the obvious benefits of increasing power transfer ratios, lowering PEC costs and improving efficiency. They are enabling a host of new solutions and opportunities, particularly when it comes to electrical systems that operate in isolation from the electrical grid.

2.1.1 The opportunity of shipboard and aircraft electrification

For example, there is a growing and informed consensus among stakeholders and technologists that future large vessel transportation system electrification will be built upon *MVDC-based power and energy delivery systems*. For the commercial ship industry, steam-powered ships have already given way to electric propulsion and, for many ship platforms, common points of MVDC interface for electric propulsion PECs are extending into MVDC distribution systems that integrate with other ship service power distribution systems, spanning large portions of ship's geography.

Along these same lines, future and existing naval ship platforms are becoming more electrified. Electric weapons and other electrified loads, requiring high energy delivery for short periods of time, are well-served by a shared MVDC bus that combines dynamically controllable energy dispatch from PEC-interfacing generation and BES systems. In naval shipboard systems there is also a trend towards the utilization of an Integrated Power System (IPS), based upon a DC Zonal Electrical Distribution System (DCZEDS), which enables the powering of all electric loads and electric propulsion systems from a small number of installed generators [10]. Furthermore, the U.S. navy's roadmap for future ship platforms includes a MVDC-based Integrated Power and Energy System (IPES), which enables dynamic dispatch of power and energy from installed generation and BES to a diverse load set through multiple reconfigurable paths throughout the ship [11]. These systems allow for the optimal use and placement of short term and long term energy storage. They also utilize modular Power Electronic Building Block (PEBB) approaches in order to integrate PECs into the entire ship electrical distribution system during the construction phase [12,13]. The PEBB approach to PECs, combined with transformational ship construction concepts that integrate an expandable and adaptable PEC-based power and energy delivery system (to the end of service life) into the ship's hull design, will, in the long term, reduce the risks and costs of new shipbuilding endeavors while providing increasingly flexible and adaptable mission capabilities [14].

For aircraft systems, increasing levels of smart electrification, associated with the concept of More Electric Aircraft (MEA) are the norm. As aircraft platform implementations shift towards electrified propulsion drives, the migration of MEA DC-based power distribution systems to higher power levels drives the need for MVDC systems [15].

A general example of an MVDC electrical architecture, for ships or aircraft, is shown in Figure 2.3a.

2.1.1.1 PEC-based protective architectures

As was discussed in relation to (Figure 2.2), the PECs must take on the power distribution functionalities of voltage transformation, galvanic isolation and current limiting. Some MVDC-based electrical systems take advantage of inherent DC-side current limiting capability within PECs to mitigate the impacts of LL, LLG, or MLG short circuit faults on the MVDC bus. The generator and BES interfacing PECs coordinate with NLSws distributed along the bus in order to locate and subsequently isolate the fault, once all current has been driven out of the NLSws. This approach is referred to as *PEC-based protection* and requires centrally managed distributed communications between the generator and BES interfacing PECs and NLSws. PEC-based protective MVDC system electrical architectures are built upon a network of DC buses spanning the length and breadth of the vessel used to deliver power and energy to downstream *in-zone* electrical loads. When a short circuit LL, LLG, or MLG fault occurs the affected MVDC bus will collapse. As a result, redundant, isolated DC buses and cross-connected feeds from both buses to critical loads are required to maintain system functionality and recoverability while faults are being isolated.

The MVDC-interfacing PECs must also be capable of controlling or blocking current flow into the MVDC bus during low impedance fault incidence. The MVDC-side PEC capacitive DC links are often directly connected across the MVDC bus and, as a result, a significant level of surge current may discharge from

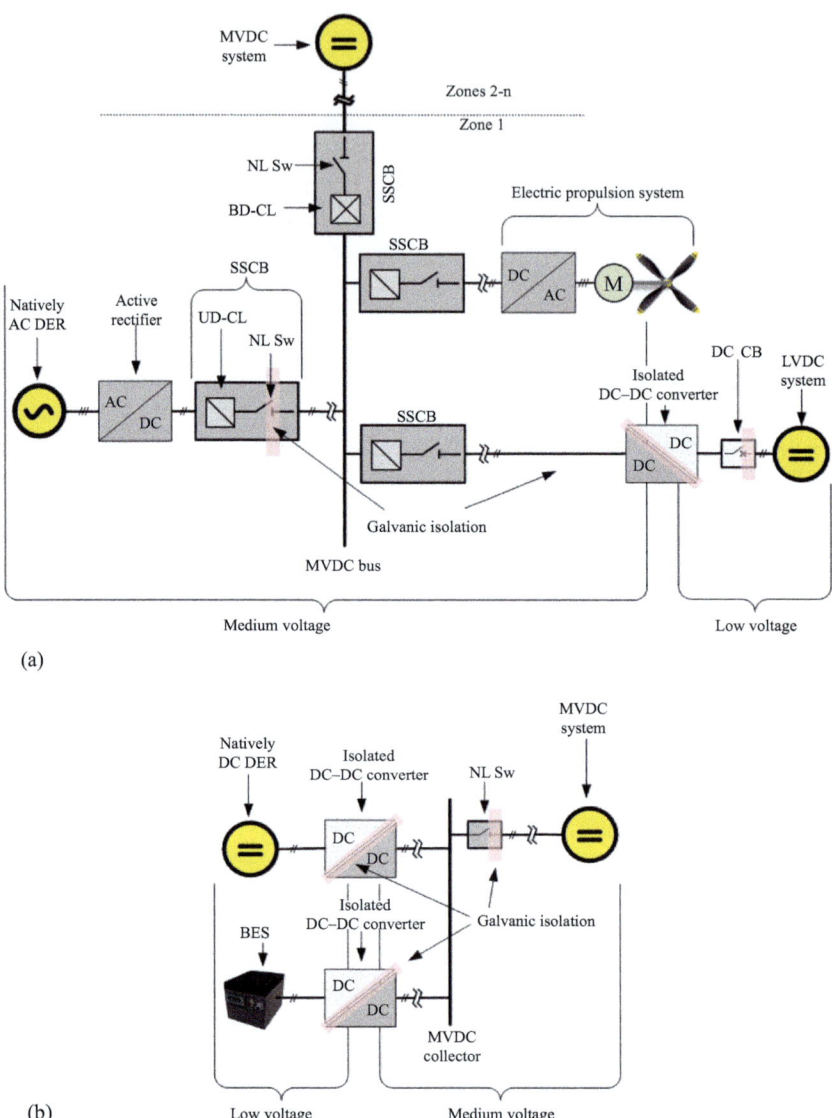

Figure 2.3   Emerging MVDC-based distribution systems: (a) MVDC-based power delivery system. (b) Shared MVDC-bus collection and distribution system

these capacitors into sudden inception short circuit LL faults. The MVDC bus conductors and NLSws must *let through* and *withstand* this discharge current (without damage) until the MVDC bus has collapsed to a significant level for the PEC active switching networks *behind* the DC links to inhibit current flow into the fault. The capacitor discharge current during fault incidence may be used by the PEC-based protection scheme to locate the fault [16,17].

2.1.1.2   Breaker-based protective architectures

Because of the networked nature of the MVDC bus, it may be advantageous to relieve the PECs of functionality associated with limiting fault currents on the MVDC bus and discriminating the location of those faults. This alternative approach to the PEC-based protective architecture is referred to as *Breaker-based protection*, and is what is represented in Figure 2.3a. Breaker-based protection requires the replacement of the NLSw with Solid State Circuit Breakers (SSCBs) which take on the power distribution functionalities of both current limiting and galvanic isolation of the fault. The SSCB generally consists of either a power semiconductor-based Unidirectional Current Limiter (UD-CL) or Bi-Directional Current Limiter (BD-CL) in series with a NLSw. The use of SSCBs in MVDC systems relieves significant burdens that are otherwise placed on the PECs and opens up the possibility of the entire range of MVDC-interfacing PEC circuit topologies. However the SSCB for MVDC systems faces many challenges that are unique and even more demanding when compared to their High Voltage DC (HVDC) counterpart, the Hybrid Solid State Circuit Breaker (HCB) (which is becoming commercially viable for HVDC transmission and multi-terminal HVDC networks) [18,19]. The MVDC SSCB is a developmental item, and, at the time of this writing, is at low technical readiness level (TRL).

2.1.2   *High renewable penetration threat to MVAC grids*

For utility interfacing applications, as the percentage of RES contribution to the grid is growing, simultaneous control of dispatchable (utility generation, co-generation, standby generation, energy storage) and non-dispatchable (wind, solar) energy is essential. For the PEC-interfaced DER, the primary function of the AC–DC converter (Figure 2.1a) or DC–DC converter (Figure 2.1b) is to maximize extraction of power or energy from the RES or to optimize energy transfer to or from the BES, while main-taining a stable DC link voltage. The primary function of the DC–AC converter is to dynamically synchronize the PEC synthesized electrical angle with the grid and control the reactive power at the point of grid interface through the power distribution com-ponents (transformer and EMCB). Due to the evolutionary way in which utility grids are built up, (i.e., without a central guiding plan), the control of energy dispatch from multiple points of source within the grid is becoming increasingly ambiguous and reliant upon distributed control concepts to achieve virtual system level control over a wide variation of time scales. The ability of the PEC to act as a buffer at the AC point of interface to voltage fluctuations and extreme events, such as transient overloads and short circuit faults, depends on the AC-side filter implementation, PEC power semi-conductor device thermal management and PEC controls adaptability. The grid-interactivity of PEC performance is difficult to specify during the procurement process and depends solely upon evolving, and slowly maturing, *quality of service* standards—aimed principally at defining power quality and ride-through parameters *from the grid-user perspective.*

A primary requirement for each DER-to-MVAC grid interfacing PEC is self-protection. During extreme events, i.e., short circuit fault incidence, fault isolation and recovery, loss of input, etc., there is a growing uncertainty regarding the resiliency of power distribution systems having a high penetration of PEC-interfacing DERs [1]. As RES contributions approach grid parity, these grids become increasingly susceptible to extreme weather events, cascading losses in grid capacity and voltage surges during event recovery [20]. Even though the aggregated monitoring and processing capabilities of the PECs present an opportunity improve grid resiliency, their primary function is power conversion. The responsibility for removal of faults and recovery from events is allocated to protective relaying functions allocated to sluggish electromagnetic/electromechanical power distribution components, which may be ill-equipped to handle the dynamics of high PEC-penetrated AC grids during extreme fault scenarios.

2.1.3   MVDC system solutions and attributes

Introduction of shared MVDC-bus collection and distribution systems within utility grids could eliminate or significantly reduce many of the inter-compatibility issues that are plaguing high PEC-penetrated MVAC grids and improve the efficiency of RES installations, such as offshore wind, wind and solar farms. They can also transmit combined RES- and BES-based energy very efficiently over long distances between geographically isolated parts of a MV distribution system. Alternatively, the performance of the MVDC-based power and energy delivery system required for the electrification of navy ships and aircraft (represented generally in Figure 2.3a) cannot be mimicked effectively using present-day approaches to integrating DERs into utility grids, which overlay PECs into conventional low frequency transformer and EMCB-based power distribution systems. This approach would eventually become unwieldy, from the standpoints of space/weight claim and costs of the the overall equipment. Even more significantly, the problems associated with high PEC-based RES penetration into utility grids would be amplified because the generation sources of mobile transportation systems are inherently weak. Therefore, the technological pull of MVDC-based transportation systems is assured.

It must be recognized that a key attribute of any MVDC-based system is that the bulk of the power and energy that is generated and used must pass through the PECs. The low-frequency isolation transformer, switchgear and the typical radial power distribution systems associated with MVAC systems are eliminated in MVDC systems. The result is a significant paradigm shift away from conventional usage of PECs within power distribution systems that may not be recognized without an understanding of the range of potential end-use configurations. While these systems present tremendous opportunities, there will also be challenges associated with realization of system compatible, sustainable and reliable MVDC-interfacing PEC solutions. It is the intention of this chapter to identify these challenges, to point to advances in technology that are making the realization of these systems possible, and to illuminate the critical areas where effort is required to bridge the *valley of death* between enabling technological advances and commercially viable solutions. To achieve this end, this chapter will frame the discussion of the design of PECs for MVDC-based electrical system around eight attributes that are at the foundation of the justification of a transition from a MVAC- to MVDC-based electrical system. Table 2.1 enumerates the attributes that justify the movement towards MVDC-based electrical distribution systems, along with example applications that utilize these attributes and the TRL of deployability of such systems into end-use applications. The attributes of Table 2.1 will be referred to in the discussion of emerging applications for MVDC-based systems. Table 2.2 provides the definition of each TRL, which will prove helpful in the discussions on emerging applications.

Table 2.1   Attributes (1–8) justifying the movement from MVAC- to MVDC-based electrical distribution systems

Attribute	Example applications	TRL
1 *Plug and play, scalable renewable energy system installations*	Utility-scale wind and solar farms	7
2 *Efficient end-to-end energy transport over long distances*	Off-shore wind, MVDC corridor	7
3 *Increased system efficiency*	Shipboard electric propulsion	8
4 *Highly dynamic energy dispatch*	MVDC Microgrids, Naval shipboard IPES	5
5 *Optimal usage of distributed short-term and long-term energy storage*	MVDC EV Charging, Naval shipboard IPES	4
6 *Improved ride-through and associated quality of service*	Common Bus VFDs, MVDC microgrids	6
7 *Highly flexible energy dispatch*	Naval shipboard IPES, MT MVDC distribution	4
8 *Dynamic re-configuration of source, load, bus interconnections*	Naval shipboard IPES, MT MVDC distribution	3

Table 2.2   Definitions of technical readiness levels (TRLs) for PEC-based systems

TRL	Description
9	System proven through successful operations
8	System completed and qualified through testing and demonstration activities (ready to procure)
7	Demonstration of a system prototype in an operational environment
6	Demonstration of system or subsystem model or prototype in the relevant environment
5	Component or breadboard validation in the relevant environment
4	Component or breadboard validation in a laboratory environment
3	Analytical and experimental critical function and/or characteristic proof of concept
2	Technology concepts and/or applications formulated
1	Basic principles observed and reported

2.2   Power electronic converter considerations

Architectures associated with various applications will be described using one line diagrams, similar to those shown in Figures 2.1, 2.2, and 2.3, in the following sections of this chapter. The PECs in these architectures are AC–DC, DC–AC, DC–DC, and isolated DC–DC converters. In all one line diagram representation of systems, light gray blocks denote Low-Voltage (LV) rated converters and dark gray blocks denote medium-voltage (MV) rated converters. This distinction is important, not only from the standpoint of the power circuit topology (i.e., single level versus multilevel) but from the insulation rating of the PEC, which incorporates cable/termination insulation dielectrics, creepages and clearances, voltage dielectric stand-off distances between energized surfaces and chassis ground and ground wall insulation dielectrics. The various converter types exhibit certain behaviors and provide certain capabilities which should be understood in order to appreciate their impacts on architectural implementations.

Figures 2.4 and 2.5 show, to begin with, the correlation between the single line symbol for these AC–DC and DC–AC PECs, respectively, and their multiple line inputs/outputs, directions of power flow control ability and input/output characteristics and capabilities. For AC–DC and DC–AC converters, energy storage is, obviously, on the DC sides. Commercially packaged power semiconductor modules having sufficient voltage and current ratings for these PECs will either pass current in both directions and block voltage in only one direction or they will pass current in one direction and block voltage in both directions. The achievement of bidirectional voltage and current blocking is only accomplished through the use of multiple modules to achieve single switch functionality, which reduces PEC efficiency, drives up cost and introduces interconnection complexities that can usually require additional snubbing and clamping components to ensure safe device operation. As a result, bidirectional switch implementations in PECs, which

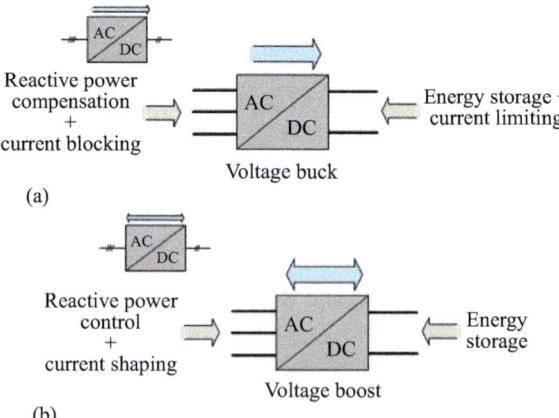

Figure 2.4   AC–DC PECs: (a) unidirectional (buck rectifier) and (b) bidirectional (boost rectifier)

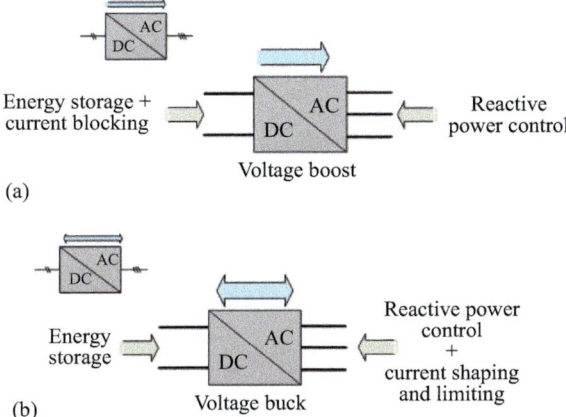

Figure 2.5   *DC–AC PECs: (a) unidirectional (boost inverter) and (b) bidirectional*
(buck inverter)

generally fall under the category of matrix converters, are not common. Consequently, buck rectifiers (Figure 2.4a) will inherently control *both* power and current flow in one direction (from AC to DC) and provide active current limiting (down to zero current) only on the DC side. On the AC side, depending upon the topology, reactive power compensation capability is limited and is highly dependent upon AC-interfacing second order (and higher) passive filters. There is a beneficial consequence of unidirectional rectification: the natural current blocking that will occur if an AC-side LL low impedance or short circuit fault occurs. As a result, this topology has full LL fault mitigation capability on the DC side and self-protective capability on the AC side.

Boost rectifiers (Figure 2.4b), on the other hand, inherently control *both* power and current in both directions (from DC to AC and vice-versa) but only limit current (down to zero current) on the AC side. Reactive power can be actively controlled on the AC side and, depending upon the topology, the current wave shape can be actively controlled so that power quality can be managed while minimizing the size in AC-side passive filtering. Because of the inherent lack of voltage blocking on the DC-side (a consequence of the reverse diodes providing bi-directional current capability on both AC and DC sides) boost rectifiers cannot control their AC-side currents when DC link energy storage capacitors are not charged above the peak rectified AC input voltage level, and are therefore dependent upon upstream switchgear to capacitor precharge and AC-side fault protection. Boost rectifiers lose control when LL low impedance or short circuit fault occurs on the DC side. When a DC-side LL short circuit fault occurs they will completely discharge their DC link capacitors into the fault, unless there is some mechanism (provided by some topologies) to block the flow of current from the DC link capacitors into the fault.

The unidirectional boost inverter of Figure 2.5a controls *both* power and current flow in one direction (from DC to AC) and provides active current limiting (down to zero current) only on the DC side. Similar to the unidirectional buck rectifier of Figure 2.4a on the AC side, and, depending upon the topology, reactive power compensation capability is limited and highly dependent upon AC-interfacing second order (and higher) passive filters. Also there is a beneficial consequence of unidirectional inversion: the natural current blocking that will occur if a DC-side LL low impedance or short circuit fault occurs. As a result, this topology has full LL fault mitigation capability on the DC side and self-protective capability on the AC side. The thyristor-based boost inverter, commonly referred to as Load Commutated Inverter (LCI), is the most commonly applied PEC topology applied to RES inter-facing PECs because the thyristor (or SCR) is well suited for MV, multi-MW applications requiring high current ratings.

In the Variable Frequency Drive (VFD) industry, common terminology for the buck inverter of Figure 2.5b is Voltage Source Inverter (VSI). In utility applica-tions, the common terminology is Voltage Source Converter (VSC). In the latter vernacular, the VSC refers to any AC-interfacing PEC capable of actively con-trolling reactive power on the AC side, whether it is the bidirectional boost rectifier (Figure 2.4b) or the bidirectional buck inverter (Figure 2.5b). The inverter controls *both* power and current in both directions (from DC to AC and vice-versa) and actively limits current (down to zero current) only on the AC side. Reactive power is actively controlled on the AC side and, depending upon the topology, the current wave shape can be actively controlled so that power quality can be managed while minimizing the size in AC-side passive filtering. Because VSC construction is inherently simpler than LCI construction and the VSC more effectively controls reactive power at the MVAC interface, there is considerable attention to the movement from LCI-based DER and grid interfacing PECs (naturally suited to utility-scale application) towards VSC-based DER and grid interfacing PECs. However the challenge is in the limitations opposed by the current and voltage limitations of the power semiconductors that are ubiquitously used in VSCs, the Silicon (Si) Insulated Gate Bipolar Transistor (IGBT) [21].

It should also be noted that the inherent lack of voltage blocking on the DC-side (a consequence of the reverse diodes providing bi-directional current cap-ability on both AC and DC sides) of the buck inverter of Figure 2.5b causes it to lose control when LL low impedance or short circuit fault occurs on the DC side. Under a DC-side short circuit they will completely discharge their DC link capa-citors into the fault, unless there is some mechanism (provided by some topologies) to block the flow of current from the DC link capacitors into the fault.

Considering DC–DC power conversion, the general solution is the bidirectional buck/boost DC–DC converter shown in Figure 2.6a. The buck/boost converter is analogous to the bidirectional AC–DC and DC–AC PECs of Figure 2.4b and 2.5b: power and current flow are controllable in either direction as long as the DC voltage on the nominally high DC voltage is higher than the nominally lower DC voltage, and vice versa. There is current limiting capability on the nominal low DC voltage side. This converter implementation will provide the best dynamic control of voltage and power over the entire nominal range. However, loss of control will occur when LL low impedance or short circuit fault occurs on the nominal high DC side and, when a DC-

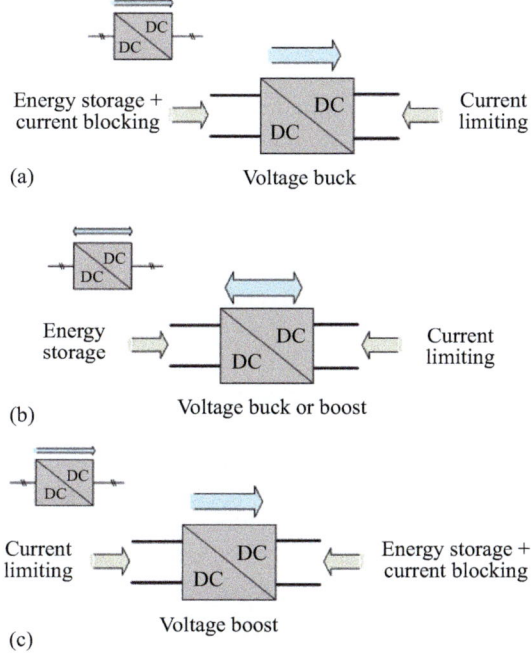

Figure 2.6 *DC–DC PECs: (a) bidirectional buck/boost converter, (b) unidirectional buck converter, and (c) unidirectional boost converter*

side LL short circuit occurs, DC link capacitor(s) (on nominal high DC side) discharge energy storage current and collapse to zero.

The selection of either unidirectional DC–DC PEC of Figure 2.6b or 2.6c is dictated by the needed block current discharge in directions opposite to the nominal current direction. For the buck converter, current blocking capability occurs on the high voltage side. For the boost converter, current blocking capability occurs on the low voltage side. The unidirectional converter is obviously connected to an energy source on the source side and a load on the side the arrows of Figure 2.6b and 2.6c are pointing to. The required current blocking capability is to ensure that energy from the load side of the converter does not regenerate back into the connected source. However, if the buck converter load voltage goes higher than the source side voltage, the associated energy will pump up the DC link capacitor voltages, unless there is some mechanism to provide for dissipating this energy. For the boost converter, blocking capability ensures that current flow from the higher voltage source into the lower voltage source is prohibited (but, for most topologies, it cannot prohibit uncontrolled flow from source to load if load voltage collapses from a low impedance or short circuit fault, and once the DC link capacitor(s) have discharged into the fault). For higher power systems, assuming conventional VSC topologies, there is no cost advantage to a unidirectional topology. The possibility of discontinuous modes of conduction under light loads introduces marked system-level disadvantages [22]. Inherently unidirectional DC–DC converter topologies will be avoided unless there is a compelling advantage, such as the introduction of soft switching operation to enable higher power density or higher efficiency. It should be noted that unidirectional AC–DC and DC–AC PECs will also have discontinuous conduction modes of operation which can be used to achieve a range of advantages [23]. The reactive power associated on the AC side provides another degree of freedom that can be utilized to avoid the voltage pump-up scenarios that plague their DC–DC converter counterparts.

The final PEC to be considered is the isolated DC–DC converter. Fundamentally, the isolated DC–DC converter comprises a DC–AC PEC and AC–DC PECs connected on either side of a medium to high frequency isolation transformer, having a switching transistor controlled fundamental frequency in the range of 0.5–50 kHz. The foundational isolated DC–DC implementation is the single phase, bidirectional configuration as shown in Figure 2.7a. The most commonly applied topology is the dual active bridge (DAB), which controls the power flow in either direction by controlling the phase displacement of actively synthesized square wave or stepped wave voltages on either side of the transformer [24]. In the MVDC system application, a primary function of isolated DC–DC PEC is to step up from the LVDC source or system to the MVDC source or system, and to dynamically control the energy dispatch. The isolated DC–DC PEC may also be used to dispatch the combined energy of the MVDC bus at a given location to a LV distribution system downstream of its LV side. For this latter scenario, the arrangement of Figure 2.7a can be considered as reversed, with voltage step-down from left to right. A primary consideration for the isolated DC–DC PEC design and implementation is achievement of galvanic isolation between primary and secondary windings and between transformer windings and core (if the core is grounded to the PEC chassis).

The isolated DC–DC PEC may be configured, by design, or controlled to operate as a unidirectional converter as indicated in Figure 2.7b. The former approach can be applied to achieve highly efficient and power dense implementation through use of a resonant DC–AC or AC–DC topology on the primary or secondary side of the transformer. The latter approach is simply a change in the way the DAB is controlled and provides a means for current limiting on either the primary or secondary fed DC buses. For example, considering the example of Figure 6, when a LL low impedance or short circuit fault occurs on the MVDC bus, fed by the dark gray, right side of the circuit, power semiconductor switching on that side of the transformer is completely inhibited. The resultant circuit on the MVDC secondary-side AC–DC converter becomes a passive AC–DC rectifier. The power semiconductor switches continue to be actively controlled on the primary side of the transformer and the current into the MVDC fault is limited by the pulse width control on the primary side. The reverse may occur if the LL fault occurs on the primary side, assuming there is a source upstream of the secondary side to supply the fault. In this operational mode, the isolated DC–DC converter is operating as a single active bridge (SAB). Consequently, the configuration of Figure 2.7a can be dynamically switched to the configuration of Figure 2.7b in

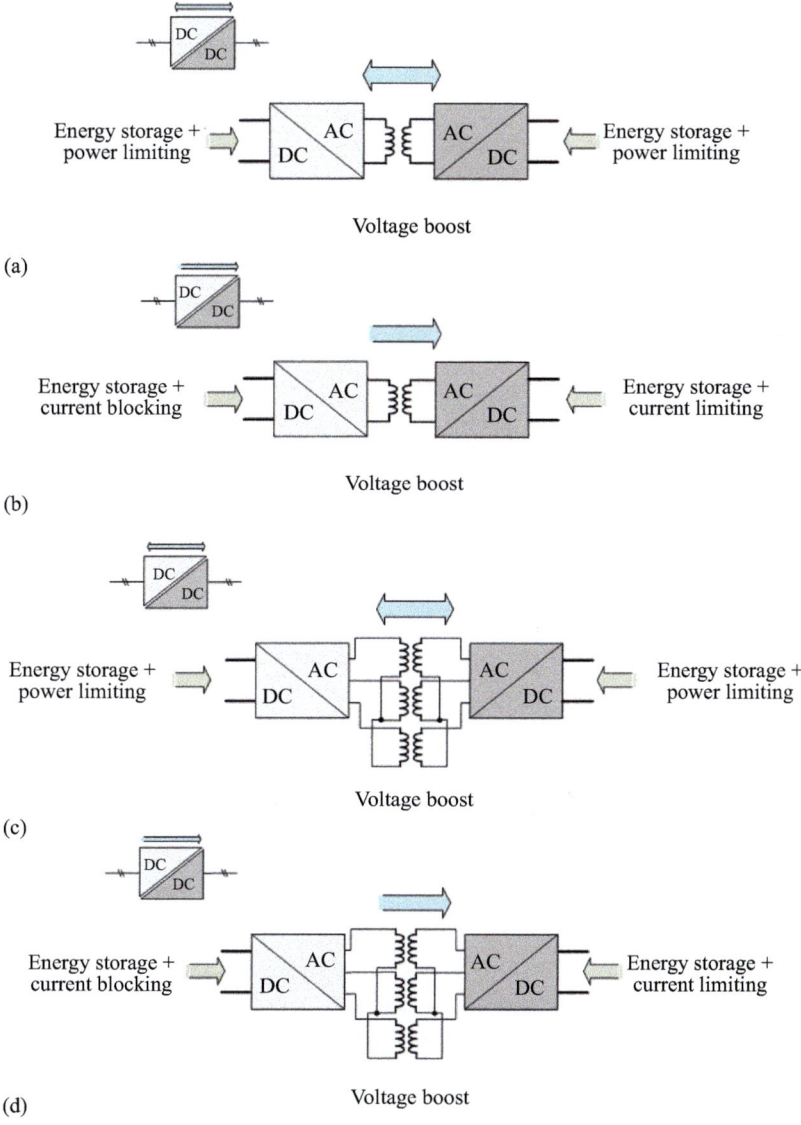

Figure 2.7   Isolated DC–DC PECs: (a) single phase bidirectional, (b) single phase unidirectional, (C) three-phase bidirectional, and (d) three-phase unidirectional

order to add currently limiting capability to the isolated DC–DC PEC and the current limiting direction (and resultant power flow direction) can be flipped actively. Active current flow can be completely blocked and LV and MV systems isolated from each other under extreme fault conditions simply by gating off all of the power semiconductor devices. This functionality promises interesting possibilities for PEC-based protective architectures. Multi-winding, multi-phase implementations of the DAB or SAB can also be realized, as demonstrated by the three-phase DAB and SAB implementations of Figure 2.7c and 2.7d [25,26].

The isolated DC–DC PEC may also be constructed from multiple lower voltage rated building blocks. In this way there is no limit to the DC voltage levels that can be achieved. One approach, building upon any of the possible configurations of (Figure 2.7), is the Input Parallel Output Series (IPOS) configuration shown in Figure 2.8a [27]. The IPOS can achieve a voltage step-up ratio of N times the voltage transformation ratio associated with its constituent building blocks, where N is the number of IPOS connected building blocks. Very high voltage step-down ratios can be achieved as well with the reverse case of Input Series and Output Parallel (ISOP) connections [28]. Here only the bidirectional case is considered, but

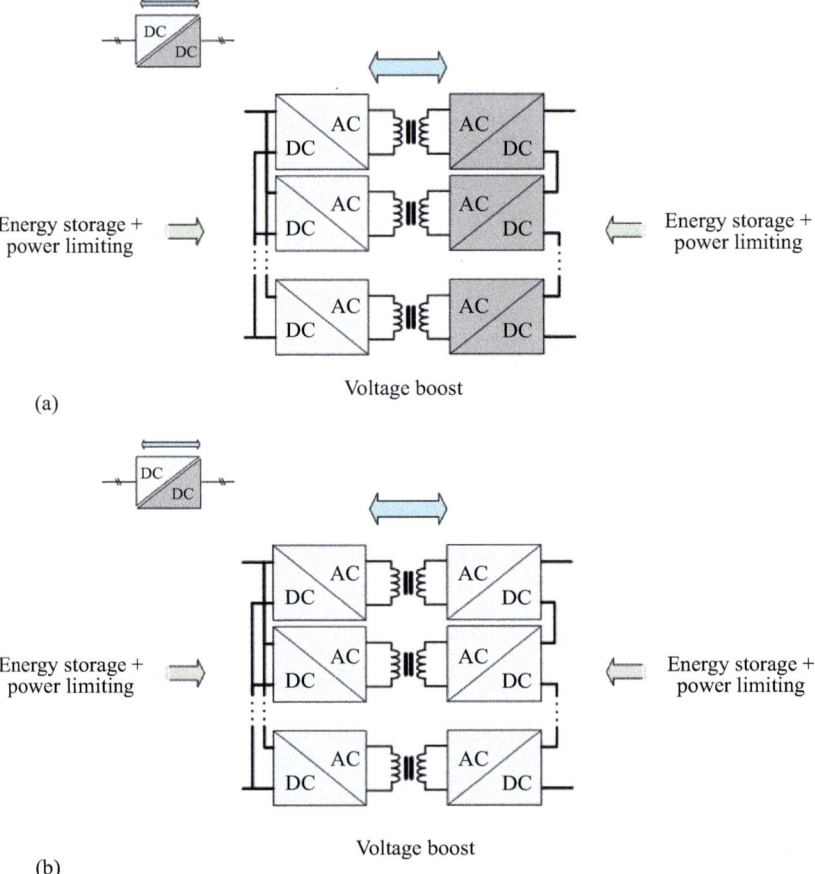

Figure 2.8   *IPOS/ISOP configurations of isolated DC–DC PECs: (a) IPOS with voltage step-up isolated DC–DC PECs and (b) IPOS with LV rated isolated DC–DC PECs*

either the IPOS or the ISOP can be either configured as unidirectional (using unidirectional building block topologies) or they can be controlled as unidirectional by transitioning from DAB to SAB mode. The IPOS or ISOP isolated DC–DC PEC may also be constructed from building blocks having LV ratings on the primary and secondary sides. This approach is shown in Figure 2.8a.

2.3   Emerging MVDC applications

In MVDC system applications, the voltage and power ranges start with solar PV systems at the low end, presently in the 1–1.5 kV voltage range with output power capacity up to 300 kilowatt (kW). MVDC transmission represents the high end operating in the range of 15–50 kV, with 30–150 megawatt (MW) capacity [29]. In MVAC systems the low-frequency transformer isolates the PEC's switching power semiconductors from medium voltage levels. However, the common requirement for PECs applied to MVDC systems is that the power semiconductors must directly connect to medium voltage levels, insulated from the direct contact across a variant MVDC bus by a large capacitor bank and/or inductive/capacitive filtering. The emergence of Wide Bandgap (WBG) power semiconductors, along with advances and experienced gained with multilevel PEC multilevel topologies, is increasing of a range of feasible PEC solutions. It is these technological advances that are paving the way for both feasible MVDC-interfacing PECs that can be applied over the entire range of applications.

Increasingly mature WBG power semiconductor device solutions, such as 6.5 kV and 10 kV Silicon Carbide (SiC) Metal Oxide Semiconductor Field Effect Transistor (MOSFET) based multi-chip power modules, promise higher efficiency, higher power density and, eventually, lower cost solutions to PECs for MVDC systems [30]. Increases in the achievable switching frequency above 10kHz, afforded by WBG power semiconductor technology, combined with increased DC voltage blocking capability of SiC MOSFETs, will lead to smaller and lower cost solutions. SiC MOSFET-based PECs will become increasingly viable as the current ratings of these modules reach the levels that are on a par with the commercially available modules utilizing Silicon IGBT and Insulated Gate Commutated Thyristor (IGCT) devices.

2.3.1   MVDC system value proposition versus technical readiness

For a given application, the MVDC system value proposition, which is outlined in Table 2.1, should be considered from the viewpoints of the stakeholders in future electrical infrastructural changes and ground-up installations and building projects, who *fund, promote and sustain* emerging applications, the end-use system developer and integrator, who must *procure* PEC (and other) equipment to meet the needs of the end-use application, and the original equipment manufacturer (OEM) of the PEC equipment, and other power electronics based equipment such as SSCB, who, in order to *supply* the needed equipment, must make business decisions based upon market opportunities, perceived risks and returns on investment.

There is justification and ample proof of feasibility for MVDC-based electrical systems applied to a range applications in the existing body of literature. The present-day challenge comes down to the viability of the MVDC-based system and justifying the internal research and development investments that the OEMs of PEC equipment must make. Even if the MVDC-based system is an undeniable "no-brainer," there are a variety of decisions that the PEC OEM must make regarding the PEC circuit topologies. In the world of power electronics-based equipment OEMs, product differentiation that distinguishes one manufacturer from another is the PEC topology selection. The MVDC system architecture, and the volt-ampere requirements of the PEC within the architecture will not only limit the range of feasible PEC topological choices but in some cases, as is illustrated in the example of the PEC- versus Breaker-based protective architectures for the MVDC power and energy delivery system of Figure 2.3a, informs the architectural implementation.

Considering Table 2.1, at first glace, *Attributes 1–3* yield quantifiable metrics in favor of the MVDC system, such as ease of installation, economies of scale and energy savings, etc.—*if and only if the associated PEC equipment is either commercially available or can be procured without the threat of obsolescence.* However, even if these justifications comprise the *only* technological push towards MVDC, it may still be debatable whether or not they will have only incremental impact, and thereby have limited life in the marketplace. The possibility of a "one off" is only one of the dilemmas that PEC OEMs face in order to ascertain whether the investments in applied research and development to emerging markets will justify the costs. There are other risks associated with the integration of the PEC into the end-use environment when there is little past experience to draw upon. For MVDC-based systems there are no governing specifications, either mandated by governmental agencies or volunteered by the efforts of industrial-academic-governmental coalitions, to ensure inter- and intra-compatibilities once the constituent parts of the MVDC system are installed into the end-use environment. The example applications associated with *Attributes 1–3* in Table 2.1 are at a relatively high TRL of 7, meaning that what is needed are qualification testing and demonstration activities of system prototypes in final installation environments, using upon Commercial Off the Shelf (COTS) PEC equipment, to move up to the next TRL. As this is accomplished with an increasing number of field installations, the commercial viability of MVDC based systems, compared to present-day MVAC based systems already at TRL 9, such as utility-scale wind and solar farms, offshore wind farms and shipboard electric propulsion, viability can be proven—and promising new markets for MVDC systems will emerge. The most obvious and significant challenges to the development of suitable PEC equipment design for these applications, *for which, also, guiding standards for full qualification testing and experiential-based practices are needed to move from TRL 7 to TRL 9,* are:

- Creepage and Clearance requirements for the internal electrical connections within the PEC and its sub-assemblies.
- Electromagnetic Compatibility (EMC) requirements applied to points of shared electrical connection and shared physical space.

- Guidance on equipment self-compatibility with its internal conducted and radiated Electromagnetic Interference (EMI) emissions and coupling paths.
- Mitigation of partial discharge failures in the presence of simultaneously high electric-field and high voltage rates of change (*dv/dt*).
- Reliability metrics and design practices for intra-equipment insulation systems.
- Systematic intra-equipment failure mode management.
- Intra-system LL and LG short circuit fault response.

Attributes 4–8 of Table 2.1 imply system-level performance capabilities and associated system inter-compatibilities. These attributes point to advantages of MVDC systems that are transformational and disruptive to the way power and energy is currently distributed and utilized. It will not be difficult for the OEM to see the need to invest when payment for non-recurring engineering associated with new equipment development and system integration is assured. However, as indicated by the low TRLs associated with these attributes, stakeholders, system developers, system integrators and PEC OEMs are exposed to significant risks.

Naval shipboard electrification presents a striking example of the TRL challenge of MVDC systems. The end-goal for the naval shipboard system really has nothing to do with the generation and distribution of power, but with justification of new mission capabilities and/or new ship construction. Nevertheless, the promise of new warfighting capabilities, which is what is important to the stakeholders (the Department of Defense and congress), is inextricably connected to the under-girding electrical power distribution system and generation. The level of existing shipboard qualified "program of record" power distribution and generation equipment upon which the electrical system can be built corresponds to the appetite the widely diverse stakeholders will have for funding new shipbuilding endeavors. A comparison between Figure 2.2a, which represents a feasible starting point for the building of a new shipboard electrical system using program of record equipment (or militarized COTS equipment), and Figure 2.3a, which represents the needed MVDC-based IPES, shows little leveraging opportunity. What drives the TRLs down for the achievement of *Attributes 4, 5, 7, and 8* needed for these systems are the MVDC voltage levels needed to achieve a system that fits within the limitations of space and hull displacement (allocated for the electrical system) *and* high dependency on low TRL technology, such SSCBs and NLSw's, having, simultaneously, sufficient power and current throughput capacity and power density. At the same time, as the electric ship research design community progresses in its understanding of power conversion and distribution requirements, there is a simultaneous need to assess the space claim *and* power density of PEC, SSCB, and NLSw topologies and ratings, and their impacts on the final electrical architecture. Once final electrical equipment requirements are determined, equipment must be procured from a limited pool of OEMs capable of building shipboard qualified equipment. OEMs must, in turn, integrate their equipment into an end-use application for which the system integrator, the shipbuilder, has a limited experience base to draw upon!

In order to achieve the benefits of *Attributes 4–8*, the requirements of the individual PECs are inextricably connected to how the MVDC-based system acts as whole. For these emerging applications (represented by the example applications in Table 2.1), if the PEC OEM waits until testing and integration to address system level interactions and inter-compatibilities, runaway non-recurring engineering costs will likely occur. The associated risks are significant compared to the TRL 7–8 level target applications of *Attributes 4–8*, for which options for which the OEMs of TRL 8–9 PEC solutions may be found or easily developed. Major challenges associated with development of highly PEC-interactive MVDC systems *for which, also, guiding standards for full qualification testing and experiential-based practices are needed move from TRL 3 to TRL 7*, are provided as follows:

- Creepage and clearance requirements for external connections to the PEC.
- System-level EMI/EMC requirements and emission limits.
- Conducted emissions versus power quality requirements at points of common MVAC interface.
- Limits of insulation approaches on power and energy through-put and power density of PEBBs and other power electronic sub-assembly implementations.
- Short circuit fault response on the MVDC side.
- Systematic intra- and inter-equipment failure mode susceptibility, vulnerability and recoverability.
- Inter-system LL, LG, LLG, and MLG short circuit fault response at MVDC and MVAC points of interface.

The following sections describe unique characteristics of emerging MVDC applications in order to shed light on PEC design challenges as they relate to these systems and the systematic considerations that will influence the selection of power circuit topology for the PEC, the focus of de-risking activities during the product development stage and the enabling technologies in which applied research and development, *by the PEC OEMs*, will pay huge dividends.

2.3.2 Utility interfacing applications

MVDC power converters have been utilized in both centralized power generations and distributed generations, including conventional non-renewable generations (internal combustion engines, gas turbines, fuel cell systems) and the emerging renewable generation units (solar power, wind power, ocean wave energy, hydro systems). Considering that MVDC power converters used for conventional (non-renewable) power generations have been discussed in other textbooks, we will mainly focus on the MVDC power converters used in the emerging utility interface applications such as solar, wind, and energy storage systems.

2.3.2.1 MVDC utility solar architecture

The conventional MVAC-based utility-scale solar power architecture, in common use today, is shown in (Figure 2.9)(a). This system consists of multiple strings of PV modules combined into arrays in combiner boxes where all of the solar energy is aggregated in parallel, is fuse protected, and is then sent through the feeder lines to the power conversion stations. The power conversion stations mainly consist of

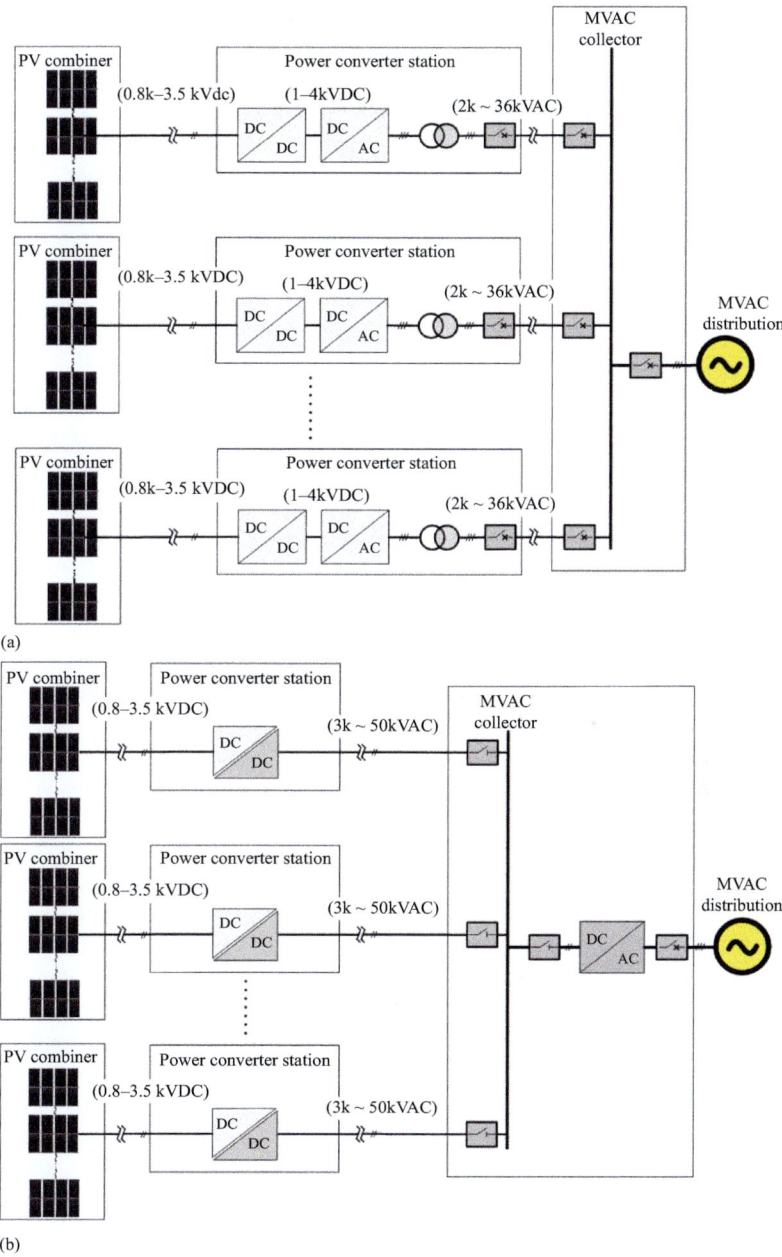

(a)

(b)

Figure 2.9   Solar power architectures: (a) MVAC with transformers and (b) MVDC with MV-PETs

solar DC–AC inverters and transformers that step up the LV system voltage to MVAC feeds of up to 35 kV, which are eventually connected to the substation and power grid through the PV combining MVAC collector and associated switchgear. These systems are built upon at 800V, 1000 V, and 1500 V DC-rated solar inverter PEC systems. Typical power ratings for the power conversion stations range from 100 kW to 10 MW. Recently, 3.5 kV DC-rated solar systems, enabled by using double glass bifacial PV modules, have emerged as a very suitable solution for higher string voltage [31], indicating a clear trend towards higher levels of MVDC.

To reduce the number of components, the related upfront cost and installation effort in the solar field, and increase efficiency, a MVDC solar power architecture, as shown in (Figure 2.9)(b), have been proposed [32], but the introduction of PEC solutions for these systems into the market place is slowed by the falling prices of conventional solar inverters [33]. MVDC utility-scale solar PV replaces the combiner boxes containing field solar inverters, transformers, and switchgears with isolated DC–DC converters capable of stepping up 1–3.5 kV string voltage to MVDC voltage levels of up to levels as high as 50 kV. Such systems will significantly reduce power cables costs and ohmic losses. It is significant to note that the value proposition of utility-scale MVDC solar PV, as well as for other applications, will be more readily realized once system-level considerations are taken into account. Such system-level considerations include the fact that more PV modules can be installed on the land area that is reserved for cables, inverters, transformers, and switchgears. The cost and footprint of auxiliary systems used for thermal management of the transformers and switchgear may also be reduced.

The need for transformer isolation in large power systems must be addressed in MVDC systems. As a result, there is great need for commercially viable MV rated, MW-scale isolated DC–DC PECs. The challenge with utility scale solar power architectures is that the PV array strings themselves are either low voltage DC (LVDC) rated or they are rated at the low end of the MVDC range. In order to ensure system-level reliability it will be necessary to transform isolated LVDC or low-end MVDC ground potentials of PV arrays from the MVDC collector, especially as MVDC levels approach the high end of 50 kV. One feasible approach to achieving a TRL 7 MVDC utility-scale solar power architecture is to integrate together the PV combiner and power conversion stage into a system at the PV array string location. With this approach lower voltage rated isolated DC–DC converters are connected in series to formulate a MVDC output feed to the MVDC collector. This approach has been proposed in [34] and is shown in (Figure 2.10). This approach is readily scalable to higher powers simply through the addition of modular integrated PV combiner and power conversion stations.

2.3.2.2   MVDC wind turbine architecture

Compared to MVDC for solar PV, MVDC power architectures for wind turbines have received considerably more attention, particularly for offshore wind power plants. Driven by the superior wind environmental conditions at deeper water, the distance of offshore wind power plants to onshore grid substations keeps increasing, i.e., beyond 60 km. Under such scenarios, the MVAC transmission is disadvantageous due to high cable cost,

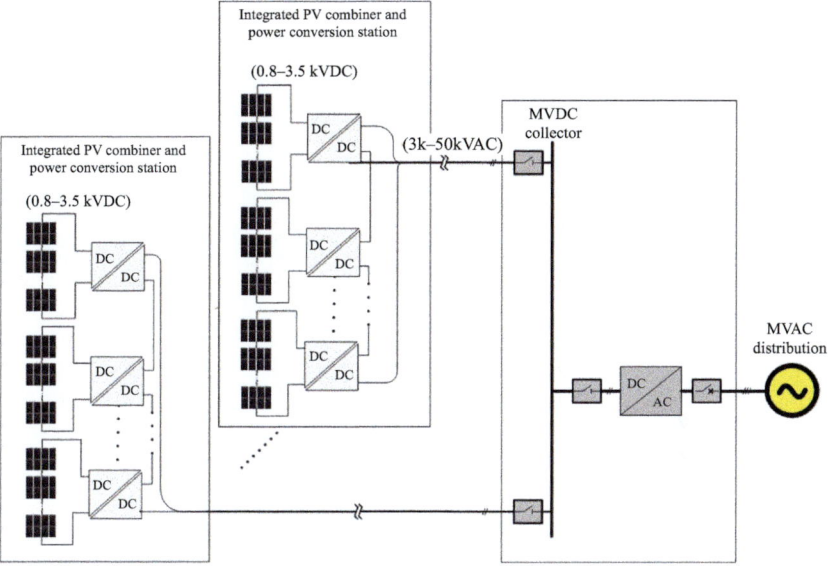

Figure 2.10   MVDC solar power architecture with series connected medium frequency isolated PV array strings

high skin effect and proximity effect losses, capacitive cable charging, and consequent needs for reactive power compensation, when compared to MVDC power transmission. The development and commercialization of multilevel VSCs, particularly the Modular Multilevel Converter (MMC) PEC topology, for HVDC systems has paved the way for viable MVDC connection long-distance offshore power integration. If the collection system of the offshore wind power plants can be configured with MVDC, a number of benefits can be achieved, including high efficiency, zero reactive power compensation, elimination of bulky low frequency, or line-frequency, (50 Hz / 60 Hz) transformers and increased power density of PECs for wind turbine energy conversion. Case studies elaborated in [35] show that adopting the MVDC offshore collection systems can yield up to 15%–20% cost saving in the total electric infrastructure, due to the elimination of the offshore substations and utilization of DC cables.

Various MVDC DC offshore wind turbine power plant configurations are shown in (Figure 2.11). For both offshore and terrestrial wind farms, the PECs are the critical apparatus to interface the wind generators and the power grid. The output voltage of the wind generators typically ranges between 690 V and 15 kV. Generator voltage output is limited by the insulation system. The power string rating from offshore turbine to shore is dictated by the turbine power rating. Currently wind turbine ratings range from 3 to 12 MW, so offshore wind and medium-voltage power converters presently represent the highest power throughput PECs applied to RES applications. Also, particularly for the larger turbine installations, both generator and PEC are designed to the same insulation level. Consistent with all of the configurations of Figure 2.11, except for Figure 2.11a, the low frequency transformer voltage step-up occurs after the PEC, or generator, switchgear, PEC, and cabling are at the same voltage rating.

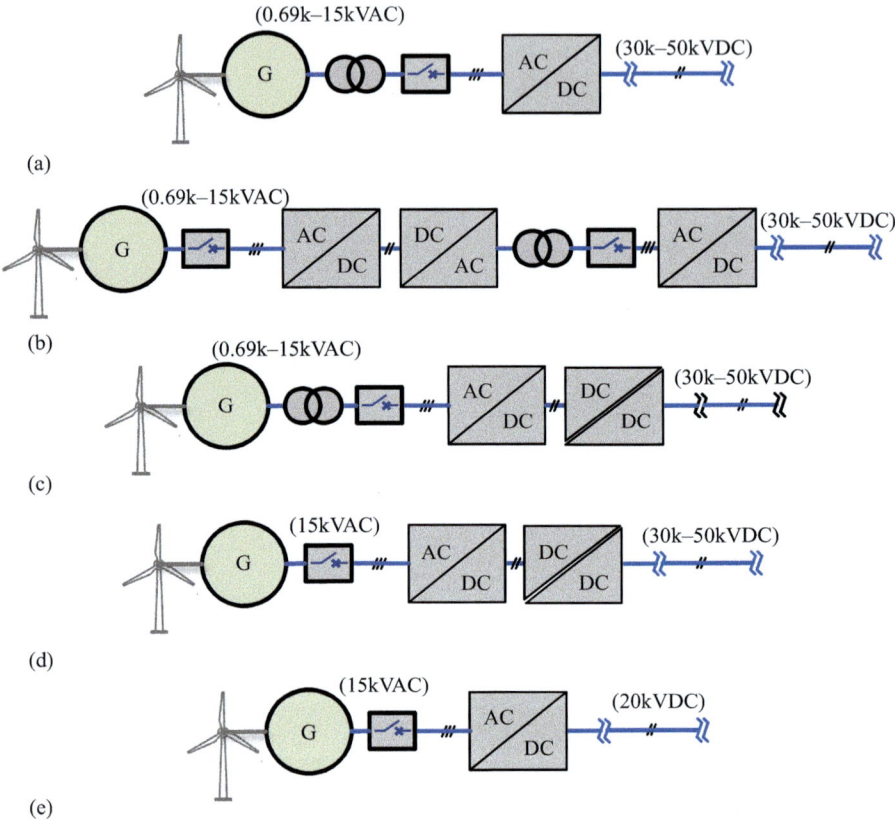

Figure 2.11   Various configurations of DC offshore wind substations:
(a) transformer-based 2-stage MVDC power converter system,
(b) transformer-based 4-stage MVDC power converter system,
(c) MVDC system based on isolated DC–DC converters, (d) MVDC
system based on non-isolated DC–DC converters, and (e) single-
stage MVDC power converter system

Figure 2.11a represents one of the most promising implementations of offshore MVDC, or even HVDC, transmission links to an onshore substation. This configuration is enabled by the MMC PEC topology, which has been deployed successfully at voltage levels as high as 500 kV in both offshore wind and grid connection installations [36]. A lower risk, transitional implementation at similar power levels but with more conventional PECs, such as offshore back-to-back LLC AC–DC and DC–AC PECs and an additional step-up transformer to DC–AC PEC in a four-stage power conversion architecture, is shown in Figure 2.11b. However, hardware complexity and low efficiency would be a major concern for this configuration. The configuration of Figure 2.11c could provide a simultaneously high power density and efficiency once high TRL MW-scale isolated DC-DC converters become commercially available. The configuration of Figure 2.11d is a non-isolated approach, similar to Figure 2.11a which could achieve a higher voltage boost for longer offshore distance applications. Figure 2.11e is the simplest power architecture, consisting only of switchgear and AC-DC converters. However, this configuration represents the low end of achievable power throughput because the MVDC bus voltage will be limited by achievable generator voltage. The most likely installation to which Figure 2.11e would apply are small-scale terrestrial wind farms. Comparatively speaking, Figure 2.11c strikes a great balance in terms of output voltage flexibility and galvanic isolation, and is presently the most common approach to MVDC that has been adopted by the wind power industry [35]. Based on such wind turbine sub-systems, large wind farms by utilizing either MVDC or MVAC collector platforms can be developed, as shown in (Figure 2.12)(a) and (b), respectively.

Certain challenges remain with MVDC offshore wind collection systems, particularly when it comes to DC protection. However, there is clear evidence from plans to deploy multi-terminal HVDC power distribution networks, such as the North Sea power hub and spoke network [37–39], that significant attention is being paid to the issues associated with, at least, HVDC grid resiliency. Commercially viable solutions are emerging [18] and TRL of these systems are expected to progress from TRL 7 to TRL 9 within the next decade. However, when it comes to MVDC, however, significant issues remain unresolved and even feasible solutions

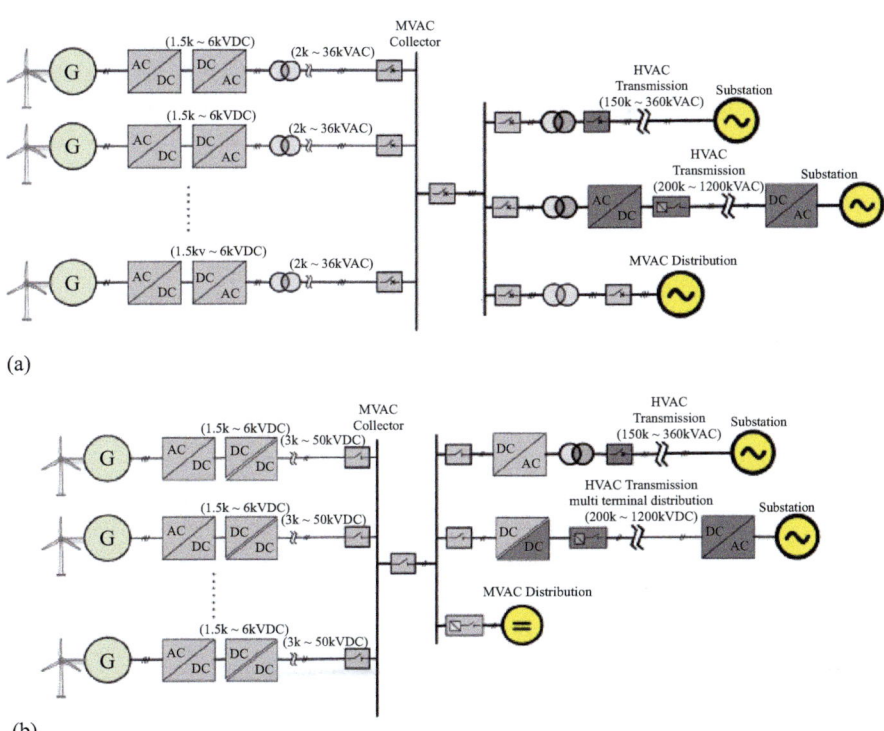

(a)

(b)

Figure 2.12   Power architectures of wind farms (a) with MVAC collector platform
and (b) with MVDC collector platform

to MVDC multi-MW system protection, which require orders of magnitude high currents compared to HVDC systems, are yet to emerge.

2.3.2.3   Utility-scale MVDC energy storage systems

Utility-scale (also called, grid-scale or large-scale) energy storage has been experiencing rapid growth in recent years, which is either paired with renewable energy sources (solar, wind, etc.) to smooth the power generations or installed in the transmission network and distribution substations to balance local supply and demand. The typical power capacity of energy storage systems ranges from hundreds of kilowatts to hundreds of megawatts. According to a recent report [40], the global energy storage installations are set to exponentially increase from 9 GW / 17 GWh deployed as of 2018–1,095 GW/2,850 GWh through 2040, as shown in (Figure 2.13).

Taking the emerging battery energy storage system (BESS) as an example, the power architectures for utility-scale BESS can be classified as transformer-based and transformerless based architectures. (Figure 2.14)(a) shows the transformer-based BESS power architecture, where a medium-voltage power inverter converts the battery DC power into AC power. This inverter can be a two-level inverter based on high-voltage semiconductor switches, or a multilevel inverter based on low-voltage semiconductor switches. Subsequently, the LVAC output is stepped up by a transformer to distribution MVAC voltage levels, such as 4 kV ~35 kV in the United States. A good example using such a power architecture is the 11 kV/200 kWh Lithium-ion battery storage system located in Norfolk, England, installed by UK Power Networks in collaboration with ABB and Durham University. The power inverter used in this project is a 3-level neutral-point-clamped inverter rated at 850 kVA [41]. This is reported as the first time an electrical energy storage system installed on an 11 kV distribution network in the UK in 2011 [41]. (Figure 2.14)(b) depicts the transformerless BESS architecture, where the battery output voltage is first boosted to medium-voltage levels through a number of isolated DC–DC boost converters connected in series, and then converted into AC voltages at distribution levels

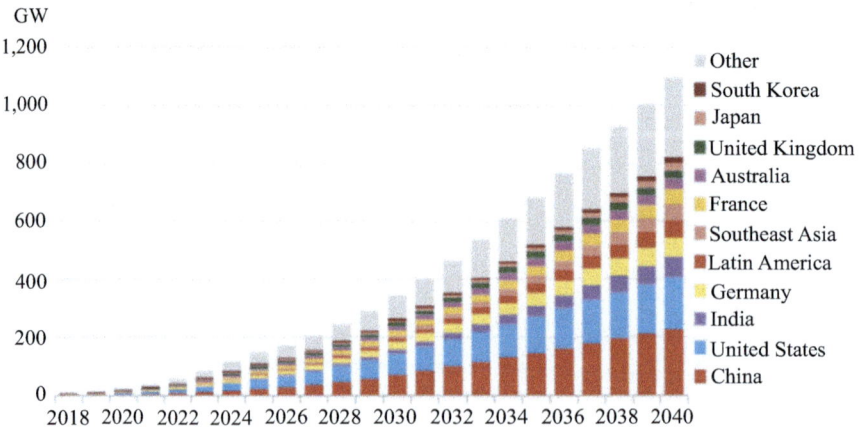

Figure 2.13   Global accumulative energy storage installations [40]

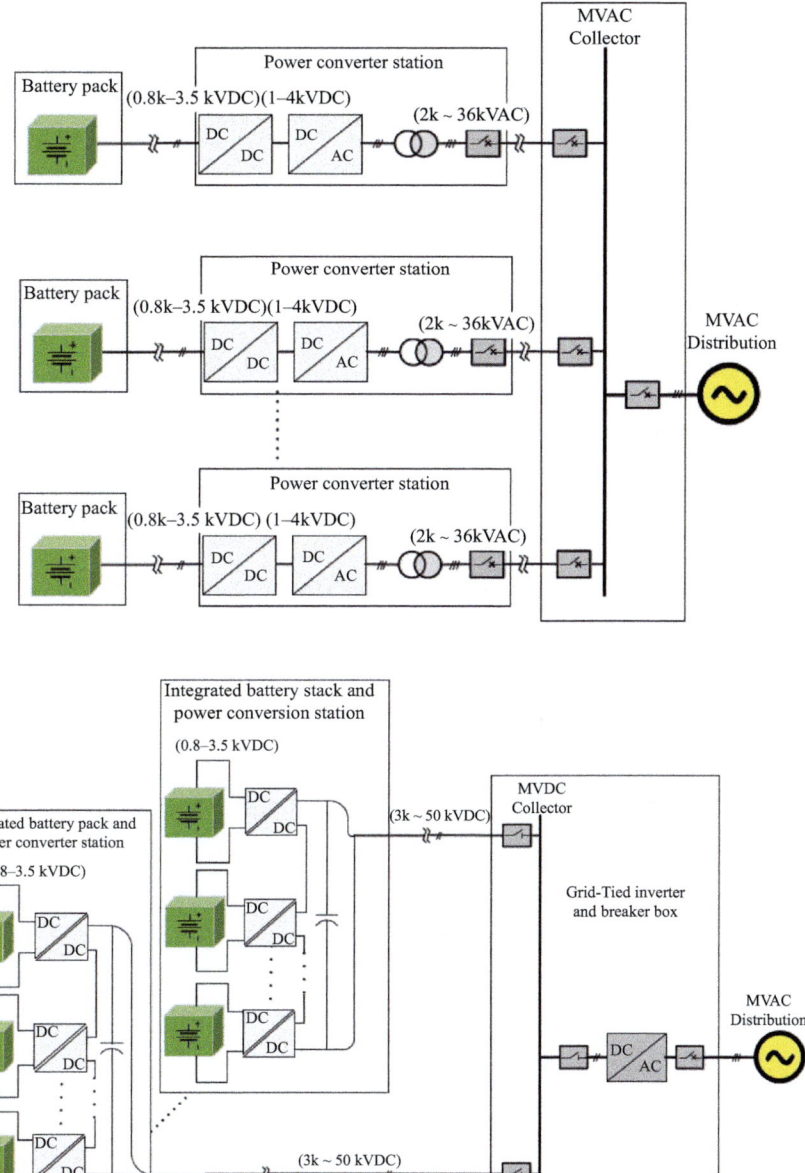

Figure 2.14   Utility-scale BESS power architecture: (a) MVAC with transformers and (b) MVDC without transformers

with a medium-voltage inverter. By comparison, one can find that the technology with (Figure 2.14)(a) has lower technical risks, although the grid-frequency transformers could make the BESS bulky and lossy. On the other hand, (Figure 2.14)(b) would be more compact and efficient, but the development of high-voltage power converters have certain risks, such as the dynamic voltage sharing among the series connection of large number of semiconductor switches.

For BESS paired with solar energy resources, there are mainly two types of power architectures in terms of power coupling connections, namely, AC coupled and DC coupled, as shown in (Figure 2.15)(a) and (b), respectively. The AC coupled configuration is the most common one utilized in existing industrial utility-scale PV systems, in which a unidirectional inverter and bi-directional inverter are connected to the PV and the battery, respectively, and the output of these inverters are connected at the AC bus of the systems after the voltage is boosted up by the transformers. One advantage with this architecture is the flexibility of retrofitting a BESS as much additional capacity as required to an existing PV system, without the need of re-wiring the DC side of the PV inverter for storage integration. However, the overall power conversion efficiency with this configuration might be lower compared to the DC-coupled architecture. This is due to the fact that the charging process with the AC coupled architecture involves two inverters and two transformers while there is only one DC/DC converter for the charging with the DC coupled architecture. Another disadvantage with the AC coupled architecture is the cost increase with the installation of the switchgear and MV transformers, which

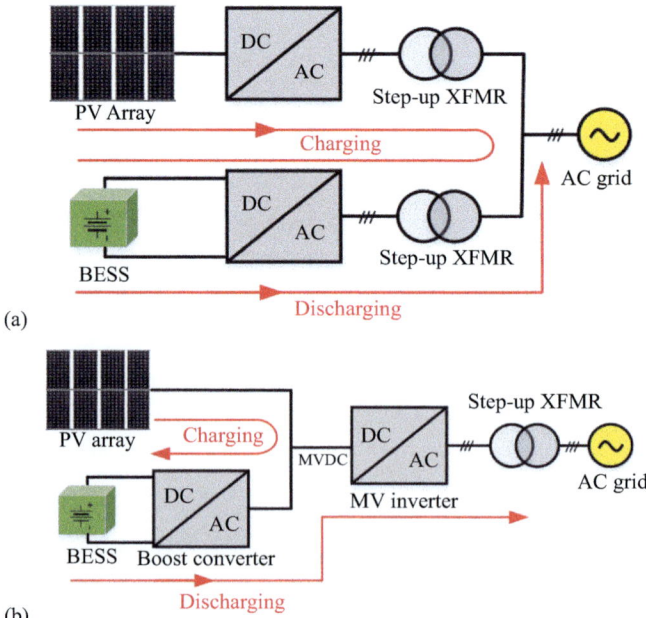

(a)

(b)

Figure 2.15   Utility-scale energy storage coupling power architectures: (a) AC coupled and (b) DC coupled

can be eliminated in DC-coupled architecture. Furthermore, for the DC coupled architecture, if the output of the PV system is curtailed due to grid restrictions or an outage due to a fault or maintenance, the curtailed energy can be recaptured by the BESS via the DC/DC converter. Nevertheless, one main drawback with the DC coupled architecture is that the capacity of the BESS is restricted by the size of the PV inverter if the BESS is to be retrofitted to an existing PV system.

To sum up, both of these power architectures have their own advantages and disadvantages, and they need to be chosen carefully depending on the comprehensive consideration of multiple factors, including cost, efficiency, utility requirements and constraints, field conditions, grid service requirements, ease of installation, communication and interoperability features, scalability, reliability, and the like.

2.3.2.4 MVDC transmission and distribution

So far, MVDC systems have been addressed from RES, BES and DER perspectives. However, there is growing interest and potential for MVDC systems as they relate to MVDC transmission of power and energy between points of generation and usage, and MVDC distribution from the load perspective. These applications are touched on briefly here.

(Figure 2.16) shows two emerging approaches to MVDC transmission. As is the case with HVDC transmission, compared to HVAC, power is transmitted over very long distances more efficiently with less loss of voltage due cable impedance and lower power dissipation. The same principle applies with MVDC versus MVAC transmission but applied more to lesser range of distance between points of generation and distribution usage. Perhaps the most viable approach to MVDC that is on the horizon and can more easily be justified in regions where conventional MVAC transmission and distribution is well established is the MVDC corridor, shown in Figure 2.16a [42–47]. An MVDC corridor between points of large or utility scale renewable generation, such as wind or solar farms, or mixed RES and BES installations enhances the transfer capacity and can be used to to increase overall distribution network power quality at points of usage. The MVDC corridor can be especially attractive for municipal grids where bulk levels of RES are installed, say, in a remote area, and there is a desire to use that energy several kilometers away to improve the power quality of a distribution grid or provide an opportunity for community usage of renewable energy. Usage of a conventional long MVAC inter-connecting feeder for the simple purpose of transmitting power and energy from a point of generation to a point of usage faces many challenges, including the need to provide phase synchronization across transmission in-feeds, interconnection between weak or unstable grids, loss of voltage due to reactive line drop and need for complicated fault protective relaying to address variation in source number, impedance and directionality caused by mixing the power electronic interface with the conventional grid. The MVDC corridor, on the other hand, provides a well-managed point of interface and transmission that, once the initial installation investment is made, avoids the need for further investment of network reinforcement at points of usage. The MVDC corridor will typically have four-quadrant PECs at each end of the link to support MVDC voltage regulation and provide reactive power support at points of interface to the MVAC grid.

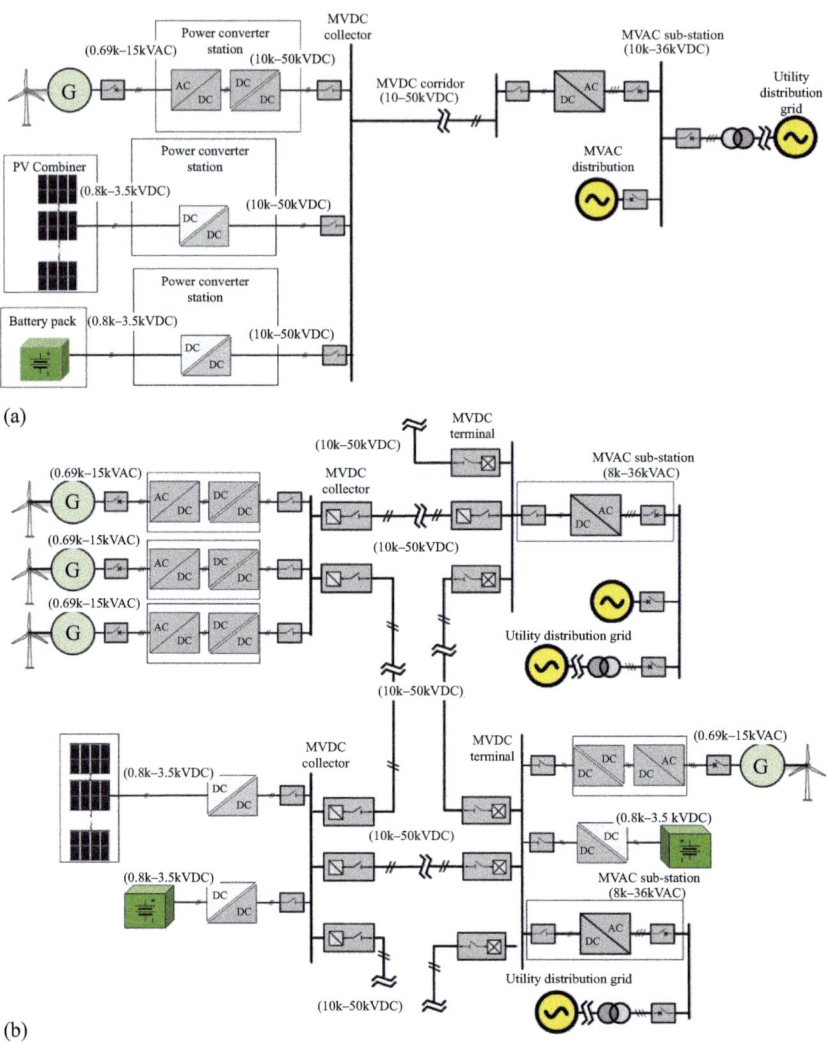

*Figure 2.16   Approaches to MVDC transmission: (a) MVDC corridor and
(b) MVDC multi-terminal transmission*

Another variant of the MVDC corridor utilizes MVDC at a substation level to provide increased power transfer capacity in order to avoid more disruptive up-rating of cables and transformers for substation reinforcement. This is achieved through what is called a "Soft Open Point" configuration where two PECs are connected back-to-back with a common MVDC bus interface [48,49]. Such an approach can improve power balancing at a heavy loaded substation in order to provide reliability and maintain short circuit levels of the existing distribution feeders in order to avoid expensive relaying upgrades or wide scale black-outs resulting from poorly planned interconnection between substations [50].

Multi-terminal HVDC (MT-HVDC) and MVDC (MT-MVDC) transmission is emerging as an increasingly viable option for providing an energy distribution network in areas where large scale changes to the energy infrastructure can be made to increase the level of RES penetration into the grid [9,37,38,51–53]. A MT-MVDC meshed transmission network is shown in Figure 2.16b. Here, the energy provided by large-scale installations of wind and solar transmitted to various points of distribution over a very wide geographical region (MT-HVDC) or between or within municipalities (MT-MVDC). The MT-MVDC network can also be an outgrowth of, expansion on, MVDC corridor.

MVDC corridor and MT-MVDC also apply to sub-sea networks [54,55]. There are a number of MVDC corridor and MT-MVDC transmission network programs in the planning stages or currently underway, including connection between the 33 kV MVAC distribution grids on the Isle of Anglesley and North Wales using a ± 27 kV MVDC corridor [46], the design of a 10 kV three-terminal DC distribution system in Zhuhai, a city in the Guangdong province of China [9,53], a four-terminal ± 15 kV DC offshore power transmission project in the Hainan Province of China [43].

Local MVDC distribution combining RES, BES and DER together at the point of usage can be accomplished through the MVDC microgrid, simplistically represented by the block diagram in Figure 2.17a [56,57]. This system could apply to, say, a small building or factory where it is practical to combine MVDC interfacing sources and loads into a common MVDC bus, with MVDC bus disconnect capability occurring at points of source or load. In practical systems, especially as the geography of the distribution grid is more spread out in, say, a large building, group of buildings (campus), military base or industrial park, a MVDC distribution network that includes multiple distribution network points and associated fault protection, will be necessary. Such a system could take the form of the ring-bus network of MVDC nanogrids or MVDC ring bus microgrid as shown in Figure 2.17b. The system of Figure 2.17b is a promising alternative to MVAC radial microgrids commonly applied to community microgrids, military bases and university campuses [3,56,58]. This meshed structure brings clusters of combined local nanogrids with co-located RES with battery or engine-driven generator energy buffers, or with combined RES and storage, into a lateral system where energy flow can be well-controlled between nodes. A demonstration project of a MVDC microgrid similar to the system of Figure 2.17b is currently under development at RWTH-Aachen University [3,58].

2.3.3 Aircraft electrification

More-electric aircraft have been proven to be a very promising approach to significantly reduce gas emissions, acoustic noise, operation cost and improve system efficiency, compared to the conventional engine based aircraft propulsion systems. For instance, the Boeing wide-body twin-engine 787 Dreamliner has a total onboard electric power of 1.45 MW, exhibiting 20% reduction of fuel and CO_2, and 60% smaller noise footprint, in comparison to Boeing 767 counterpart [59]. From

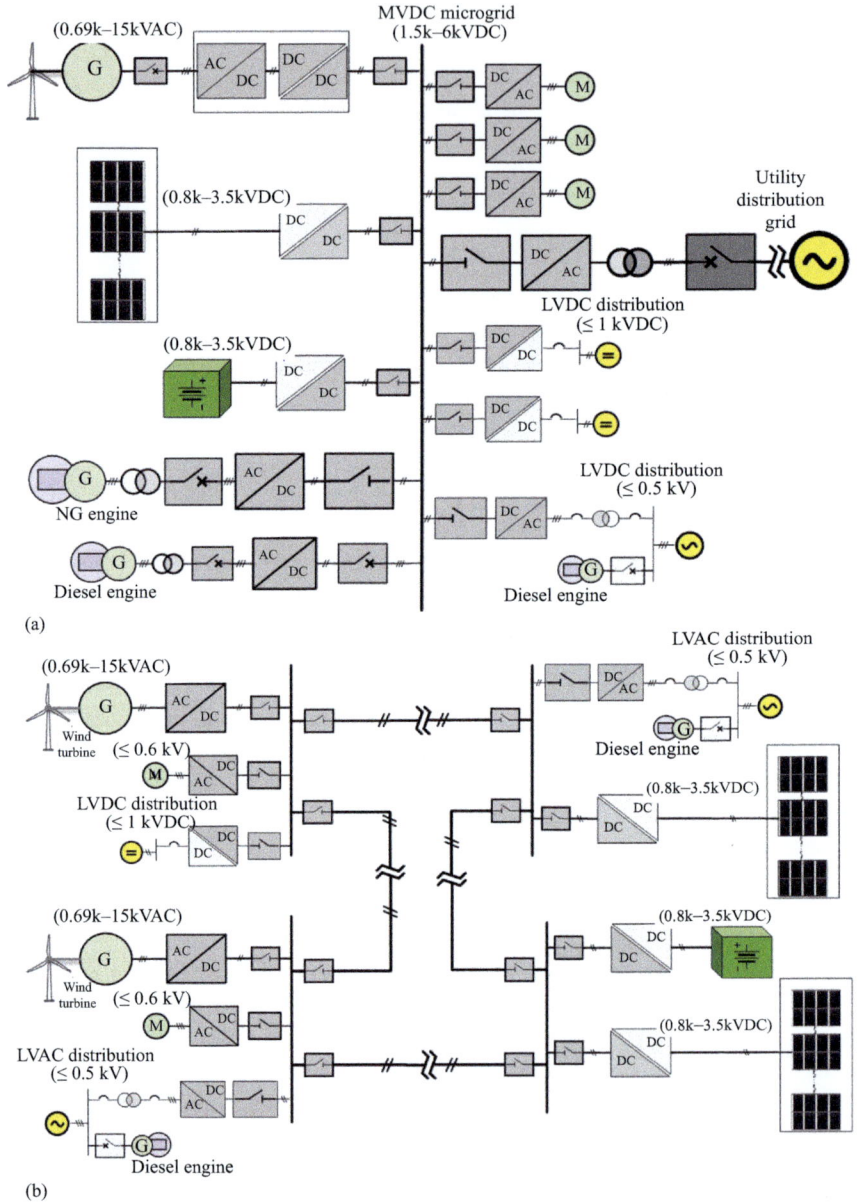

(a)

(b)

Figure 2.17   MVDC Microgrids: (a) building MVDC microgrid and (b) MVDC ring bus campus or military-based microgrid

the level of electrification, there are four types of aircraft propulsion systems, namely, turbo-electric, series hybrid, parallel hybrid mode, and all electric, as illustrated in (Figure 2.18). The typical ratings of the propulsion power ranges from 100 kW for air taxis to tens of megawatts for large wide-body airplanes.

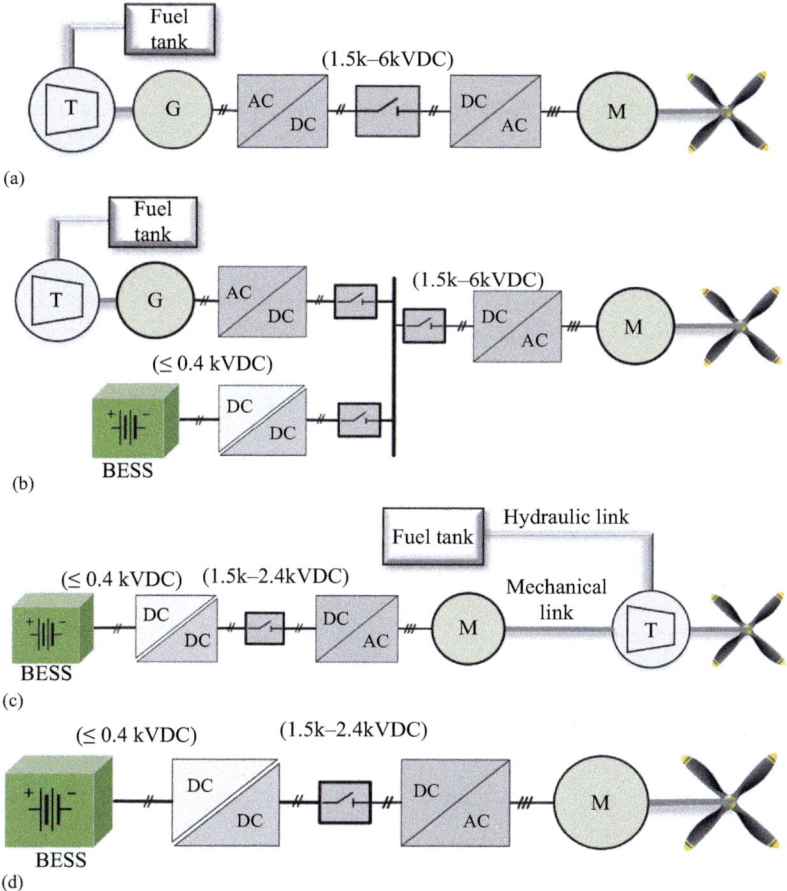

Figure 2.18 *Various power architectures of aircraft propulsion systems:*
(a) turbo-electric propulsion system, (b) hybrid-series propulsion,
(c) hybrid-parallel propulsion, and (d) all-electric propulsion

Constrained by the limited energy density and reliability concerns with the present battery technologies, hybrid-electric propulsion (either series or parallel) seems to be more feasible for regional large-capacity air transportation, where the on-board electric power is typically in the range of megawatt scales. To effectively enhance the power density and efficiency of aircraft MW-scale electrical systems, MVDC transmission and distribution has been regarded as an enabling technology that can dramatically improve the system power density and efficiency. Specifically, MVDC has the following advantages compared to the existing low-voltage DC (LVDC) or low-voltage AC (LVAC) for aircraft electrical systems:

- **Lower cable weight and cost.** For DC transmission systems containing only the positive conductor and the ground, a single conductor can be used, while three-phase conductors are required in the AC transmission systems. The adoption of MVDC transmission may lead to a significant reduction of cable weight and cost, especially for wide-body large-capacity airplanes.

- **Elimination of skin effect and proximity effect losses.** In traditional aircraft AC transmission systems, typically high frequency (e.g., 400~800 Hz) is used to reduce the size of magnetic components such as electric machines or transformers. Consequently, skin effect and proximity effect losses are significant, but such losses can be eliminated in DC transmission in aircraft electrical systems.
- **Lower corona effect.** Corona effects tend to be less significant on MVDC conductors than for MVAC counterparts. In a MVAC system, due to the higher voltage and faster switching with the modern semiconductor devices, the increased dv/dt poses more challenges on the insulation design of the electric apparatus, such as stator windings of electric machines, laminated high-voltage bus bar designs of power converters, and the like.

In the present more-electric aircraft systems, such as Boeing 787 and Airbus 380, 270 V is the dominant DC-bus high voltage rectified from engine-driven generators. If such DC-bus voltage can be boosted to 1000 V, 2000 V, or above, the systematic power density and efficiency will be dramatically increased. Aviation manufacturers have already started developing MVDC aircraft powertrain systems. For instance, Airbus and Rolls Royce have jointly launched the E-Fan X flight project for commercial aircraft with serial hybrid-electric powertrain, where the MVDC distribution is based on 3 kV [60]. General Electric Corporation and NASA have been jointly developing hybrid-electric aircraft propulsion based on 2.4 kV SiC power inverter [61]. An aircraft MVDC distribution system based on hybrid-electric propulsion is shown in (Figure 2.19)(a). Specifically, multiple electric power paths either coming from the turbo engines or batteries are aggregated, resulting a MVDC common bus typically ranging at 1.5–3 kV. Various DC loads (28 V, 270 V, etc.), AC loads (115 V, 230 V, and so on), and the propulsors are powered through different power converters connected to the MVDC common bus. Likewise, a turbo-electric propulsion based aircraft MVDC distribution system is shown in (Figure 2.19)(b). As can be seen, in both of these two power architectures, bulky transformers are eliminated, and the cable weight and systematic ohmic losses will be dramatically reduced accordingly. Particularly, the recently commercial availability of high-efficiency high-voltage SiC MOSFETs (e.g., 1.7 kV, 3.3 kV, and so on) further promote the development of MVDC systems for hybrid-electric or all-electric aircraft [61].

Nevertheless, there are a few key challenges to be addressed during the development of MVDC systems for electric aircraft, which are summarized as follows:

- **Increased insulation stress at high altitude.** According to Paschen's Law (the Paschen's curves are shown in (Figure 2.20) [62]), for a given clearance between conductors, the higher the altitude, the lower the breakdown voltage. This is mainly due to the fact that air is less dense as an insulator when the

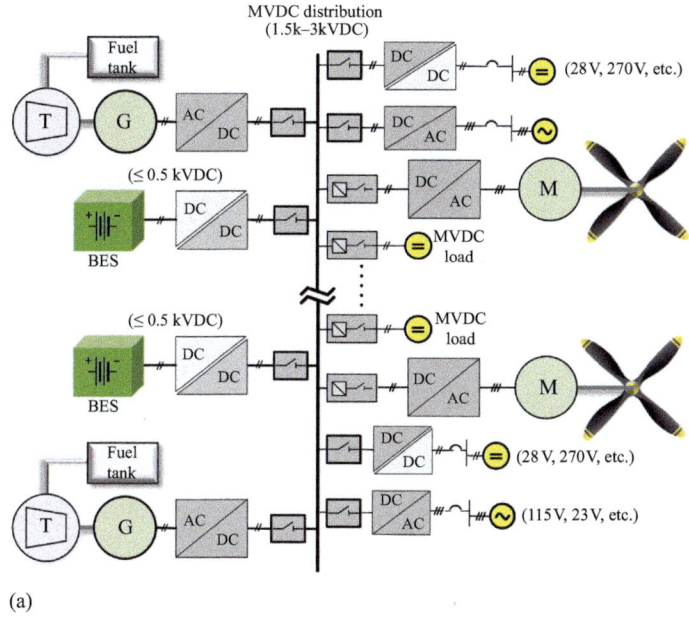

(a)

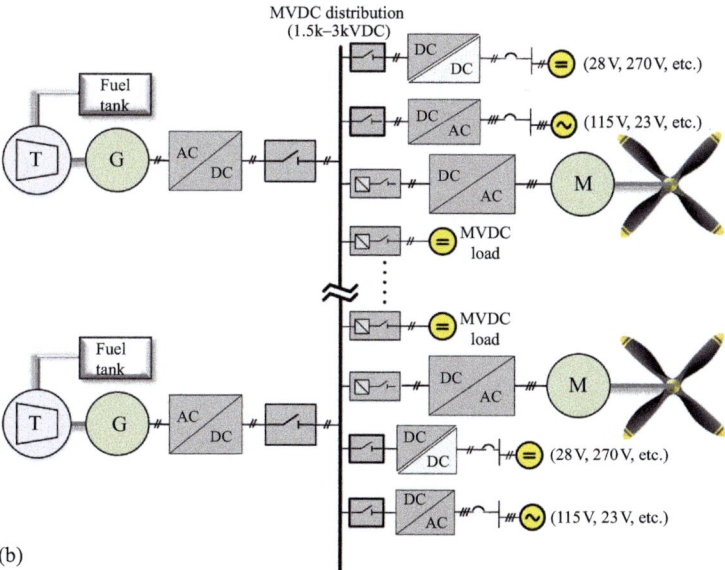

(b)

Figure 2.19 Aircraft MVDC distribution system based on various powertrain architectures: (a) hybrid-electric (b) turbo-electric

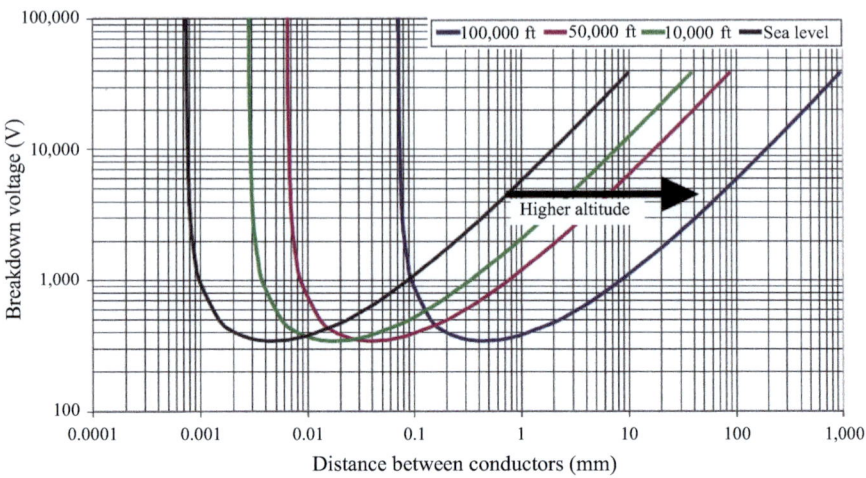

*Figure 2.20   Paschen's curves: breakdown voltage vs. conductor distance and
          altitude [62]*

altitude increases. For instance, for a 10-mm distance between conductors, the
breakdown voltage for such a clearance will be 30 kV DC at sea level, which
decreases to only 1.2 kV at 50,000 ft. This poses a challenge to the clearance
and creepage design in aircraft MVDC systems at high altitude, which will
reduce the volumetric power density of aircraft power apparatus.

• **Fault protection in MVDC systems.** Fault protection in DC systems is more
challenging than that in AC systems due to the unavailability of a natural zero
crossing in DC current which is required for contacts separation in circuit
breakers. However, intense research has been conducted to develop various
solutions for MVDC systems, including hybrid circuit breakers and solid-state
circuit breakers, which are detailed in [63]. Typically, the hybrid breakers have
very low on-state conduction losses, but the fault interruption speed could be
slow. On the contrary, the solid-state breaker is fast in fault isolation, but the
on-state losses can be significant, especially for MVDC where a number of
semiconductor switches will be connected in series to withstand the high DC
voltage.

• **Cosmic ray impact on high-voltage semiconductor devices at high altitude.**
It is reported that intensive cosmic rays at high altitude may damage semiconductor devices since the high-energy terrestrial neutrons might cause a
temporary short between the device terminals resulting in gate rupture or
burnout due to increased gate field or parasitic bipolar turn-on, respectively. As
the altitude increases, the associated neutral flux also approximately increases
exponentially. According to [64,65], for the same reliability target, the higher
the DC-bus voltage of power electronic systems, the lower utilization of the
blocking voltage rating of the semiconductor devices. A mathematical model
has been developed to calculate the cosmic ray related failure rate of

semiconductor devices by accounting for the three important factors, including blocking DC-bus voltage, junction temperature, and altitude levels, as shown in (2.1):

$$\lambda = C_3 \cdot \exp\left(\frac{C_2}{C_1 - V_{DC}}\right) \cdot \exp\left(\frac{25 - T_j}{47.6}\right) \cdot \exp\left[\frac{1 - \left(1 - \frac{h}{44300}\right)^{5.26}}{0.143}\right] \quad (2.1)$$

where V_{DC} and T_j are the blocking voltage in Volts and junction temperature in degree Celsius of the semiconductor devices, respectively; h refers to the altitude in meters above sea level. C_1, C_2, and C_3 refer to the module-based constants related to the blocking voltage ratings, junction temperature, and altitude levels, respectively, which can be obtained from the semiconductor manufacturers.

Experimental results on the emerging SiC MOSFETs at sea level and room temperature show that the 1.7 kV / 2 kA SiC MOSFETs have to derate to 1.1 kV, in order to meet the failure rate of 100 FIT [66]. For aircraft MVDC systems, to meet the reliability requirements at high altitude, high-voltage (e.g., 3.3 kV) semiconductor devices may have to be used, which may exhibit relative higher device losses than low-voltage devices (e.g., 1.2 kV and 1.7 kV). Optimization should be conducted by targeting both high efficiency and high reliability for aircraft high-altitude applications.

2.3.4 MVDC commercial shipboard electric propulsion systems

Steam and gas powered plants using reciprocating engines have been the principal means for ship propulsion throughout the nineteenth and twentieth centuries. For large ships, the steam pipe fitter played a crucial role in ship construction and operation. In the twentieth century, with the advent of steam turbines and liquid natural gas internal combustion engines, combined with marine reduction gears, the construction and operation of ship propulsion systems shifted towards mechanical drive systems. Ships also require electrical power generation to support personnel and passengers and, increasingly, to support a host of hull mechanical and electrical loads throughout the ship, such as pumps, auxiliary loads that provide coolant to main mechanical systems, hoists and capstans. Although the concept of electric ship propulsion was first introduced by the U.S. Navy in 1922 [67], large-scale adoption of electric propulsion has been driven by global initiatives to reduce the greenhouse gases produced by the commercial shipping industry, which, if conventional mechanical propulsion were to continue, would, by 2050, be responsible for a fifth of global carbide emissions. The result would be a rise in air pollution to as high as 250% from container and cruise ships, oil tankers and cargo vessels alone [68,69]. As of 2014, maritime shipping was reportedly responsible for 2%–3% of global carbon dioxide emissions [70] and is predicted to increase to as high as 17% by 2050 [71].

Table 2.3 shows the types of commercial vessels. There are a wide range of vessel types that make up the commercial shipping industry as shown in Table 2.3.

Table 2.3   Commercial vessel types versus propulsion power and SMCR

Ship type	Propulsion (MW)	Engine type*	SMCR (MW)
Fishing vessels	0.2–6	D, LNG	0.3–8
Off-shore vehicles	1–10	D, LNG	1–13
Roll on/roll off	1–25	RD, GT	2–19
High-speed craft	0.5–10	D, LNG	1–21
Bulk carrier	1–30	RD	1–22
Ferries	0.04–40	RD, GT, LNG	1–34
Tankers	3–50	RD, ST	2–44
Specialty vessels	1–40	D, GT, ST, NST, LNG	1–72
Passenger ships	4–44	D, GT, ST, LNG	2–80
Container ships	5–100	RD	3–86

*Diesel (D), Liquefied Natural Gas (LNG), Reciprocating Diesel (RD), Gas Turbine (GT), Steam Turbine (ST), Nuclear-power Steam Turbine (NST).

In most cases, the vessel type is defined by the cargo that it ships. The exceptions to this rule are ships that are defined by their special purpose or mission. Examples of these exceptions to cargo ships are fishing vessels (trawlers, purse seiners, fish processing), tugs, tenders, cable layers, research vessels and icebreakers. The ship power requirements, both mechanical and electrical, define its capabilities, sea distance and carbon footprint. Each vessel implementation must have sufficient engine capacity to meet a Specified Maximum Continuous Rating (SMCR) as defined by its mission and requirements. SMCR is derived from the ship propulsion requirements, as shown in Table 2.3, and is based upon the propeller power and speed under heavy propeller running, taking into account both sea and engine margin. This defines the continuous operating point of the engine. Propulsion power is the power that must be delivered to the propeller shaft(s) taking into account mechanical losses. For some ships the propulsion power exceeds the SMCR, meaning that propulsion power may be delivered to the shaft at a higher level than its continuous duty range to achieve higher speeds for short periods of time, but not continuously.

All ships require some level of installed electrical generation for human support (lights, heating, ventilation, air conditioning, refrigeration), maneuvering and navigation (bow thruster, radar, echo sounder), and auxiliaries (pumps, hoists, capstans, hoists). These electrical power systems usually operate from a separate smaller engine (in addition to the main propulsion engine(s)) that is dedicated to the generation of electric power. For ships having a high human support load requirement, i.e., passenger ship, or additional mission load requirements, i.e., specialty vessels, the installed engine SMCR may also include significant reserve to account for electrical power generation. It should be noted that, in the ship-building industry, SMCR is typically applied vessels where the movement of cargo is the primary purpose (i.e., container ships, tankers, bulk carrier, roll on/roll off) and, therefore, relates directly to the propulsion shaft. However, for the purpose of illustrating the transition from auxiliary power requirements to full electric propulsion the SMCR values in Table 2.3 include electrical generation power as well. In mechanically driven ships, the power train is designed to maximize efficiency, as it relates to fuel usage, at the SMCR operating point. If the SMCR operating condition were the only consideration, then the argument for transition to electric propulsion would not be compelling. In fact, the efficiency of power usage falls off considerably at speeds other than the SMCR operating point. It is the need for efficient fuel usage over a wide variation of ship speeds that justifies the transition from mechanical to electric propulsion.

(Figure 2.21) shows two conventional implementations of electric propulsion systems. In both cases, MVAC power generation is added to the ship's electrical distribution system to accommodate the considerable electrical power required by propulsion. The electrical systems of (Figure 2.21) show that the LVAC electrical power generation for the auxiliary loads remains separate from the MVAC propulsion power. Figure 2.21a shows a direct replacement of the mechanical system with an electric propulsion system. This is a notional layout for a large vessel, where the total required propulsion power is divided up between four (or more) propellers, with two motors per propeller shaft. In general, depending upon the ship propulsion requirements, propulsion power is accomplished with one or more propulsion shafts. Usually, more than one motor is ganged up on one shaft because of throughput limitations of the PEC(s) associated with the VFD functionality. As we will see, for the shipboard application, PEC power ratings fall outside of the range where commercial solutions are readily available. One way to address this limitation is through various approaches that enable the aggregation of power from multiple lower power PECs. Shipboard electrical systems will typically be redundant on port and starboard sides in order to enhance survivability. The MVAC cross-feeds of Figure 2.21b ensure that the ship can make it back to shore if the propulsion system is lost on one side of the ship.

In a mechanical propulsion system, engine power is either directly connected to each shaft or it is coupled to the engine through a gear box. The latter approach is required for a slow moving large vessel, say, a container ship, tanker or bulk carrier, where a long stroke Reciprocating Diesel (RD) engine is used in order to enable high torque at low speeds. The replacement of mechanical with electric propulsion eliminates the need for gear boxes and the RD engine by allowing direct torque control of the motor shaft by the motor(s). With electric propulsion, the determination of the propulsion power capability per shaft starts with the installed generation capability, or drives the need to increase installed generation. Since the engines can now operate at a single speed, where power production can be maximized, the engines in shipboard installations are more appropriately considered as prime movers, as opposed to propellers. Diesel, liquefied natural gas (LNG) engines are commercially available for marine applications, having power ratings from 450 kW to 12 MW with speeds ranging from 600 rpm to 1800 rpm. Gas Turbine (GT) engines are available for marine applications, having ratings from 3 MW to 42 MW with speeds ranging from 3,000 rpm to 15,000 rpm. A summary of commercially available marine generators with their power, voltage and speed ranges is shown in Table 2.4. Considering this information regarding marine grade

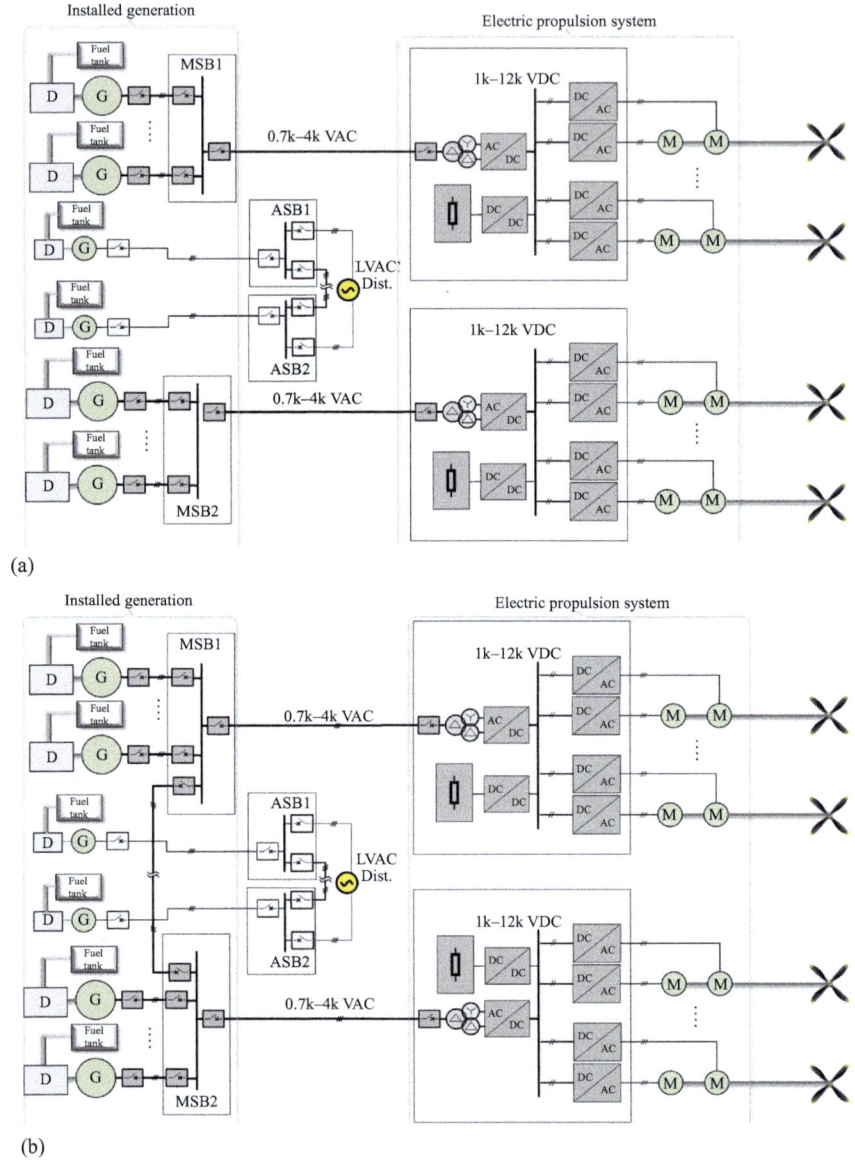

Figure 2.21   *Shipboard electric propulsion system: (a) direct replacement of mechanical propulsion with electric propulsion and (b) direct replacement with MVAC cross-feeds between propeller drives*

engine and generator commercial offerings [72], if a gear box is to be avoided, matching generator to engine rpm limits the power range of an individual engine/generator to 12 MW. Taking this consideration into account, a hypothetical and feasible electric propulsion system, matching the layout of Figure 2.21a, could be

Table 2.4   Commercially available generators for marine applications

Manufacturer*	Power range (MVA)	Voltage range (kV)	Speed range (rpm)
ABB	1–50	1–15	500–1,800
Siemens	0.38–20	0.4–13.8	500–1,800
GE	2.5–45	0.4–15	300–1,200

*Data represent market survey in 2018 [72].

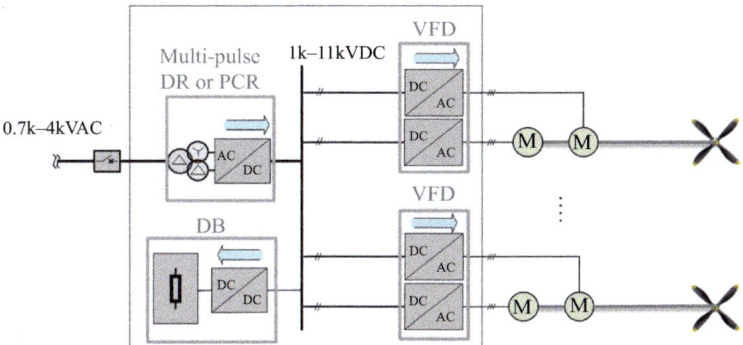

Figure 2.22   High TRL shipboard electric propulsion drive

implemented with four 4.2 kV MVAC, 12 MW diesel-fed generators as prime movers. Taking into account efficiencies, the propulsion power capability would be 40 MW. Such a system would be suitable for electric propulsion of bulk carrier, icebreaker, tanker or mid-size carrier ship.

In any electric propulsion implementation, the AC power produced by the generator(s) dedicated to this purpose is first converted to DC through an AC–DC PEC as shown in (Figure 2.22). The turbo-electric aircraft propulsion system of Figure 2.19b appears to be similar to (Figure 2.22), but there is very important distinction. In aircraft systems the power train between a turbo-electric generator and a propulsion shaft is an end-to-end transformation of power from prime mover to propeller. As the propulsion power need grows, additional turbo-electric generator driven propellers are added. Furthermore, the increased wingspan associated with the larger plane brings with it additional real estate for the propulsion system. Therefore, the build-up of electric propulsion in aircraft is modular. Ship propulsion is very different. Some specialized vessels may be built up in a modular fashion like the aircraft. However, for most ships, electrification of propulsion brings with it an increased payload to accommodate additional PECs and switchgear. The combination of hull design for large ships and increased payload from electric propulsion equipment does not improve the aerodynamics of the ship. Additionally, the need for a large propeller shaft limits the available or convenient real estate for placement of the generation needed for propulsion.

For shipboard propulsion electrification, the power throughput requirements for the AC–DC PEC are typically in the tens of MW range, as they must be matched to the marine generator ranges of Table 2.4. One or more generator(s) may be paralleled into a common MVAC bus by combining switchgear, as shown in (Figure 2.21) according to the propulsion power need and to achieve redundant propulsion feed capability in the event of a loss of generation. At the time of this writing, commercially available AC–DC PECs are limited to around 12 MW and, due to power semiconductor device limitations, the MVAC distribution voltage feed to the PEC is usually limited to less than 4 kV. Custom systems may be developed to handle higher MVAC voltage levels but such systems are very costly and generally impractical for commercial vessels, since it is the propulsion load that drives the need for MVAC distribution in the first place. The story is very different for naval vessels where the the requirement to build the system upon program of record switchgear equipment at 4.2 kV and 15 kV levels and, as a result, drives the need for development for high voltage rated PECs. For commercial ship electrification, the most common AC-DC PEC implementation, shown in (Figure 2.22), utilizes a low frequency transformer having two three-phase winding sets (one delta-connected and one wye-connected) on the secondary. The AC-DC PEC has a six-phase input and is typically a twelve-pulse (12-pulse) Rectifier consisting of two six-pulse Diode Bridge Rectifier (DBR) units. This is sufficient for MVAC systems, however a 12-pulse Phase Controlled Rectifier (PCR) may be utilized to provide inrush control during start-up.

Simplistically, the electric propulsion system of (Figure 2.22) is a drop-in replacement for mechanical propulsion. For the MVAC-based electric propulsion, the components shown in (Figure 2.22) are co-located, typically in the same cabinet or group of adjacent cabinets. The key required sub-components of the electric propulsion system are the AC–DC PEC, one or more VFDs and a means for handling regenerative power from the VFDs, which will inevitably occur when the ship speed is decelerating. The most common approach to regenerative energy management is to dissipate the energy using a resistive Dynamic Brake (DB). The DB consists of a buck DC–DC PEC implementation feeding a bank of resistors. The number of VFDs connected to the DC link, with associated propulsion shafts and DBs, may vary. Notionally, as the propulsion power requirement grows, the system of (Figure 2.22) is duplicated. MVAC-side switchgear is part of an external MVAC distribution system, as shown in (Figure 2.21). Installations of electric propulsion in commercial ships represented by the systems of (Figure 2.21) are arguably at TRL 9, per the definitions of Table 2.2.

It should be noted that for the power distribution architectures of (Figure 2.21), the generation for ship propulsion and generation for all other auxiliary and hull mechanical and electrical loads are isolated from each other. Future trends in shipboard electrification are moving towards the concept of an IPS, shown in (Figure 2.23), where generation for propulsion and all other electrical loads are integrated together onto the same MVAC bus [73]. Comparing Figures 2.21 and 2.23, the transition from implementations of (Figure 2.21) to the IPS of Figure 2.23a simply requires MVAC to LVAC transformer feeds from the MVAC

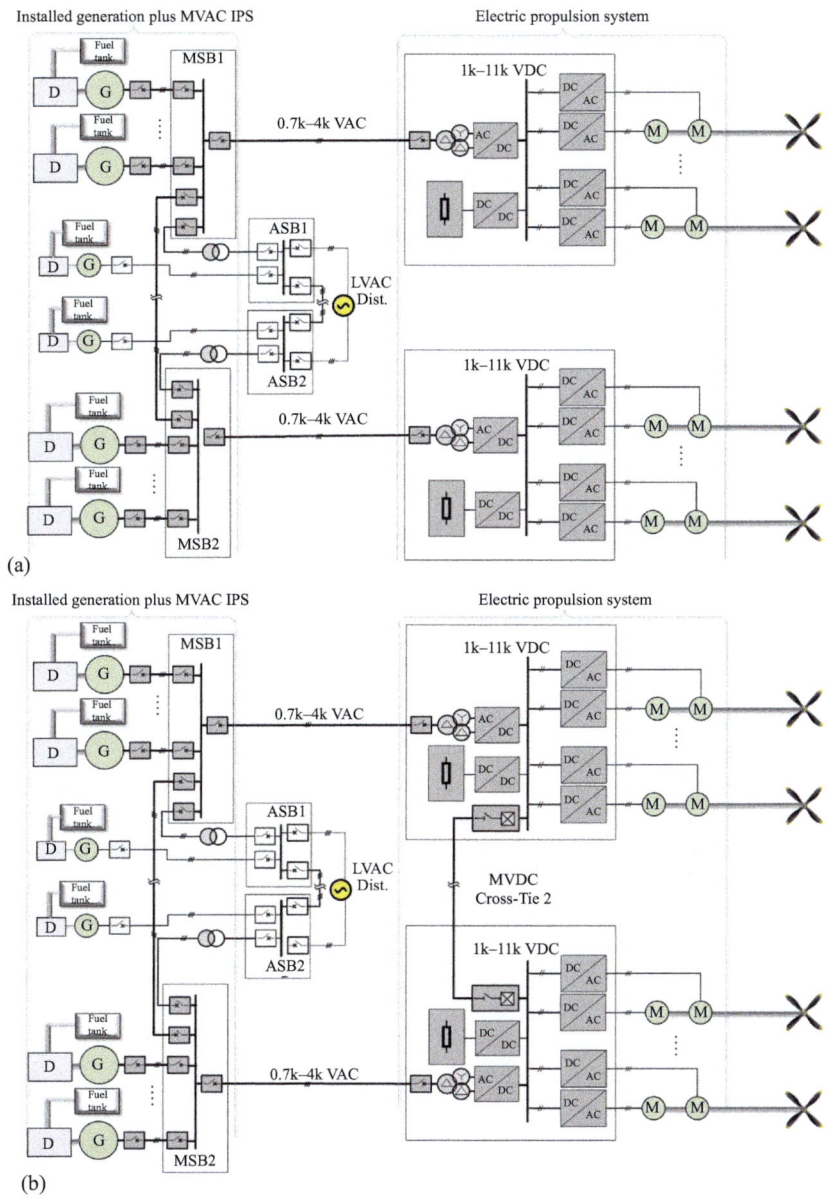

Figure 2.23 Shipboard Integrated Power System: (a) High TRL MVAC IPS (b) MVAC IPS with MVDC cross-feeds between propeller drives

main breaker panels MSB1 and MSB2 to the LVAC auxiliary breaker panels of ASB1 and ASB2. The MVAC cross-feeds, introduced in Figure 2.21b are present within the MVAC IPS, but MVDC cross-ties between propulsion cabinets may be added to enhance survivability, as shown in Figure 2.23b making the cross-tie connection with SSCBs internal to the propulsion system cabinets. IPS adoption has driven significant change within the commercial shipbuilding industry. Specifically, as a result of IPS adoption, the All Electric Ship (AES) has become the recognized standard for modern cruise liners and has become the design framework for modern military vessel electrical distribution systems [74–78].

The move towards AES has led to the recognition of significant benefits of MVDC distribution in future shipboard implementations for both commercial and military vessels. It is both feasible and viable to implement a MVDC distribution bus simply by converting the internal DC link of the MVAC electric propulsion system of (Figure 2.22) to a share the MVDC bus. In this way, the AC–DC PEC can be located in a different part of the ship that the electric propulsion, say in close proximity to the installed generation. Such a system is shown in (Figure 2.24). For MVDC-based electric propulsion, the MVDC cable interconnections between rectification and propulsion are more vulnerable to damage, so DC disconnection capability is required at the the rectification and propulsion sides of the MVDC bus. This can be accomplished quite simply using NLSws. In the event of a LL short circuit fault, for example, the MVDC bus can be folded back and supply current from the rectifier driven to zero either through phasing back the thyristors in the multi-pulse PCR or opening the MVAC-side circuit breakers, for the multi-pulse implementation made up of one or more six-pulse DBR units. The implementation of the AC–DC PEC must take into account MVDC bus voltage regulation and stability considerations [79,80]. The simplest, cheapest and least problematic implementation of the AC–DC PEC is the DBR unit based multi-pulse rectifier, with voltage regulation handled by a generator dedicated to the AC–DC PEC feed. However, low cost and simplicity must be traded off against very poor fault response. The multi-pulse PCR is very effective at managing faults, because output voltage can be folded back actively to drive current to zero in the NLSws so the fault can be isolated in an orderly and coordinated fashion. However, stability, light load discontinuous conduction and dynamic voltage control considerations introduce significant challenges that must be addressed during design and system integration.

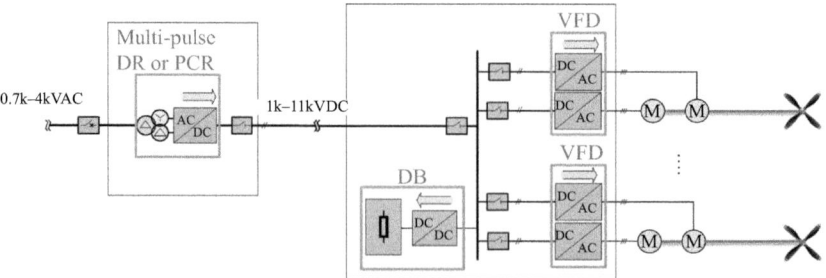

Figure 2.24   High TRL shipboard electric propulsion drive components for shipboard MVDC system

A notional high TRL implementation of MVDC IPS is shown in (Figure 2.25). Significant benefits to MVDC IPS have been identified in the literature. The most significant benefits are summarized as follows:

- Generator location can be decoupled from propulsion shaft placement.
- Use of more power dense high speed generators with direct rectification into the electrical distribution system.
- Reduction in cable size.
- Elimination of bulky low frequency transformers.
- Increased fuel efficiency on the generation side.
- Flexibility in design of the electrical power system.
- Ease of integration of energy storage at high level feeds within the electrical distribution system.

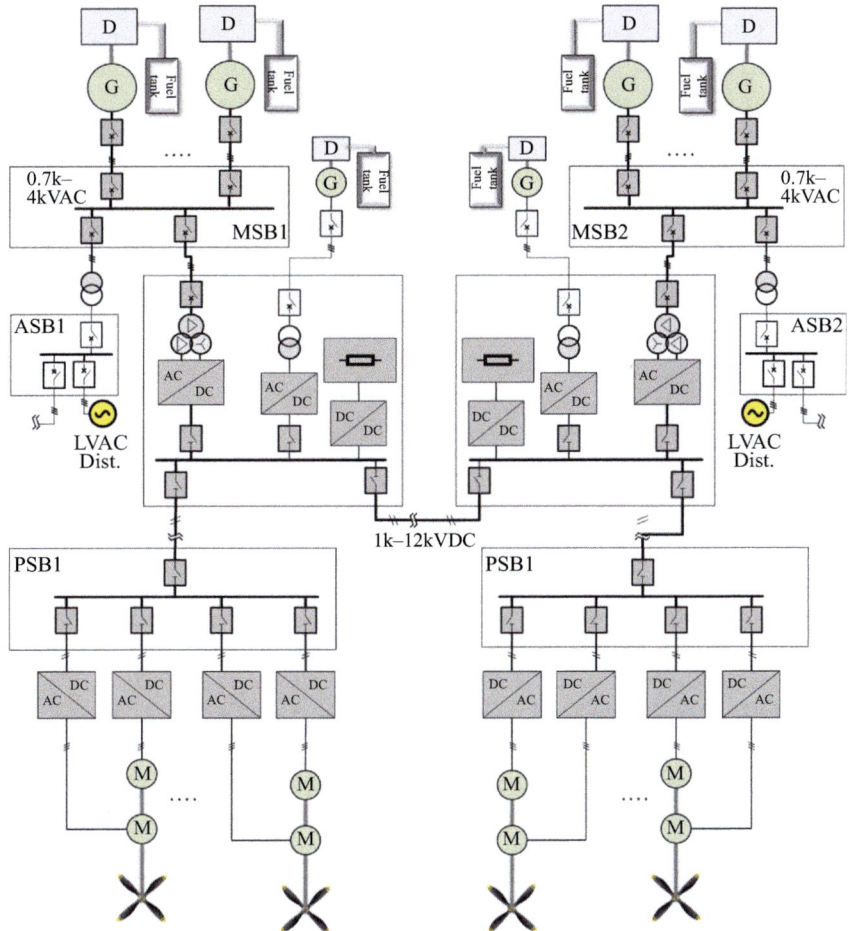

Figure 2.25   High TRL shipboard IPS implementation with MVDC propulsion bus

- Potential for optimal usage of regenerative energy from electric propulsion to improve fuel usage efficiency.
- Potential for flexible and autonomous redirection of energy as a means of enabling increased load diversity.
- Potential for flexible and autonomous fault isolation and recovery.
- Use of energy storage to improve power availability.
- Improvement in overall system efficiency.

As a result, a number of initiatives began in the years between 2010 and 2015 to move forward the development of MVDC-based ship electrification [81] for both commercial [77,82] and military ships [11,83]. One development of note was the redesign of Royal Princess at Fincantieri shipyard of Monfalcone, Italy in 2012 to incorporate MVDC. Although this design was never realized physically, it represented what would have been the largest AES with a MVDC electrical distribution system to the present date. The total installed generation was 78 MVA/62.4 MW with an 11 kV MVDC distribution bus [77]. These MVDC distribution power and voltage levels provide a good representation of what can be feasibly achieved at the date of this book publication.

Widespread deployment of shipboard MVDC system faces the following challenges:

- All power and energy flows through PECs.
- High required power rating of generator interfacing AC–DC PECs versus lack of commercial options.
- High power throughput rating of isolated DC–DC PECs versus lack of commercial options.
- High interconnection densities.
- MVDC bus protection coordination.

As has already been pointed out, the required high power rating of generator interfacing AC–DC PECs is a key challenge that must be addressed. Conventional MVAC based implementations of AC–DC PECs, (Figure 2.22), rely upon a low frequency delta-wye transformer in order to parallel two lower rated PECs and achieve some voltage boost for the internal DC link bus. It is feasible to apply the same approach to MVDC systems, as indicated in (Figure 2.24), but the MVDC bus requirements are more demanding than the single purpose internal DC link bus of the MVAC-fed electric propulsion drive system of (Figure 2.22). The most obvious challenge is the extra real estate required by the low-frequency transformer, other challenges exist, the most significant of which is the uni-directionality of all subsystems and the need for a DB to manage regenerative energy. As a result, such a system cannot utilize the regenerative energy to improve fuel usage efficiency and it is challenging to integrate energy storage for the purpose of capturing regenerative energy for later usage. An electric propulsion system that takes full advantage of the MVDC bus structure is shown in (Figure 2.26). The associated MVDC IPS implementation is shown in (Figure 2.27).

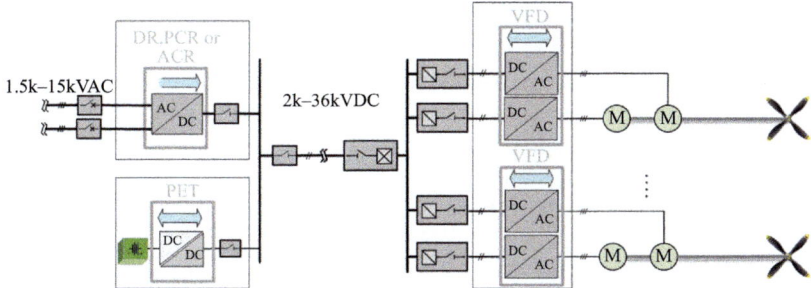

Figure 2.26 Low TRL shipboard MVDC-fed, fully regenerative electric propulsion drive system

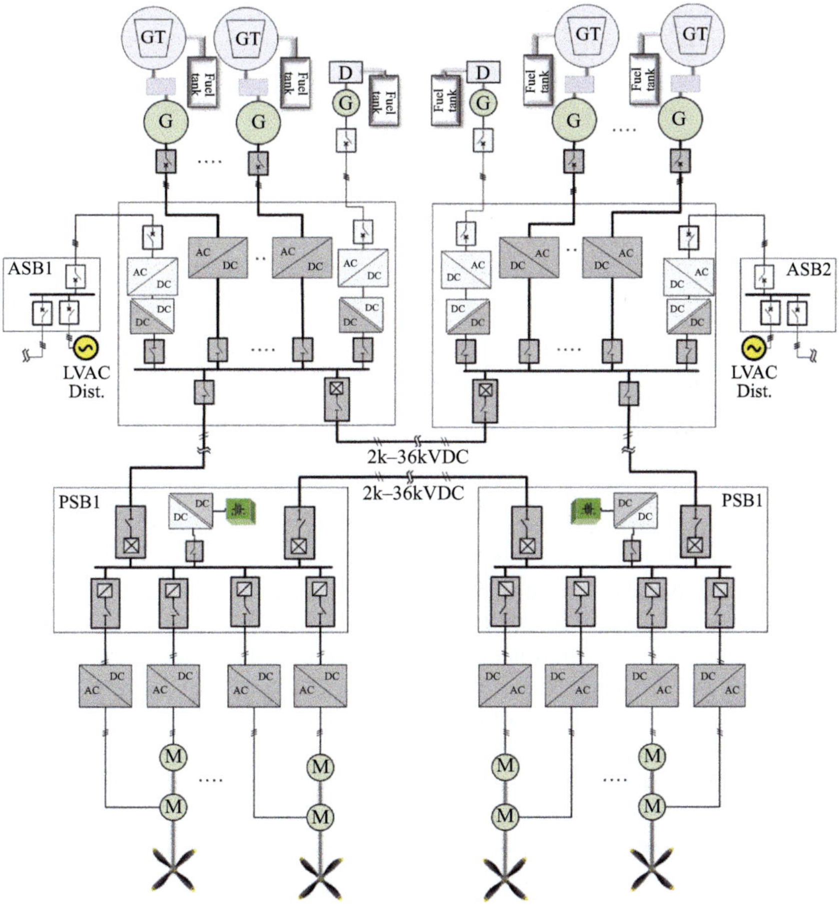

Figure 2.27 Low TRL shipboard MVDC-based IPS implementation

The challenges associated with achieving the full benefit of MVDC base ship electrification are highlighted by comparing (Figure 2.27) with (Figure 2.25). The system of (Figure 2.27) is intended to enable electrification of the entire range of commercial vessels represented in Table 2.3. The D, GT, NST and LNG engines will be used in future MVDC ship platforms. The ability to process higher power through the generator interfacing AC–DC PEC is a key enabler to deployment of electric propulsion in all ship platforms. Referring to (Figure 2.25), conventional MVAC distribution components, associated with MSB1 and MSB2, still represent a significant portion of the MVDC IPS. Since, in the MVDC based system nearly all of the power flows through PECs, continuation of the high TRL architecture of (Figure 2.25) is at cross purposes with the MVDC potential benefits of higher power density, efficiency and flexibility because, essentially, PECs are simply overlaid on top of conventional MVAC distribution equipment.

The goal of future shipboard electrification would be to replace conventional power distribution components, i.e., EMCBs and low frequency transformers, with PECs. For example, the system of (Figure 2.27) only utilizes the EMCB as an isolating switch between each generator and its associated AC–DC PEC, otherwise, all EMCBs, switchgear and associated power panels and low-frequency transformers are eliminated. Key features of the MVDC-based electric propulsion system are shown in (Figure 2.26). The AC–DC PEC may interface to the MVAC generator directly through multiple three-phase lines, which removes some of the burden and risk associated with paralleling directly AC–DC PECs to achieve higher throughput power. Energy can be recovered from the electric propulsion system regeneration to a BES through an isolated DC–DC converter, as opposed to utilizing the DB to dissipate this energy. Similarly, isolated DC–DC converters can be utilized to interface a common MVDC bus to the low voltage distribution system (LVDS), eliminating the need for bulky low frequency MVAC–LVAC transformers. The SSCB plays a clear role in enabling dynamic disconnection of, say, faulted VFDs from the common MVDC bus and distribution of MVDC bus energy throughout the system. Because long MVDC cable runs may exist, SSCBs having bi-directional fault current interruption capability are required and associated MVDC fault protective relaying. Achieving the vision of MVDC-based shipboard electrification presents a number of technological developmental challenges. A significant challenge is the development of PECs capable of handling multi-MW throughput at high levels of voltage interface. This challenge is amplified when SSCBs are part of the system. Commercially viable, MVDC rated SSCBs capable of operating continuously at 1,000–3,000 A are needed. These SSCBs must be capable of operating at high efficiency must have the capability of reliably interrupting 10s of thousands of amperes.

There are a number of ways that the high power throughput challenge of the AC–DC PEC can be addressed, that are either incremental or transformational. Since, as of this writing, commercially available AC–DC PECs and switchgear are limited to 4 kV voltage ratings, an incremental approach to achieving higher AC-DC PEC power throughput has been to address the problem via multi-phase generation [84–86]. The approach is illustrated in (Figure 2.28) for a GT engine

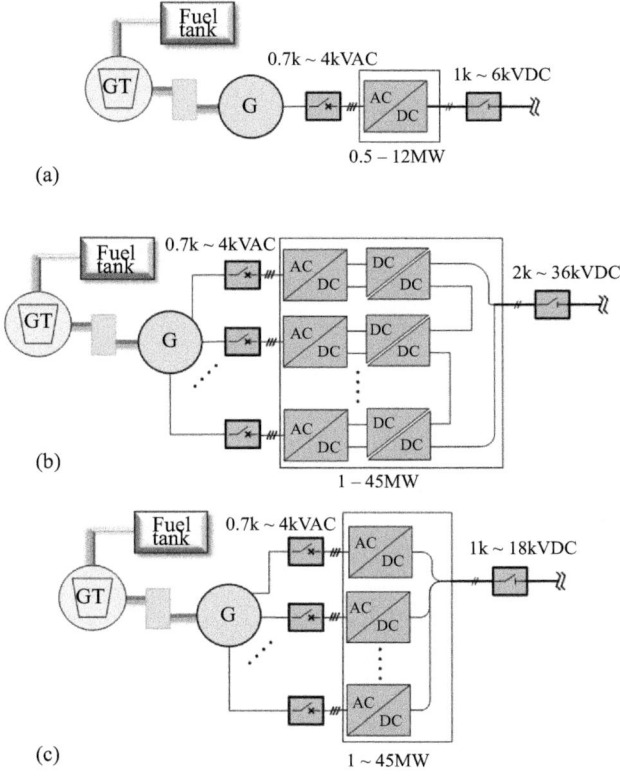

Figure 2.28 *MVDC power generation. (a) Matched MV generator with gas turbine engine prime mover to AC–DC PEC. (b) Series connection of MV multi-phase generator with gas turbine engine prime mover to isolated AC–DC PECs. (c) Parallel connection of MV multi-phase generator with gas turbine engine prime mover to AC–DC PECs*

driven system. Figure 2.28a shows a system that can be implemented using presently available COTS generator and PEC, which limits the achievable MVDC bus voltage to 6 kV and power throughput to 12MW. The AC–DC PEC may be accomplished through a DR, PC or Active Controlled Rectifier (ACR) that utilizes PWM control to dynamically regulate the MVDC bus and, in combination with MVAC side filtering (internal to the ACR) controls the MVAC side power quality. Figure 2.28b and 2.28c shows the achievement of high voltage and power rated systems through series or parallel connections of AC–DC PEC modules. In the paralleled case, Figure 2.28c, AC–DC PEC modules are isolated from each other on the MVAC through separate isolated sets of three-phase windings within the multi-phase generator. In the series case, Figure 2.28b, isolation must be added on the MVDC side through separate PECs. Similar approaches may be applied to achieving propulsion motor interface off of higher voltage grids through DC–AC PECs as shown in (Figure 2.29).

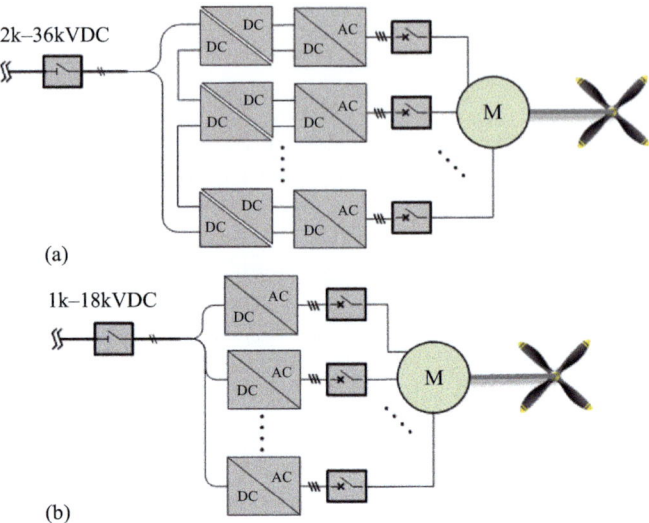

(a)

(b)

Figure 2.29   Approach to higher MVDC bus-fed VFD implementations. (a) Series connection of MV multi-phase propulsion motor through isolated DC–AC PECs. (b) Parallel connection of MV multi-phase propulsion motor through DC–AC PECs

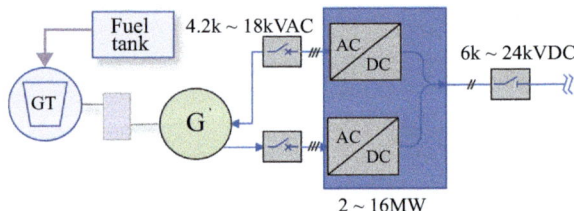

Figure 2.30   MVDC power generation matched to a six-phase MV generator (with gas turbine engine prime mover) utilizing a six-leg, multilevel, SiC MOSFET based AC–DC PEC

More transformational approaches to high power throughput AC–DC PECs are represented generally in (Figure 2.30). With the emergence of WBG power semi-conductors and advances in multilevel PEC topologies, it is possible to operate PECs at higher switching frequencies and at higher voltages per each three-phase generator connection. These state-of-the-art technologies enable dynamic energy and the control of power at MV levels as high as 36 kV, while maintaining rea-sonable levels of power density and space claim [87–90].

2.3.5   Navy shipboard MVDC electrical distribution systems

Compared to commercial ships, the electrification of naval shipboard systems overall has proceeded at a slow and measured rate. While naval ship electrification is motivated by some of the same factors as commercial ships, i.e., the movement away from steam based systems and the need for increased fuel economy and reduced environmental impact, the principal motivation for navy shipboard electrification is to ensure survivability of personnel and mission support systems. The design philosophy for naval power systems is expressed by the Naval Sea Systems Command (NAVSEA) Design Practices and Criteria Manual, Electrical Systems for Surface Ships as follows:

"The primary aim of the electric power system design will be for survivability and continuity of the electrical power supply. To ensure continuity of service, consideration shall be given to the number, size and location of generators, switchboards, and to the type of electrical distribution systems to be installed and the suitability for segregating or isolating damaged sections of the system."

At the same time, naval ships have employed electrical power systems for over a hundred years. Norbert Doerry, in reference [91], provides an excellent history of the naval shipboard stages of electrification. According to Doerry, "Virtually all U. S. naval ships generate and distribute 3 phase 60 Hz electrical power at 440 Volts or 4,160 Volts. The governing power quality interface standards are MIL-STD-1399 section 300 and section 680. Traditionally the U.S. Navy used radial distribution systems and provided vital loads with alternate sources of power from different switchboards. During the 1990s, the Navy introduced zonal distribution systems in response to the ever increasing number of vital loads. In a radial system, vital loads are provided alternate sources of power via longitudinal feeder cables from different switchboards. In a zonal system, vital loads are provided alternate sources via much shorter transverse feeder cables from port and starboard switchboards. By replacing long feeder cables with short feeder cables, zonal distribution systems reduced cost and weight" [91]. The zonal distribution system that emerged in the 1990s set the stage for emerging naval shipboard LVDC and MVDC distribution systems [10,92–94].

DC electrical distribution on navy ships became a serious consideration for new ship designs in the 1990s. This shift in the way ship electrical systems were designed and implemented came about because of a significant growth in the ratio of shipboard electronic loads (i.e., instrumentation, communications, mission loads, etc.) to purely fixed frequency electrical loads (i.e., heaters, pumps, fan coil units, etc.).

The Navy is currently facing an unprecedented challenge to develop new ship designs under compressed schedules that incorporate emerging technologies for high power energy conversion in order to enable smaller ship designs with a very high degree of electrification, high energy directed weapons and survivability enabled by high power, high power density PECs [11,83,95,96]. The Navy is already building shipboard DCZEDS-based IPS platforms that incorporate a highly flexible and survivable power delivery system that enables a significantly larger level of installed load compared to the installed generation capability. The DCZEDS enables multiple power delivery paths throughout the system. In the context of DCZEDS, a *zone* comprises both the functionalities of electrical zones of protection (akin to land-based electrical protective systems) and the geographical zones of protection for equipment arranged into areas of the ship that are partitioned by bulkheads and decks.

With present-day capabilities, using program of record equipment and/or through procurement of developmental equipment installations, existing ship platforms are being adapted to increasing levels of electrification and new ships are being built that include hybrid electric or electric propulsion systems requiring multi-megawatt (4–20 MW) power throughput capability. So far, such systems are built upon under girding MVAC distribution systems in order to mitigate the extreme risk associated with the integration of PEC-based equipment, given present-day experience of naval shipbuilders and a limited number of shipboard qualified PEC OEMs. These systems are also enabling incremental insertion of directed energy weapons, such as lasers, fed from the MVAC distribution system. Directed energy weapons loading to the electrical system manifests as single or a limited number of repetitive, very high rate of change energy pulses, having a volt-ampere draw that may far exceed the steady-state installed generation capacity within the ship. For MVAC fed high energy loads, load demand will exceed any direct line to portions of the installed generator capacity. As a result, the integration of directed energy weapon loads into MVAC distribution systems will rely heavily upon localized BESs to deliver the excess energy required. There is only a limited possibility within the IPS-based ship electrical system to dynamically share energy between the IPS, electric propulsion, hybrid electric propulsion and directed energy weapons.

Future shipboard systems must fully integrate propulsion and electric weaponry with all other electrical load demands in the ship to enable future naval warship demands for increased deployment of directed energy weapons throughout the ship. Naval vessels having a MVDC based electrical distribution system will be necessary in order to enable large scale deployment directed energy weapons and to achieve the navy's vision for future shipboard capabilities [95,96]. The MVDC based IPES, shown in Figures 2.31 and 2.32, is the way forward for these future electrified ships. (Figure 2.31) shows a MVDC electrical distribution architecture that utilizes PEC-based protection. (Figure 2.32) utilizes Breaker-based protection. Architectures shown in these figures effectively enable the delivery of power and energy to any part of the ship and optimization of use and placement of BESs throughout the system. Note that the systems of Figures 2.31 and 2.32 are notional and represent one way in which power and energy flow paths between the MVDC system, MVDC loads and medium frequency transformer isolated portions of the LVDS. The high energy pulsed load represented in these systems is the Electromagnetic Rail Gun (EMRG).

Using the accepted taxonomy of the electric ship research design community, MVDC-based IPES enables dynamic, bi-directional power and energy transfer between distributed generation and energy storage and all shipboard electrical loads using power electronics-based *Modules*. Fundamental MVDC interfacing Modules perform the following principal functions:

- The *Power Generation Module* (*PGM*) converts chemical energy to MVDC electrical energy
- The *Power Conversion Module* (*PCM*) provides energy conversion between MVDC and the LVDC

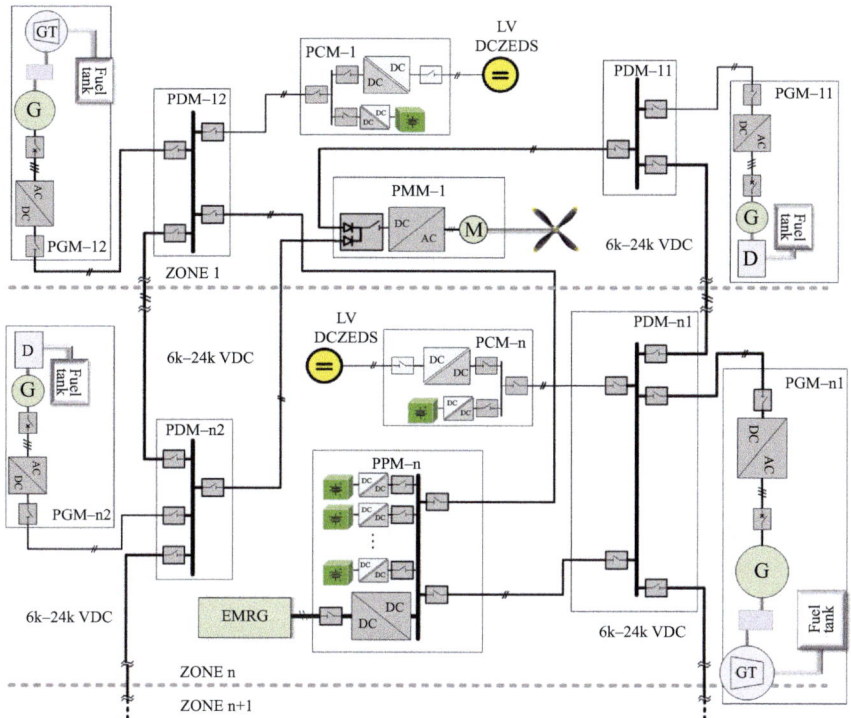

Figure 2.31   MVDC IPES with PEC-based protective architecture

- The *Power Distribution Module* (*PDM*) distributes and directs power and energy flow and includes fault isolation capability
- The *Propulsion Motor Module* (*PMM*) converts MVDC electrical energy to mechanical propulsion energy
- The *Pulsed Power Module* (*PPM*) interfaces the MVDC system to high energy pulsed loads

MVDC-based electrical distribution in naval vessels is motivated by the same benefits and challenges associated with MVDC in commercial vessels, that were outlined in Section 2.3.4. The AC–DC PEC associated with the PGM, in the naval ship taxonomy, is referred to as the PGM-Rectifier. Total installed generation capacity on the order of 100MW will be required for next generation IPES platforms. The generators themselves are distributed in zones throughout the ship, as represented in Figures 2.31 and 2.32, but this does not necessarily mean that there will be more PGM-Rectifiers with lower power ratings. There are also constraints in large ship designs as to where the the generator can be installed, according to the location of engine rooms and the need to co-locate exhaust stacks with the generators that span multiple decks. Just as will be the case for future commercial shipboard electrification, the GT engines associated larger

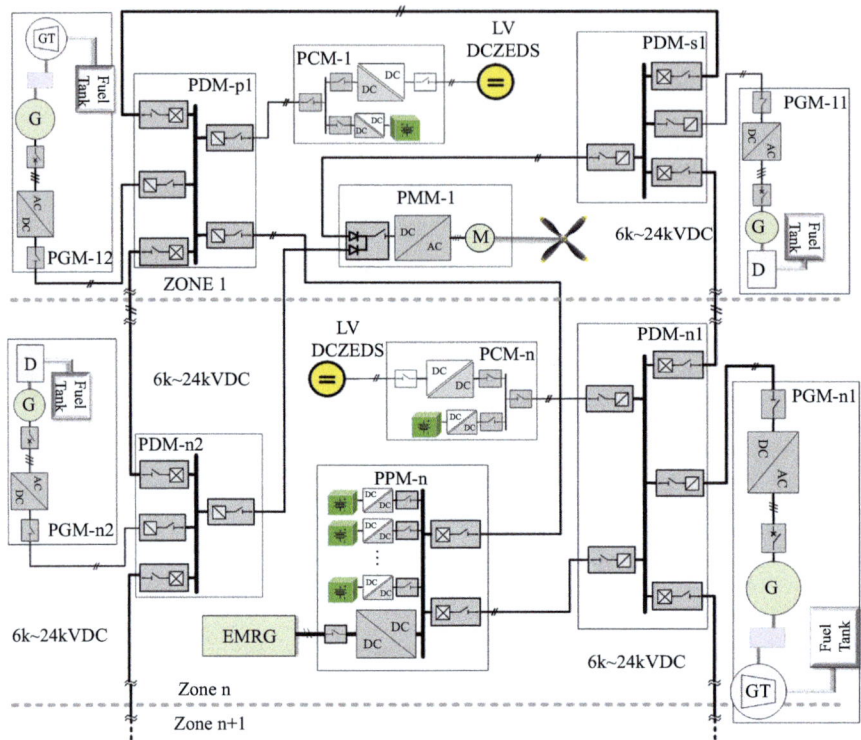

Figure 2.32   MVDC IPES with breaker-based protective architecture

main generators in naval vessels (represented by PGM-12 and PGM-n1 in Figures 2.31 and 2.32) will come from the same limited supply base of marine engine OEMs and fall in the 2.5–80 MW power range. The naval vessel MVDC IPES architecture will require direct generation and power conversion (through the PGM-Rectifier) off of these generator and, as a result, the PGM-Rectifier faces the exact same power throughput challenge as was outlined in detail in Section 2.3.4.

Furthermore, MVDC distribution in naval ships is motivated by much more than just electrification of propulsion. The PGM-Rectifier and high energy, multi-MW loads, such as PMM and PPM, may be special high power PEC implementations procured from OEMs having the capability of designing and manufacturing large, MV-rated PEC cabinets for the shipboard environment. But there is a need to develop a large number of modular, multi-use Lowest Replaceable Units (LRUs) from which scalable systems can be built within PDMs and PCMs. This new concept for multi-platform system building for navy ships has motivated significant investment by the Office of Naval Research and NAVSEA in the development of PEBBs [12,15,97–102], NLSw's [103] and SSCBs [104,105].

2.4   Technical requirements and challenges on MVDC power converters

This section will present the technical requirements and challenges from the MVDC applications to the power converter circuit topologies, and serves as a transition from all the aforementioned MVDC applications to the specific power converter topologies. The intent is to provide a general evaluation guideline for the various MVDC converter topologies which will be presented in Chapters 4 and 5 of this book. Specifically, technical requirements on both the input and output of the MVDC power converters, the challenges on switching devices, and the performance requirements from the system levels are briefly discussed.

2.4.1   Input and output requirements

2.4.1.1   Harmonic distortions

As is well known, the switching of the MVDC converters may cause significant harmonic distortions to interfacing electric machines or to grid. Harmonic components around the switching frequency of the converter will enter into the machines or grid, unless a passive filter, such as an inductive-capacitive-inductive (LCL) filter, is designed to attenuate such harmonic distortions in the line voltage and current. Otherwise, these harmonic distortions may cause substantial winding losses to the machines or transformers, cause nuisance tripping or other system inter-compatibility issues, such as ringing with system resonances, due to conducted EMI. For utility interfaced MVDC applications, existing standards and guidelines such as IEC 61000 and IEEE Standard 519-2014 can be used when designing the converter and filter topologies [106].

2.4.1.2   Overvoltage induced by high *dv/dt*

In a machine-connected MVDC systems, such as wind turbine generator systems, or aircraft and shipboard propulsion motor-drive systems, the dv/dt stress might be considerable due to the high voltage levels and fast switching of modern semiconductor devices such as Silicon Carbide (SiC) MOSFETs. The high differential-mode dv/dt combined with the characteristic impedance mismatching between the machine, cables and the MVDC converter, very likely induce high-frequency overvoltage spikes on the machine terminals with an amplitude of more than twice the converter DC-bus voltage [107]. Such high-frequency voltage spikes, if not mitigated, may eventually damage the insulation of the machine stator windings, especially the first few winding turns closing to the converter side. This phenomenon becomes more severe if there are long cables (e.g., > 50 ft) connected between the machine and the MVDC converter. For instance, the recent trend of commercial wind turbine generator systems is to install the MVDC converters on the bottom of the tower for easy access and maintenance convenience, requiring long cables interconnected between the generator in the nacelle and the MVDC converter located on the bottom of the tower, which will possibly incur overvoltage issues in the generator, as illustrated in (Figure 2.33). Therefore, to mitigate the

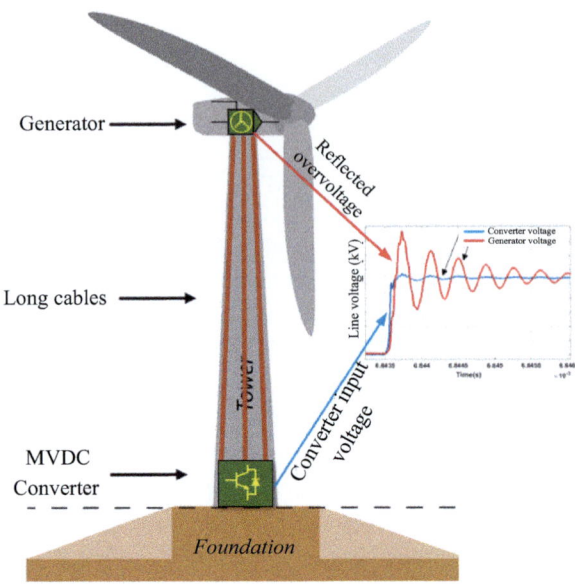

Figure 2.33   Reflected overvoltage transients in wind turbine generator connected to MVDC converter with long power cables

insulation stress on the generator stator windings, passive dv/dt filters are applied. These stresses may also be reduced through the use of special front-end converter topologies, such as the Current Source Converters (CSC) or multilevel VSC-based active rectifiers that produce staircase line voltage waveforms. From a PEC topology perspective, two-level CSCs and VSCs based on series connected power modules are feasible, but the high dv/dt in the line voltage will be a drawback, requiring a dv/dt or sinewave filter installed at the output of the converter, although such filters typically are very lossy [108]. VSC implementations of these PECs will require complex voltage balancing gate drive circuits and low inductance mono-lithic buses for devices connections, so each implementation is a custom design. CSCs are well-suited for higher current rated devices, such as the Gate Turn-Off Thyristor (GTO), and do not demand the extreme measures required by VSCs with series devices to manage device interconnections. A range of multilevel VSC topologies may be considered as an alternative which can be exploited reduce PEC size and weight and eliminate the need for dv/dt passive filters and voltage balancing snubber circuits.

2.4.1.3   Common-mode voltage stress

Common-mode voltage stress from MVDC converters can be severe due to the higher voltage level and the fast switching speed with new-generation semi-conductor devices. For back-to-back MVDC converters, the peak value of the common-mode voltage can be as high as the total DC-bus voltage when two opposite zero vectors are used by the rectifier and inverter simultaneously [109]. If

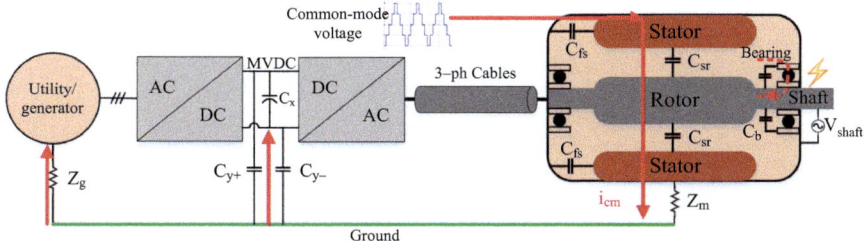

Figure 2.34 Common mode voltage stress in MV motor-drive systems

not mitigated, the common-mode voltage will cause dramatic EMI issues to the whole system, tripping the microcontrollers, gate drivers, or other sensitive devices. Particularly, in electric propulsion motor-drive applications, the high common-mode voltage will induce substantial bearing current through the parasitic capacitors between the motor frame and the stator windings, causing premature bearing failures by electrical discharge occurring through the lubricant [110]. For example, the common-mode dv/dt from an IGBT based MVDC inverter can easily reach 5 kV/us, and the typical parasitic capacitance of an electric machine would be as much as 10 nF. As a result, a total common-mode current of 50A (peak value) can be induced in the system. If the emerging SiC MOSFET devices are used to configure the MVDC converter, the common-mode current can be much higher due to the high dv/dt during the switching. The common-mode voltage stress in a MV machine-drive system is depicted in (Figure 2.34). For CSCs and multilevel VSCs, the common-mode voltage is typically much lower due to the more sinusoidal output voltage waveforms. In many cases, a passive common-mode filter or special PWM strategies can be developed to mitigate the common-mode voltage stress [111].

2.4.2   Switching devices

2.4.2.1   Emerging SiC MOSFET modules

Driven by the requirements of high efficiency and high power density in many applications such as transportation electrifications and renewable energies, SiC MOSFETs have been experiencing rapid development in both packaging technologies and voltage/current levels. At present, 1.7 and 3.3 kV half-bridge SiC MOSFET modules are already commercially available in the market, and SiC modules with higher voltage ratings are under development and possibly will be released to market soon, including 4.5, 6.5, 10 and 15 kV. Various packages of 1.7 and 3.3 kV half-bridge SiC MOSFET modules are shown in (Figure 2.35). Compared to the conventional silicon IGBT counterparts, SiC MOSFET modules can withstand higher voltage, higher temperature, and higher switching frequency, as shown in (Figure 2.36) [112]. Higher device breakdown voltage (i.e., 3.3 kV ratings and above) combined with low loss fast switching capability of SiC MOSFET modules enables direct connection of PECs to MVAC and MVDC grids.

(a)   (b)

(c)   (d)

*Figure 2.35   Half-bridge SiC MOSFET modules: (a) GE 1.7 kV / 450 A in
EconoDual package, (b) Cree/Wolfspeed 1.7 kV / 225 A in 62 mm
package, (c) Hitachi 3.3 kV / 800 A in nHPD 2 package, and
(d) Mitsubishi 3.3 kV / 750 A in LV100 package*

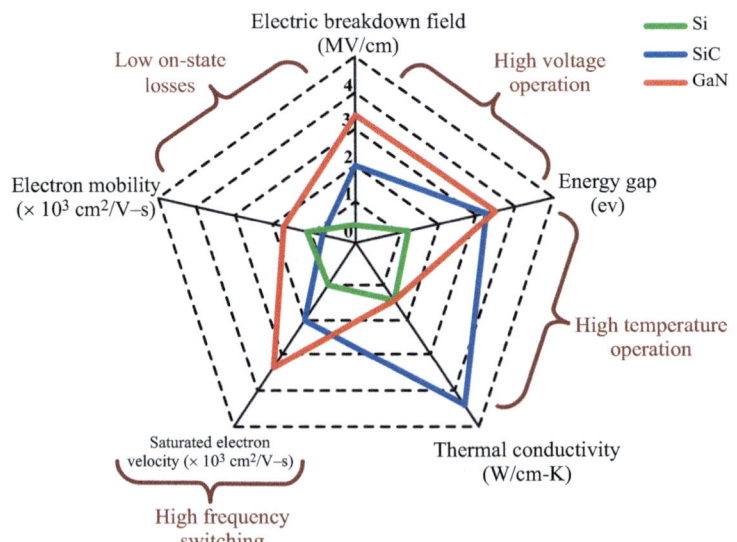

Figure 2.36   Comparison of Si, SiC and GaN semiconductor devices [112]

This capability will have a transformative impact on medium voltage distribution paradigms by enabling dynamic energy and power control of energy with significant reductions in through-put losses and equipment space-claim at voltage/power levels not achievable with present-day technology. The availability of high performing and high TRL MVDC interfacing PECs to broader applications will increase as the market price of these devices decreases.

The integration of SiC into larger systems introduces challenges due to some undesired penalties associated with high efficiency switching of power semiconductors, and never before encountered packaging challenges, which require special attention during the design process:

- *Increased EMI.* Specifically, the high common-mode dv/dt resulting from the fast switching coupled with the module-to-heatsink stray capacitance as well as the gate driver transformer primary-secondary capacitance ($C * dv/dt$) will generate significant common-mode current during switching, which may trip the system microcontrollers and saturate the common-mode filter inductors. Under such a scenario, common-mode EMI filters need to be properly designed to attenuate the high dv/dt.

- *Increased surge voltage stress.* The high differential-mode dv/dt generated from the fast-switching of high-voltage SiC MOSFETs, if not mitigated properly, will cause significant surge voltage on the load terminals, especially if there are long cables interconnected between the SiC based MVDC inverters and the loads. If the load is an electric machine, the first few winding turns closer to the inverter side might be subject to severe winding insulation stress, in which the high-frequency voltage spikes might double the DC-bus voltage. As mentioned in Section 2.4, this is mainly caused by the short rise time of the SiC switches and the mismatch of surge impedance among the inverters, cables, and machines. To reduce the high differential-mode dv/dt output from the inverter, there are a few mitigation approaches to be considered depending on the specific applications: (1) design dv/dt RLC filter or sinewave filter at the inverter output; (2) slow down the switching speed with active or passive dv/dt control; (3) integrated motor-drive solution, where no power cables are needed between the inverter and the motor; (4) soft-switching techniques; (5) use multilevel VSC or CSC topologies to obtain more sinusoidal output voltage waveforms to work against the MVAC-side interface.

- *Weaker short-circuit withstanding capability.* For conducting the same amount of current, the required die size for SiC MOSFETs will be much smaller than that for Silicon devices. Nevertheless, such high-density SiC chips also indicate shorter thermal time constant and weaker short-circuit capability. The short-circuit withstanding time for SiC MOSFETs is typically shorter than 5 μs, while the standard short-circuit withstanding time for Si IGBTs is normally 10 μs. This may pose a challenge on the short-circuit protection due to the requirement of fast fault detection speed.

- *Higher turn-off voltage overshoot.* Due to the large di/dt of SiC MOSFETs during switching, the turn-off voltage overshoot could be considerable, especially in MVDC applications where the parasitic inductance of the high-voltage SiC modules and the commutation loops in MVDC converters might be significant. For instance, with a typical di/dt of 20 kA/us for SiC MOSFETs, the turn-off voltage overshoot can reach 1 kV superposed on the DC-bus voltage if the parasitic inductance of the commutation loops is 50 nH ($V = L * di/dt$). This requires very dedicated design of the bus bars and circuit layout of the converters. Under some scenarios, advanced modulation strategies can be adopted for MVDC converters where all the commutation occurs within the semiconductor modules to avoid the large commutation loops, if it is too challenging to further reduce the parasitic inductance of the commutation loops [113]. Additionally, active overvoltage clamping techniques can be utilized for the gate driver design, in case of any potential overvoltage failures in the SiC modules.

- Trade-offs between the thermal conducting capability of insulating substrate between the module direct bonded copper and baseplate and dielectric voltage withstand capability. Power electronic modules are mounted on heatsinks which are usually at chassis ground potential. If SiC modules are applied where MVDC system level matches its nominal voltage rating (i.e., 6 kVDC for the 10 kV SiC MOSFET), then its corresponding module isolation voltage is sufficient (i.e., 15 kV for 10 kV rated devices). However, if devices are effectively placed in series to enable LL voltage operation in excess of a single device nominal rating, then the potential of a device within the stack with respect to chassis will increase to a level corresponding to the system LG voltage. If the heat sink is not sufficiently isolated from chassis then dv/dt induced displacement current through module-to-heatsink stray capacitances, combined with high mixed mode (low frequency and high frequency) electric fields within the insulating substrate of the module, will lead to increased incidence of module partial discharge failure.

2.4.2.2   Power electronics building blocks

The PEBB is a generic concept that integrates power semiconductor devices, gate drivers, controls, filters, sensors, protections, and thermal management into building blocks with defined functionality, covering electrical, mechanical, and thermal domains. The PEBB is particularly useful when it comes to the implementation of PECs for MVDC connected systems. The PEBB should be modular and scalable to stack up to configure high-power electrical systems in the power range of hundreds of kW to hundreds of MW. The PEBB concept was originally proposed by the U.S. Office of Naval Research (ONR) in 1997 for shipboard applications with the main motivation of driving down the power electronics cost [114], which has evolved for broad applications in the past decades, including medium-voltage or high-voltage motor drives, power quality, renewable energy, distributed generation, energy storage, aircraft and shipboard propulsion. Compared to the conventional monolithic designs, the PEBB approach has the potential advantages of reduced cost, power losses, weight, size, and easier maintenance.

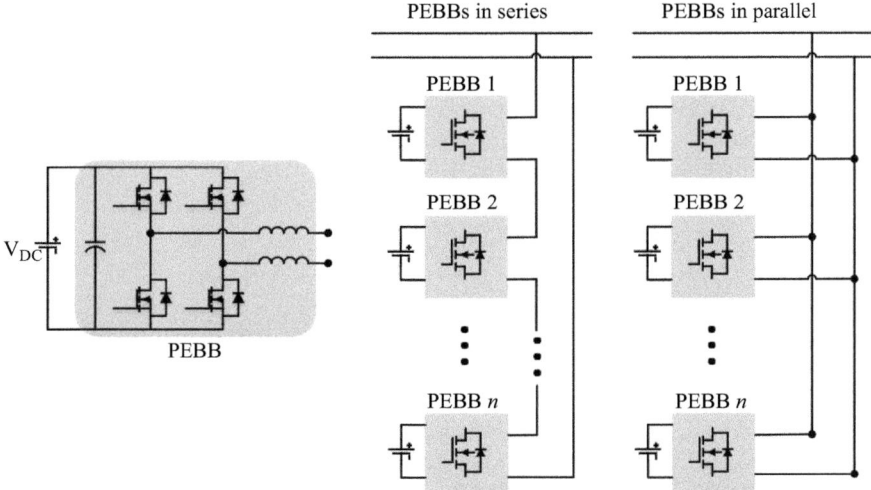

Figure 2.37   PEBBs in series or parallel connection

The internal power converter topology of a PEBB can be a half bridge, full bridge, two-level three-phase inverter, or multilevel converters. Series and parallel connections of the PEBB can be conducted to meet the high voltage and high current requirements of the systems, as shown in (Figure 2.37). In [115], a full-bridge PEBB with rated DC-bus voltage of 1 kV was developed based on 1.7 kV SiC MOSFET modules, demonstrating an efficiency of 98% at 100 kW with 100 kHz. Aiming at utility-scale solar inverter applications, 1.5 kV PEBB based on 1.7 kV/450 A SiC MOSFET modules has been introduced in [116], with a demonstrated CEC efficiency of 99% at MW-scale power.

The first scalable PEBB demonstration based on 10 kV/240 A SiC MOSFET modules was developed to constitute a compact 13.8 kV to 265 V AC/AC solid-state power substation, where the PEBBs were connected in series at the input and in parallel at the output to share the high input voltage and high output current, respectively [115]. Significant size and weight reduction of the solid-state power substation system were demonstrated. Since then, there has been significant progress made in the packaging of the 10 kV into a 6 kV PEBB as attention has been paid to many of the design challenges [102,117,118].

To enable broader penetration of PEBBs in various applications, the future trend of the PEBBs may include smart multi-timescale control architectures, advanced packaging and integration technologies to yield more compact and robust designs, higher reliability and efficiency, multi-domain modeling, simulation, and optimization to reduce the cost [119].

It is also important to consider the impact of system level voltage on the insulation coordination of the PEBBs. For example, stacking lower voltage rating PEBBs to achieve a higher voltage rated system is perfectly acceptable from a LL, differential mode electrical voltage stress perspective. However, the LG voltage stresses on any one of the building blocks will be the same as the system level LG voltage stress. Emerging solutions are the additional voltage dielectric stand-off distances around the individual PEBBs and use of conductive and dielectric field grading at PEBB electrical interconnections with other LRUs (such as high frequency transformers and inductors) to ensure adequate isolation between any part of the PEBB that connects electrically to the system voltage [120–123]. Isolation and insulation coordination methods must also account for the ground wall isolation of the thermal management system.

2.4.2.3   Switching frequency

From the harmonic distortion perspective, high switching frequency is preferred for the MVDC converters due to the required compliance to the grid code for utility interfaced systems or due to the high power density requirement for high-speed propulsion motor-drive systems. In many applications, high switching frequency is recommended for the MVDC converters to reduce the size of magnetic components, such as filter inductors, power transformers, and electric machines. However, high switching frequency will inevitably cause substantial switching losses of the converters and increase the required thermal dissipation capacity of the cooling equipment, even configured with the emerging low-loss 1.7 or 3.3 kV high-voltage SiC MOSFET modules. The switching frequency with the modern medium-voltage IGBT/IGCT modules and the SiC MOSFET modules is typically constrained no more than 10 and 30 kHz in MVDC applications, respectively. Thus, systematic trade-off studies considering all these critical factors is very necessary to seek the optimal value of the switching frequency for the MVDC converters.

2.4.2.4   Series connection

Currently, the highest voltage blocking ratings for commercial Silicon IGBT and IGCT modules is 6.5 kV, and the highest voltage blocking ratings for the SiC modules is 3.3 kV up to now [124]. Although SiC MOSFET modules with higher blocking voltages including 6.5, 10 and 15 kV have been under development by semiconductor manufacturers, it may take at least a few more years to become commercialized into the market [125,126]. To meet the high voltage levels with MVDC converters, such as 20–30 kV, series connection of the semiconductor switches will be unavoidable to develop the MVDC converters. However, it is very common to observe unequal dynamic or static voltage sharing among series-connected semiconductors due to mismatched gate driver performance or uneven semiconductor parameters. Such uneven voltage sharing among the series-connected devices can lead to overvoltage failures, if no balancing techniques are available. Common voltage balancing techniques include passive snubber circuits, controllable RC snubber circuits, active miller clamping techniques, and so forth [127]. On the other hand, from the circuit topology perspective for MVDC applications, instead of adopting two-level converters based on a number of series-connected switches, multilevel converters is preferred due to the avoiding of large number of semiconductor switches in series, higher apparent switching frequency, and other benefits.

2.4.2.5   Parallel connection

Connecting semiconductor devices in parallel is a common practice in MVDC applications when the current rating of a single device is insufficient. It is well known that asymmetrical current distribution during static and dynamic modes among the semiconductor devices in parallel has been a challenge, constraining the number of devices that can be connected in parallel. Any variations of the device characteristics, modules, gate driver circuits, converter circuit impedance, and cooling homogeneity can cause asymmetrical current distribution.

The present Si IGBTs and SiC MOSFETs have a positive temperature coefficient (PTC), which are beneficial for parallel connection since larger on-state resistance at higher temperature will in turn naturally reduce the current distribution if one device shares more current. However, most anti-parallel diodes have a negative temperature coefficient (NTC) in their rated current range, leading to more unbalanced current distribution than the devices with PTC characteristics, particularly for MVDC converters operating in low power factor.

From the topological perspective, paralleling two-level or multilevel VSCs as building blocks is another approach to meet the high power requirements in MVDC applications. Paralleling converters also have other benefits, including modularity, easy maintenance, harmonic reduction through interleaving, and fault tolerance with (N+1) configuration.

2.4.3   System requirements

2.4.3.1   Efficiency

In almost every MVDC applications, there are strict requirements on the converter efficiency. For instance, in electric aircraft and shipboard propulsion systems, the efficiency of the converter basically determines the mileage range. For MVDC converters used in solar PV, wind turbines, and battery energy storage applications, there are rigorous efficiency requirements and standards for the converters to comply before release to the market. From a systematic viewpoint, the efficiency is determined by both the hardware configurations and the firmware including PWM and control strategies. The hardware configuration incorporates the semiconductor devices, power converter topologies, filter topologies, and the like. The system efficiency is also closely related to the operating points such as the converter switching frequency, power factor, and modulation index. It should be noted that, not all the MVDC systems are mostly operated at full load during their mission profiles. For instance, large uninterruptible power supplies (UPS) typically operate at light load around 20%–30% in data center applications, and aircraft propulsion systems are normally operated around 30% of the full power during the cruising mode. Hence, a systematic efficiency optimization considering the specific mission profiles is necessary. For MW-scale MVDC converters, the converter efficiency generally can reach above 90%, depending on the semiconductor devices, circuit topologies, conductor material, modulation and control strategies, ambient temperature, as well as cooling conditions.

2.4.3.2   Power density

Among the aforementioned applications, the transportation industry has more rigorous requirements on the power density of the MVDC converters due to the limited space and allowable weight on board, compared to other scenarios such as the outdoor wind farm or solar farm applications where the weight and physical volume is not so sensitive. The power density of MVDC converters is determined by a number of factors, including semiconductor and filter losses, thermal management, dielectric insulation requirements, and mechanical design. The present gravimetric power density (i.e., specific energy) of MVDC converters is approximately in the range of 5–15 kW/kg, and the volumetric power density is in the range of 4–10 kW/L. Generally, the higher the efficiency of the converter, the higher the power density. However, higher reliability design of the MVDC converter may decrease the power density, due to design factors such as excessive electrical design margins, redundant backup power modules or phase legs as well as the associated extra semiconductor losses and cooling design, which will be detailed next.

2.4.3.3   Reliability

Reliability of MVDC converters has always been a concern due to the utilization of a large number of semiconductor devices and other passive components such as film capacitors, especially for safety-critical applications such as airplanes, shipboard systems, power grid, and large data centers. Any single fault in the converter circuits may lead to a severe malfunction, systematic collapse, or even a disaster. These faults can be device-level faults such as open-circuit switching faults and short-circuit switching faults, or converter-level faults such as line voltage unbalance, line-to-ground faults, DC-link shoot-through faults, and a few others. For instance, overvoltage and overcurrent are two common failure causes in MVDC converters if electrically designed or operated inappropriately; dramatic decrease of power cycling lifetime with MVDC inverters may occur by high junction temperature swings during heavy-duty low-speed (e.g., a few Hertz) operation. To ensure high reliability of MVDC converters, in addition to having sufficient design margins on the active and passive power components, dedicated online prognosis, diagnosis, and fault-tolerant operation strategy is also of high necessity to avoid the failures and minimize the downtime cost. A hierarchic diagram of a high-reliability design paradigm is shown in (Figure 2.38). For the fault prognosis and diagnosis, they are typically sensory data-based software algorithms embedded in the microcontrollers, unless ultra-fast detection speed (e.g., a few microseconds) is required such as semiconductor short-circuit faults whose detection is generally integrated in the gate driver circuits for prompt actions, since the semiconductor short-circuit withstanding time is generally no longer than 10 µs. Fault-tolerant design of MVDC converters can be classified into two categories. One category is achieved based on the inherent redundant switching states of the converter topologies, which is feasible in multilevel converters and other topologies. Such an approach does not require any redundant hardware components, but it might be challenging to achieve full-power ratings during post-fault operation unless the power components are

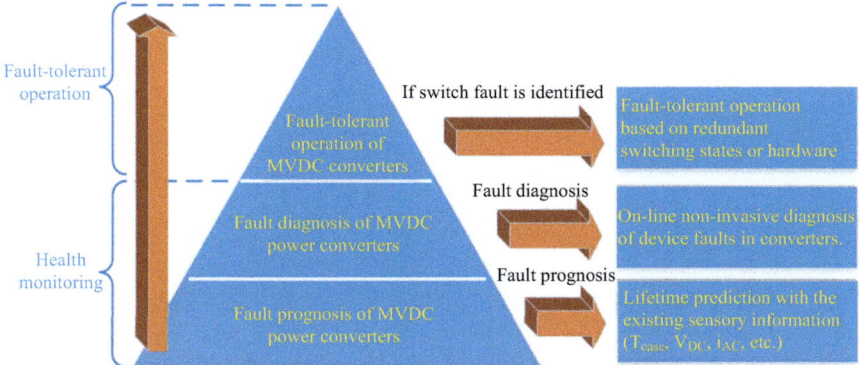

Figure 2.38 A hierarchic diagram of a high-reliability design paradigm for MVDC converters

oversized. The second category of approaches is redundant hardware-based fault-tolerant design, including adding redundant power modules, phase legs, or even an additional identical converter for backup purpose. Obviously, such redundant hardware design will significantly decrease the efficiency and power density at an increased system cost, although this approach can better achieve full power ratings during fault-tolerant operation. Thus, reliability design of MVDC converters requires optimization of the holistic systems, with comprehensive consideration of the specific reliability requirements, efficiency, cost, and power density.

2.5 Conclusions

This book chapter has provided an extensive review of emerging MVDC applications and how their viability and distribution architecture implementations are influenced by emerging power electronic device technologies and multilevel circuit topologies. The realization of MVDC systems represents an extremely rich area for research, particularly when it comes to pushing technologies, such as WBG power semiconductors and packaging of increasingly commercially available multi-chip modules into larger systems. While any MVDC compatible PEC can be designed and built for a specific application, the real challenge comes to viability is in the maturing of truly plug and play, modular approaches to the build-up of PEC solutions across a wide application space. This is especially challenging to the MVDC environment where there is little experience with regard to practicalities such as impacts of this environment to creepage and clearance constraints (at various levels of electrical interface), reliability, resiliency, sustainability and EMC/EMI. It is expected that significant future research will apply to insulation coordination, partial discharge and conducted internal and intra- and inter-system EMC at system and equipment levels. In short, significant attention is being given to the complexities associated with the packaging and controls of the modular *cyber-physical* structure of MVDC system compatible PECs in order to make future MVDC systems viable.

References

[1]   B. Kroposki, B. Johnson, Y. Zhang, V. Gevorgian, P. Denholm, B.-M. Hodge, and B. Hannegan. Achieving a 100% renewable grid: Operating electric power systems with extremely high levels of variable renewable energy. *IEEE Power and Energy Magazine*, 15(2): 61–73, 2017.

[2]   G. F. Reed, B. M. Grainger, A. R. Sparacino, R. J. Kerestes, and M. J. Korytowski. Advancements in medium voltage DC architecture development with applications for powering electric vehicle charging stations. In *2012 IEEE Energytech*, pages 1–8. IEEE, 2012.

[3]   F. Mura and R. W. De Doncker. *Preparation of a medium-voltage DC grid demonstration project*. E. ON ERC, 2012.

[4]   R. Yu, Y. Chen, J. Pan, R. W. Vesel, *et al.* Generic reliability evaluation method for industrial grids with variable frequency drives. *Energy and Power Engineering*, 5(4B): 83–88, 2013.

[5]   M. Stieneker and R. W. De Doncker. Medium-voltage DC distribution grids in urban areas. In *2016 IEEE 7th International Symposium on Power Electronics for Distributed Generation Systems (PEDG)*, pages 1–7. IEEE, 2016.

[6]   G. Wang, G. Konstantinou, C. D. Townsend, J. Pou, S. Vazquez, G. D. Demetriades, and V. G. Agelidis. A review of power electronics for grid connection of utility-scale battery energy storage systems. *IEEE Transactions on Sustainable Energy*, 7(4): 1778–1790, 2016.

[7]   Y. Rongrong and C. Yao. Common DC bus concept in power plant auxiliary system: Part ii economic analysis. In *IECON 2017—43rd Annual Conference of the IEEE Industrial Electronics Society,* pages 1839–1843. IEEE.

[8]   M. Stieneker, B. J. Mortimer, A. Hinz, A. Müller-Hellmann, and Rik W. De Doncker. Mvdc distribution grids for electric vehicle fast-charging infrastructure. In *2018 International Power Electronics Conference (IPEC-Niigata 2018-ECCE Asia)*, pages 598–606. IEEE, 2018.

[9]   J. Chen, Y. Liu, L. Qu, Q. Song, Z. Yu, R. Zeng, Z. Yuan, Z. Nie, and B. Zhao. Main loop design of Zhuhai three-terminal DC distribution system. In *2019 4th IEEE Workshop on the Electronic Grid (eGRID)*, pages 1–5. IEEE, 2019.

[10]   J. G. Ciezki and R. W. Ashton. A technology overview for a proposed navy surface combatant DC zonal electric distribution system. *Naval Engineers Journal*, 111(3):59–69, 1999.

[11]   J. Kuseian. *Naval power systems technology development roadmap. Electric Ships Office, PMS*, 320, 2013.

[12]   T. Ericsen. Power Electronic Building Blocks—A systematic approach to power electronics. In *Power Engineering Society Summer Meeting, 2000. IEEE*, volume 2, pages 1216–1218. IEEE, 2000.

[13] T. Ericsen, Y. Khersonsky, P. Schugart, and P. Steimer. PEBB-power electronics building blocks, from concept to reality. 2006.

[14] J del Á Ferrandis, J. Chalfant, C. M. Cooke, and C. Chryssostomidis. Design of a power corridor distribution network. In *2019 IEEE Electric Ship Technologies Symposium (ESTS)*, pages 284–292. IEEE, 2019.

[15] J. Benzaquen, J. He, and B. Mirafzal. Toward more electric powertrains in aircraft: Technical challenges and advancements *CES Transactions on Electrical Machines and Systems*, 5(3):177–193, 2021.

[16] R. Cuzner and A. Jeutter. Dc zonal electrical system fault isolation and reconfiguration. In *Electric Ship Technologies Symposium, 2009. ESTS 2009. IEEE*, pages 227–234. IEEE, 2009.

[17] M. W. Rose and R. M. Cuzner. Fault isolation and reconfiguration in a three-zone system. In *2015 IEEE Electric Ship Technologies Symposium (ESTS)*, pages 409–414. IEEE, 2015.

[18] R. Derakhshanfar, T. U. Jonsson, U. Steiger, and M. Habert. Hybrid HVDC breaker–a solution for future HVDC system. *Cigre,* pages 24–29, 2014.

[19] J. Häfner and A. Hassanpoor. HVDC hybrid circuit breaker with snubber circuit, 2014. US Patent 8,891,209.

[20] E. O'Neill-Carrillo, I. Jordan, A. Irizarry-Rivera, and R. Cintron. The long road to community microgrids: adapting to the necessary changes for renewable energy implementation. *IEEE Electrification Magazine*, 6(4): 6–17, 2018.

[21] A. Fernández-Guillamón, K. Das, N. A. Cutululis, and Á. Molina-Garca. Offshore wind power integration into future power systems: overview and trends. *Journal of Marine Science and Engineering*, 7(11): 399, 2019.

[22] R. M. Cuzner, A. R. Bendre, J. D. Widmann, K. A. Stonger, S. M. Peshman, J. S. Carlton, and J. A. Fischer. Considerations when diode auctioneering multiple DC buses in a non-isolated DC distribution system. In *2011 IEEE Electric Ship Technologies Symposium*, pages 277–282. IEEE, 2011.

[23] Z. Zhao, M. Xu, Q. Chen, J.-S. Lai, and Y. Cho. Derivation, analysis, and implementation of a boost–buck converter-based high-efficiency pv inverter. *IEEE Transactions on Power Electronics*, 27(3): 1304–1313, 2011.

[24] M. N. Kheraluwala, R. W. Gascoigne, D. M. Divan, and E. D. Baumann. Performance characterization of a high-power dual active bridge DC-to-DC converter. *IEEE Transactions on Industry Applications*, 28(6): 1294–1301, 1992.

[25] S. P. Engel, N. Soltau, H. Stagge, and R. W. D. Doncker. Dynamic and balanced control of three-phase high-power dual-active bridge DC–DC converters in DC–grid applications. *IEEE Transactions on Power Electronics*, 28(4): 1880–1889, 2012.

[26] T. Jimichi, M. Kaymak, and R. W. D. Doncker. Design and experimental verification of a three-phase dual-active bridge converter for offshore wind turbines. In *2018 International Power Electronics Conference (IPEC-Niigata 2018-ECCE Asia)*, pages 3729–3733. IEEE, 2018.

[27] C.-M. Young, J.-W. Siao, W.-S. Yeh, and S.-J. Cheng. An input-parallel and output-series-parallel phase-shift full-bridge converter with maximum power point tracking for wind turbine. In *2013 International Conference on Renewable Energy Research and Applications (ICRERA)*, pages 133–136. IEEE, 2013.

[28] H. Fan and H. Li. High-frequency transformer isolated bidirectional DC–DC converter modules with high efficiency over wide load range for 20 kva solid-state transformer. *IEEE Transactions on Power Electronics*, 26(12): 3599–3608, 2011.

[29] M. I. Rycroft. Medium and low voltage DC networks: an emerging alternative to AC. *EE Publishers*, 2017.

[30] H. Tang and A. Q. Huang. A design investigation of A 1 MVA SiC medium voltage three phase rectifier based on isolated dual active bridge. In *2018 IEEE Applied Power Electronics Conference and Exposition (APEC)*, pages 2409–2416, San Antonio, TX, USA, 2018. IEEE.

[31] U. A. Yusufoglu, T. M. Pletzer, L. J. Koduvelikulathu, C. Comparotto, R. Kopecek, and H. Kurz. Analysis of the annual performance of bifacial modules and optimization methods. *IEEE Journal of Photovoltaics*, 5(1): 320–328, 2015.

[32] E. A. Gunther. The state of medium voltage DC architectures for utility-scale PV. *PVTech*, 2018.

[33] E. Wesoff. *First Solar Postpones its Move to a Medium-Voltage DC Utility Solar Architecture. Greentech Media*, 2017.

[34] H. Wang, X. Huang, Y. Wang, and H. Xu. Series-connected PV MVDC converter for large scale pv system. In *2019 10th International Conference on Power Electronics and ECCE Asia (ICPE 2019-ECCE Asia)*, pages 1246–1251. IEEE, 2019.

[35] J. Pan, S. Bala, M. Callavik, and P. Sandeberg. Platformless DC collection and transmission for offshore wind. In *11th IET International Conference on AC and DC Power Transmission*, pages 1–6, 2015.

[36] F. Martinez-Rodrigo, D. Ramirez, A. B. Rey-Boue, S. De Pablo, and L. C. Herrero-de Lucas. Modular multilevel converters: control and applications. *Energies*, 10(11):1709, 2017.

[37] T. M. Haileselassie, M. Molinas, T. Undeland, *et al.* Multi–terminal VSC–HVDC system for integration of offshore wind farms and green electrification of platforms in the North Sea. In *Nordic Workshop on Power and Industrial Electronics (NORPIE/2008), June 9–11, 2008, Espoo, Finland.* Helsinki University of Technology, 2008.

[38] D. V. Hertem and M. Ghandhari. Multi-terminal VSC HVDC for the European supergrid: obstacles. *Renewable and sustainable energy reviews*, 14(9): 3156–3163, 2010.

[39] A. Ray, K. Rajashekara, H. Krishnamoorthy, *et al.* Novel HVDC power transmission architectures for subsea grid. In *Offshore Technology Conference*, 2019.

[40] BloombergNEF. Energy storage outlook 2019 report. https://about.bnef.com/. accessed: 2020-04-20.

[41] A. B. B. Website. DynaPeaQ energy storage system: a UK first. https://new. abb.com/facts/references/reference-dynapeaq-a-uk-first. accessed: 2020-04-21.

[42] J. M. Bloemink and T. C. Green. Benefits of distribution-level power electronics for supporting distributed generation growth. *IEEE Transactions on Power Delivery*, 28(2): 911–919, 2013.

[43] G. Bathurst, G. Hwang, and L. Tejwani. MVDC-the new technology for distribution networks. 2015.

[44] Q. Qi, C. Long, J. Wu, K. Smith, A. Moon, and J. Yu. Using an MVDC link to increase dg hosting capacity of a distribution network. *Energy Procedia*, 142: 2224–2229, 2017.

[45] J. Yu, K. Smith, M. Urizarbarrena, N. MacLeod, R. Bryans, and A. Moon. Initial designs for the angle DC project; converting existing AC cable and overhead line into DC operation. 2017.

[46] T. Joseph, J. Liang, G. Li, A. Moon, K. Smith, and J. Yu. Dynamic control of MVDC link embedded in distribution network: case study on angle-DC. In *2017 IEEE Conference on Energy Internet and Energy System Integration (EI2)*, pages 1–6. IEEE, 2017.

[47] A. Shekhar, T. B. Soeiro, L. Ramrez-Elizondo, and P. Bauer. Weakly meshing the radial distribution networks with power electronic based flexible DC interlinks.

[48] W. Cao, J. Wu, N. Jenkins, C. Wang, and T. Green. Operating principle of soft open points for electrical distribution network operation. *Applied Energy*, 164: 245–257, 2016.

[49] C. Wang, G. Song, P. Li, H. Ji, J. Zhao, and J. Wu. Optimal siting and sizing of soft open points in active electrical distribution networks. *Applied energy*, 189: 301–309, 2017.

[50] J. Barron and M. Zaveri. Power restored to Manhattan's West Side after major blackout. *The New York Times*, 2019.

[51] B. K. Johnson, R. H. Lasseter, F. L. Alvarado, and R. Adapa. Expandable multiterminal DC systems based on voltage droop. *IEEE Transactions on Power Delivery*, 8(4): 1926–1932, 1993.

[52] L. Xu, B. W. Williams, and L. Yao. Multi-terminal DC transmission systems for connecting large offshore wind farms. In *2008 IEEE Power and Energy Society General Meeting-Conversion and Delivery of Electrical Energy in the 21st Century*, pages 1–7. IEEE, 2008.

[53] L. Qu, Z. Yu, Q. Song, Z. Yuan, B. Zhao, D. Yao, J. Chen, Y. Liu, and R. Zeng. Planning and analysis of the demonstration project of the MVDC distribution network in Zhuhai. *Frontiers in Energy*, 13(1): 120–130, 2019.

[54] K. Rouzbehi, A. Miranian, A. Luna, and P. Rodriguez. Towards fully controllable multi-terminal DC grids using flexible DC transmission systems. In *2014 IEEE Energy Conversion Congress and Exposition (ECCE)*, pages 5312–5316. IEEE, 2014.

[55] A. Escobar-Mejia, Y. Liu, J. C. Balda, and K. George. New power electronic interface combining DC transmission, a medium-frequency bus and

an AC-AC converter to integrate deep-sea facilities with the AC grid. In *2014 IEEE Energy Conversion Congress and Exposition (ECCE)*, pages 4335–4344. IEEE, 2014.

[56] D. Kumar, F. Zare, and A. Ghosh. Dc microgrid technology: system architectures, AC grid interfaces, grounding schemes, power quality, communication networks, applications, and standardizations aspects. *Ieee Access*, 5: 12230–12256, 2017.

[57] C. Yuan, M. A. Haj-Ahmed, and M. S. Illindala. An MVDC microgrid for a remote area mine site: protection, operation and control. In *2014 IEEE Industry Application Society Annual Meeting*, pages 1–9. IEEE, 2014.

[58] F. Mura and R. W. D. Doncker. Design aspects of a medium-voltage direct current (MVDC) grid for a university campus. In *8th International Conference on Power Electronics-ECCE Asia*, pages 2359–2366. IEEE, 2011.

[59] Boeing Online Document. Boeing 787 dreamliner. www.boeing.com/commercial/787/. Accessed: 2019-09-18.

[60] Airbus. E-fan x: A giant leap towards zero-emission flight. www.airbus.com/innovation/. Accessed: 2020-04-20.

[61] D. Zhang, J. He, and D. Pan. A megawatt-scale medium-voltage high efficiency high power density SiC+Si hybrid three-level anpc inverter for aircraft hybrid-electric propulsion systems. *IEEE Transactions on Industry Applications*, 55(6): 5971–5980, 2019.

[62] I. Cotton, R. Gardner, D. Schweickart, D. Grosean, and C. Severns. Design considerations for higher electrical power system voltages in aerospace vehicles. In *2016 IEEE International Power Modulator and High Voltage Conference (IPMHVC)*, pages 57–61, 2016.

[63] X. Pei, O. Cwikowski, D. S. Vilchis-Rodriguez, M. Barnes, A. C. Smith, and R. Shuttleworth. A review of technologies for MVDC circuit breakers. In *IECON 2016—42nd Annual Conference of the IEEE Industrial Electronics Society*, pages 3799–3805, 2016.

[64] ABB Application Note. Failure rates of IGBT modules due to cosmic rays. https://library.e.abb.com/. Accessed: 2019–10–18.

[65] A. Akturk, J. McGarrity, N. Goldsman, D. J. Lichtenwalner, B. Hull, D. Grider, and R. Wilkins. The effects of radiation on the terrestrial operation of sic MOSFETS. In *2018 IEEE International Reliability Physics Symposium (IRPS)*, pages 2B.1–1–2B.1–5, 2018.

[66] A. Bolotnikov *et al.* Overview of 1.2kV-2.2kV SiC MOSFETs targeted for industrial power conversion applications. In *2015 IEEE Applied Power Electronics Conference and Exposition (APEC)*, pages 2445–2452, 2015.

[67] W. M. McBride. *Technological change and the United States Navy, 1865–1945*, volume 27. JHU Press, 2000.

[68] R. Becker. Marpol 73/78: An overview in international environmental enforcement. *Georgetown International Environmental Law Review*, 10:625, 1997.

[69] M. Julian. Marpol 73/78: the international convention for the prevention of pollution from ships. *Maritime Studies*, 2000(113): 16–23, 2000.

[70] T. W. P. Smith, J. P. Jalkanen, B. A. Anderson, J. J. Corbett, J. Faber, S. Hanayama, E. O'keeffe, S. Parker, L. Johansson, L. Aldous, *et al.* Third IMO greenhouse gas study 2014. *International Maritime Organization*, 327, 2014.

[71] M. Cames, J. Graichen, A. Siemons, and V. Cook. Emission reduction targets for international aviation and shipping. *Directorate General for Internal Policies; European Parliament–Policy Department A; Economic and Scientific Policy: Bruxelles, Belgium*, 2015.

[72] U. Javaid, F. D. Freijedo, D. Dujic, and W van der Merwe. MVDC supply technologies for marine electrical distribution systems. *CPSS Transactions on Power Electronics and Applications*, 3(1): 65–76, 2018.

[73] D. H. Clayton, S. D. Sudhoff, and G. F. Grater. Electric ship drive and power system. In *Conference Record of the 2000 Twenty-Fourth International Power Modulator Symposium*, pages 85–88. IEEE, 2000.

[74] J. S. Thongam, M. Tarbouchi, A. F. Okou, D. Bouchard, and R. Beguenane. All–electric ships – a review of the present state of the art. In *2013 Eighth International Conference and Exhibition on Ecological Vehicles and Renewable Energies (EVER)*, pages 1–8. IEEE, 2013.

[75] G. Sulligoi. All electric ships: present and future after 20 years of research and technical achievements. 2011.

[76] A. Vicenzutti, D. Bosich, G. Giadrossi, and G. Sulligoi. The role of voltage controls in modern all–electric ships: toward the all electric ship. *IEEE Electrification Magazine*, 3(2): 49–65, 2015.

[77] D. Bosich, A. Vicenzutti, R. Pelaschiar, R. Menis, and G. Sulligoi. Toward the future: the MVDC large ship research program. In *2015 AEIT International Annual Conference (AEIT)*, pages 1–6. IEEE, 2015.

[78] Z. Jin, G. Sulligoi, R. Cuzner, L. Meng, J. C. Vasquez, and J. M. Guerrero. Next-generation shipboard DC power system: introduction smart grid and DC microgrid technologies into maritime electrical Networks. *IEEE Electrification Magazine*, 4(2): 45–57, 2016.

[79] G. Sulligoi, D. Bosich, G. Giadrossi, L. Zhu, M. Cupelli, and A. Monti. Multiconverter medium voltage DC power systems on ships: constant-power loads instability solution using linearization via state feedback control. *IEEE Transactions on Smart Grid*, 5(5): 2543–2552, 2014.

[80] G. Sulligoi, A. Vicenzutti, V. Arcidiacono, and Y. Khersonsky. Voltage stability in large marine-integrated electrical and electronic power systems. *IEEE Transactions on industry applications*, 52(4): 3584–3594, 2016.

[81] IEEE Standards Association et al. IEEE recommended practice for 1 kV to 35 kV medium voltage DC power systems on ships. *IEEE Std 1709TM–2010*.

[82] Y. Khersonsky and G. Sulligoi. New IEEE & IEC standards for ships and oil platforms. In *2014 IEEE Petroleum and Chemical Industry Technical Conference (PCIC)*, pages 191–199. IEEE, 2014.

[83] N. Doerry. Next generation integrated power systems (NGIPS) for the future fleet. In *IEEE Electric Ship Technologies Symposium*, volume 150, pages 200–250, 2009.

[84] G. Sulligoi, A. Tessarolo, V. Benucci, M. Baret, A. Rebora, and A. Taffone. Modeling, simulation, and experimental validation of a generation system for medium-voltage DC integrated power systems. *IEEE Transactions on industry applications*, 46(4): 1304–1310, 2010.

[85] M. Aizza, D. Bosich, S. Castellan, R. Menis, G. Sulligoi, and A. Tessarolo. Coordinated speed and voltage regulation of a DC power generation system based on a wound field split-phase generator supplying multiple rectifiers. 2012.

[86] A. Tessarolo, S. Castellan, R. Menis, and G. Sulligoi. Electric generation technologies for all-electric ships with medium-voltage DC power distribution systems. In *2013 IEEE Electric Ship Technologies Symposium (ESTS)*, pages 275–281. IEEE, 2013.

[87] A. Kadavelugu, S. Bhattacharya, S.-H. Ryu, E. V. Brunt, D. Grider, A. Agarwal, and S. Leslie. Characterization of 15 kV SiC n-IGBT and its application considerations for high power converters. In *2013 IEEE energy conversion congress and exposition*, pages 2528–2535. IEEE, 2013.

[88] S. Madhusoodhanan, K. Hatua, S. Bhattacharya, S. Leslie, S.-H. Ryu, M. Das, A. Agarwal, and D. Grider. Comparison study of 12 kV n-type SiC IGBT with 10 kV SiC MOSFET and 6.5 kV Si IGBT based on 3L-NPC VSC applications. In *2012 IEEE Energy Conversion Congress and Exposition (ECCE)*, pages 310–317. IEEE, 2012.

[89] J. E. Huber and J. W. Kolar. Optimum number of cascaded cells for high-power medium-voltage AC–DC converters. *IEEE Journal of Emerging and Selected Topics in Power Electronics*, 5(1): 213–232, 2016.

[90] A. Q. Huang. Wide bandgap (WGB) power devices and their impacts on power delivery systems. In *2016 IEEE International Electron Devices Meeting (IEDM)*, pages 20–1. IEEE, 2016.

[91] N. Doerry, J. Amy, and C. Krolick. History and the status of electric ship propulsion, integrated power systems, and future trends in the us navy. *Proceedings of the IEEE*, 103(12): 2243–2251, 2015.

[92] J. G. Ciezki and R. W. Ashton. Selection and stability issues associated with a navy shipboard DC zonal electric distribution system. *IEEE Transactions on Power Delivery*, 15(2): 665–669, 2000.

[93] S. D. Sudhoff, S. Pekarek, B. Kuhn, S. Glover, J. Sauer, and D. Delisle. Naval combat survivability testbeds for investigation of issues in shipboard power electronics based power and propulsion systems. In *IEEE Power Engineering Society Summer Meeting*, volume 1, pages 347–350. IEEE, 2002.

[94] N. Doerry. Zonal ship design. *Naval Engineers Journal*, 118(1): 39–53, 2006.

[95] L. J. Petersen, M. Ziv, D. P. Burns, T. Q. Dinh, and P. E. Malek. *US Navy efforts towards development of future naval weapons and integration into*

an *All Electric Warship (AEW)*. In *IMarEST Engine as a Weapon (EAAW) Int. Symp. UK*, 2011.

[96] N. Doerry and J. Amy. MVDC shipboard power system considerations for electromagnetic railguns. In *Proc. 6th DoD Electromagn. Railgun Workshop*, 2015.

[97] P. K. Steimer. Power electronics building blocks-a platform-based approach to power electronics. In *Power Engineering Society General Meeting, 2003, IEEE*, volume 3, pages 1360–1365. IEEE, 2003.

[98] C. DiMarino, I. Cvetkovic, Z. Shen, R. Burgos, and D. Boroyevich. 10 kv, 120 a sic mosfet modules for a power electronics building block (PEBB). In *2014 IEEE Workshop on Wide Bandgap Power Devices and Applications*, pages 55–58. IEEE, 2014.

[99] I. Cvetkovic, Z. Shen, M. Jaksic, C. DiMarino, F. Chen, D. Boroyevich, and R. Burgos. Modular scalable medium-voltage impedance measurement unit using 10 kV SiC MOSFETs PEBBs. In *Electric Ship Technologies Symposium (ESTS), 2015 IEEE*, pages 326–331. IEEE, 2015.

[100] J. Wang, Z. Shen, I. Cvetkovic, N. R. Mehrabadi, A. Marzoughi, S. Ohn, J. Yu, Y. Xu, R. Burgos, and D. Boroyevich. Power electronics building block (PEBB) design based on 1.7 Kv SIC MOSFET modules. In *Electric Ship Technologies Symposium (ESTS), 2017 IEEE*, pages 612–619. IEEE, 2017.

[101] J. Wang, R. Burgos, D. Boroyevich, and Z. Liu. Design and testing of 1 kv h-bridge power electronics building block based on 1.7 Kv SIC MOSFET module. In *2018 International Power Electronics Conference (IPEC-Niigata 2018-ECCE Asia)*, pages 3749–3756. IEEE, 2018.

[102] H. Song, J. Wang, Y. Xu, R. Burgos, and D. Boroyevich. A high-density single-turn inductor for a 6 Kv sic-based power electronics building block. In *2020 IEEE Applied Power Electronics Conference and Exposition (APEC)*, pages 1127–1134. IEEE, 2020.

[103] A. Challita and M. Uva. High speed medium voltage direct current isolation device. In *ASNE Electric Machines Technology Symposium (EMTS)*, 2014.

[104] J. Langston, K. Schoder, M. Sloderbeck, M. Steurer, and A. Rockhill. Testing operation and coordination of DC solid state circuit breakers. In *IECON 2018—44th Annual Conference of the IEEE Industrial Electronics Society*, pages 3445–3452. IEEE, 2018.

[105] L. Qi, P. Cairoli, Z. Pan, C. Tschida, Z. Wang, V. R. Ramanan, L. Raciti, and A. Antoniazzi. Solid-state circuit breaker protection for DC shipboard power systems: breaker design, protection scheme, validation testing. *IEEE Transactions on Industry Applications*, 56(2): 952–960, 2019.

[106] IEEE recommended practice and requirements for harmonic control in electric power systems. *IEEE Std 519-2014 (Revision of IEEE Std 519-1992)*, pages 1–29, 2014.

[107] J. He, G. Y. Sizov, P. Zhang, and N. A. O. Demerdash. A review of mitigation methods for overvoltage in long-cable-fed PWM AC drives. In *2011 IEEE Energy Conversion Congress and Exposition*, pages 2160–2166, 2011.

[108]   J. He, C. Li, A. Jassal, N. Thiagarajan, Y. Zhang, S. Prabhakaran, C. Feliz, J. E. Graham, and X. Kang. Multi-domain design optimization of dv/dt filter for sic-based three-phase inverters in high-frequency motor-drive applications. *IEEE Transactions on Industry Applications*, 55(5): 5214–5222, 2019.

[109]   A. M. De Broe, A. L. Julian, and T. A. Lipo. Neutral-to-ground voltage minimization in a PWM-rectifier/inverter configuration. In *1996 Sixth International Conference on Power Electronics and Variable Speed Drives (Conf. Publ. No. 429)*, pages 564–568, 1996.

[110]   J. A. Oliver, G. Guerrero, and J. Goldman. Ceramic bearings for electric motors: eliminating damage with new materials. *IEEE Industry Applications Magazine*, 23(6): 14–20, 2017.

[111]   H. Chen and H. Zhao. Review on pulse-width modulation strategies for common-mode voltage reduction in three-phase voltage-source inverters. *IET Power Electronics*, 9(14): 2611–2620, 2016.

[112]   P. Roussel. Sic market and industry update. In *International SiC Power Electronics Applications Workshop*, 2011.

[113]   J. He, D. Zhang, and D. Pan. PWM strategy for MW-scale SiC+Si active NPC converter in electric aircraft propulsion applications. *IEEE Transactions on Industry Applications*, pages 1–9, 2020.

[114]   T. Ericsen, Y. Khersonsky, P. Schugart, and P. Steimer. PEBB – power electronics building blocks, from concept to reality. In *2006 3rd IET International Conference on Power Electronics, Machines and Drives – PEMD 2006*, pages 12–16, 2006.

[115]   T. Ericsen, R. Raju, R. Burgos, D. Boroyevich, and S. Beermann-Curtin. Advances in sic-based power conversion for shipboard electrical power systems. In *2015 IEEE 3rd Workshop on Wide Bandgap Power Devices and Applications (WiPDA)*, pages 341–346, 2015.

[116]   M. Harfman Todorovic, R. Datta, L. Stevanovic, X. She, P. Cioffi, G. Mandrusiak, B. Rowden, P. Szczesny, J. Dai, and T. Frangieh. Design and testing of a modular sic based power block. In *PCIM Europe 2016; International Exhibition and Conference for Power Electronics, Intelligent Motion, Renewable Energy and Energy Management*, pages 1–4, 2016.

[117]   C. DiMarino, M. Johnson, B. Mouawad, J. Li, R. Skuriat, M. Wang, Y. Tan, G. Q. Lu, D. Boroyevich, and R. Burgos. Fabrication and characterization of a high-power-density, planar 10 Kv SIC MOSFET power module. In *CIPS 2018; 10th International Conference on Integrated Power Electronics Systems*, pages 1–6. VDE, 2018.

[118]   J. Yu, R. Burgos, and D. Boroyevich. Emi study on control implementation in PEBB-based converter. In *2018 IEEE 19th Workshop on Control and Modeling for Power Electronics (COMPEL)*, pages 1–5. IEEE, 2018.

[119]   D. Boroyevich. Building block integration in power electronics. In *2010 IEEE International Symposium on Industrial Electronics*, pages 3673–3678, 2010.

[120]   A. Christe, E. Coulinge, and D. Dujic. Insulation coordination for a modular multilevel converter prototype. In *2016 18th European Conference on*

Power Electronics and Applications (EPE'16 ECCE Europe), pages 1–9. IEEE, 2016.

[121] T. Guillod, F. Krismer, and J. W. Kolar. Electrical shielding of mv/mf transformers subjected to high DV/DT PWM voltages. In *2017 IEEE Applied Power Electronics Conference and Exposition (APEC)*, pages 2502–2510. IEEE, 2017.

[122] T. Batra, G. Gohil, A. K. Sesham, N. Rodriguez, and S. Bhattacharya. Isolation design considerations for power supply of medium voltage silicon carbide gate drivers. In *2017 IEEE Energy Conversion Congress and Exposition (ECCE)*, pages 2552–2559. IEEE, 2017.

[123] Q. Chen, R. Raju, D. Dong, and M. Agamy. High frequency transformer insulation in medium voltage SIC enabled air-cooled solid-state transformers. In *2018 IEEE Energy Conversion Congress and Exposition (ECCE)*, pages 2436–2443. IEEE, 2018.

[124] N. Soltau, E. Wiesner, R. Tsuda, K. Hatori, and H. Uemura. Impact of gate control on the switching performance of a 750a/3300v dual SIC-module. In *2018 20th European Conference on Power Electronics and Applications (EPE'18 ECCE Europe)*, pages P.1–P.7, 2018.

[125] J. W. Palmour, L. Cheng, V. Pala, E. V. Brunt, D. J. Lichtenwalner, G. Wang, J. Richmond, M. O'Loughlin, S. Ryu, S. T. Allen, A. A. Burk, and C. Scozzie. Silicon carbide power MOSFETS: breakthrough performance from 900 v up to 15 kv. In *2014 IEEE 26th International Symposium on Power Semiconductor Devices IC's (ISPSD)*, pages 79–82, 2014.

[126] K. Vechalapu, S. Bhattacharya, E. Van Brunt, S. Ryu, D. Grider, and J. W. Palmour. Comparative evaluation of 15–Kv SIC MOSFET and 15–Kv SIC IGBT for medium–voltage converter under the same DV/DT conditions. *IEEE Journal of Emerging and Selected Topics in Power Electronics*, 5(1): 469–489, 2017.

[127] N. Y. A. Shammas, R. Withanage, and D. Chamund. Review of series and parallel connection of IGBTs. *IEE Proceedings–Circuits, Devices and Systems*, 153(1): 34–39, 2006.

Chapter 3

Restructuring the existing medium voltage distribution grids using DC systems

Aditya Shekhar[1] and Pavol Bauer[1]

3.1 Introduction

Typical distribution networks (DN) are predominantly AC and radial in nature, as shown in Figure 3.1. The voltage of a long distance high voltage (HV) network is stepped down to a medium voltage (MV) level at a substation located a few tens of kilometers outside an urban area. Power demand of the downstream network is delivered to an inner city substation using multiple parallel operating 3-phase AC links. Considering the critical function of this distribution link, adequate redundancy is employed to maintain the required power capacity during (n−1) contingencies, which refers to the operating condition with single component failure in the system.

3.1.1 Emerging challenges

Figure 3.1 highlights the emerging grid components such as electric vehicle chargers, traction systems and distributed generation resources like photovoltaics and wind farms. With mass deployment of such energy resources and high power loads, the existing grid infrastructure must be upgraded to meet the changing needs.

- Sustainable energy transition is leading to greater electrification. For example, it is anticipated that electrical energy demand will rise by at least 2–3 times by 2050 with increasing share of electric vehicles (EVs) and heat pumps [1]. Depending on the consumption patterns, this will translate to significant increase in peak power demand that can overload the critical AC power links in the system. In such systems, reinforcing the infrastructure with adequate redundancy for enhanced capacity during (n−1) contingencies becomes imperative.
- Dispersed and variable nature of renewable energy resources leads to power mismatch between generation and consumption even while the energy

[1]DC Systems, Energy Conversion & Storage group in the Department of Electrical Sustainable Energy, Delft University of Technology, Delft, The Netherlands

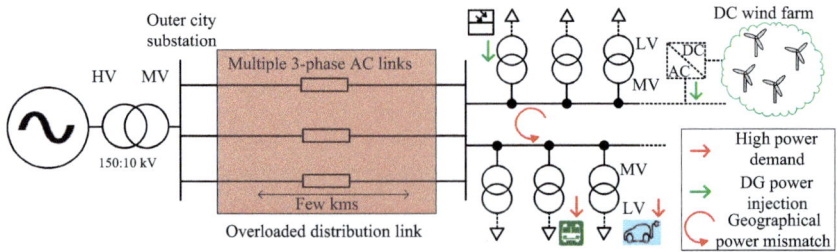

Figure 3.1   A typical medium voltage AC distribution network

dependence on the HV network is reduced. The radial structure of the MV distribution network further exacerbates the geographical mismatch associated with local pockets of excess power generation and consumption. In such operational scenarios, load balancing with efficient and controlled power redirection within the network is necessary.

- Direct integration and energy sharing between these components has led to the development of microgrid substructures capable of autonomous operation. The architecture of the point of common coupling between these microgrids and the main grid must handle operational challenges such as bidirectional power flows and protection aspects.

3.1.2   AC distribution network expansion

In response to growth in load, the Distribution Network Operators (DNOs) determine an optimal expansion plan to meet the future power demands. With network conductors as the main asset category, the decision on infrastructure reinforcement is a multi-objective consideration of trade-offs in investment cost, energy loss and reliability [2]. The three main constraints that must be satisfied during the problem formulation for expansion planning are (i) radial operation of the AC network, (ii) node voltage and branch current limits during normal operation, and (iii) adequate redundancy capacity during $(n-1)$ contingencies. Some operational flexibility to these constraints is introduced by considering the short duration overload capacity of the infrastructure and the possibility of reconfiguration using normally open tie-lines in the DN. According to [3], the hosting capacity of a radial DN downstream of a 11 kV, 10 MVA substation can be improved with optimal conductor sizing. The study defined a feeder reinforcement index corresponding to annual cost of energy losses relative to the investment cost and identified that the branch conductors closest to the substation needed to be upgraded with additional conductor cost compared to the base system. While re-conductoring and/or adding new multi-circuit AC branches is widely used by the DNOs for network capacity augmentation, almost 80% of the incurred costs are toward installation procedure (related to digging and routing) as compared to that of the actual cable material [4]. Further, the socio-economic difficulties can limit the practical viability of large-scale digging and installation of underground cable infrastructure, particularly in heritage cities like Amsterdam in The Netherlands [5]. Different solutions have emerged to address the need for deferring the DN capacity investments:

- **Energy Storage System (ESS)** can be a non-wire alternative to achieve peak demand shaving to prevent overloads in the network upstream of the installation node. For example, Deboever et al. [6] investigated the increase in the required energy and power capacity of the ESS as a function of current clipping objective including the impact of load growth for an actual 27.6 kV, 15 MVA feeder. The study suggested that a 15% peak shaving of current overload can be achieved with a 16.6 MWh, 2.5 MW ESS. A cost–benefit analysis for different 20 kV DN suggested that the target ESS cost of 60–80 €/kWh can be considered as a competing solution to traditional network reinforcement [7]. The paper anticipated a decline in future storage costs, use of second life batteries and increased penetration of EVs as possible scenarios where ESS based capacity reinforcement of secondary substations can be considered as an attractive solution.
- **Distributed Energy Resources (DERs):** While the variable nature of generation with DERs can cause grid congestion, their optimal location, sizing and operation in coordination with DN planning can be beneficial in addressing load growth. For example, an integrated approach of installing DERs along with upgrading tie-lines, conductor and transformer can offer 10% savings over a 20 year investment horizon [8]. The assumption is that the installed DER is able to supply power during peak load requirements in the system. A dynamic improvement of inter-area power transfer capability is suggested in [9] for under-utilized grid infrastructure under low wind conditions. However, some regulations prohibiting the DNOs from owning generation plants may lead to potential inefficiencies in supply infrastructure because the role of DERs in deferring investments of network expansion maybe overlooked as a consequence [10]. A distributed ownership of resources can, therefore, necessitate some degree of coordination between different stakeholders. The concept of prosumer based on clustered micro-generators could be used, especially with autonomous transaction with bitcoins/smart-meters.
- **Smart Loads:** Controllable power electronics assisted loads can offer some flexibility to address the supply challenges. Smart charging of electric vehicles and intelligent heating systems can shape the consumption patterns to relieve and/or support the DN during peak demand hours. It is observed that while the EV hosting capacity of a test DN can be maximized within the defined chargeable region with electricity price based demand response (DR), the capacity violations cannot be completely prevented [11].

A combination of aforementioned solutions along with use of Information and Communication Technologies (ICT) has led to the development of Demand Side Management (DSM) concept for smart grids [12]. Recognizing the limitation of market based DR in completely alleviating congestion, it is mixed with binding physical DR requests of load shedding for grid relief.

3.1.3 DC technology based solutions

The role of DC technologies in enabling smart grids is three-fold: (i) higher infrastructure efficiency and power density as compared to AC; (ii) potential for greater inter-connectivity as compared to conventional radially-constrained AC DN in power redirecton and improving availability; (iii) efficient integration because most emerging grid resources like ESS, DERs and EVs have intermediary DC conversion stage. This understanding has led to research initiatives in the control, protection and architectural design of DC distribution and microgrid systems [13]. However, the possibility of large-scale deployment of multiterminal DC grids at medium voltage level by DNOs is impeded by the cost of ownership and operational maturity with the existing AC infrastructure [14]. Therefore, the subsequent sections introduce working solutions for restructuring the existing AC grids with an objective of increasing the DC based distribution technologies in these systems.

This chapter emphasizes the role of DC technologies in achieving grid transition as an important step to realize sustainable energy transition. The discussed principles are applicable for MV levels of 10–66 kV, short link lengths space in between a few tens of kilometers and distribution system power capacities in the range of a few MWs. The proposed concepts have potential practical applications, for example in Siemens approach with 'MVDC Plus' [15].

3.2 Refurbishing AC infrastructure for DC operation

This section discusses the conventional solutions for AC DN planning to address the challenge of load growth. The potential power capacity gains by converting the AC infrastructure for DC operation is presented and the research opportunities on this topic are highlighted.

3.2.1 Capacity enhancement with DC operation

The existing AC conductor infrastructure can be refurbished to operate under DC conditions by AC/DC converters on either side of the link as shown in Figure 3.2.

The refurbishment strategy is proposed for operating branches as well as normally open tie-lines which can potentially reduce future overloads in the network [16]. Capacity enhancement can be achieved without installation of additional link conductors because DC operation has inherently higher power density as compared to AC as indicated in Figure 3.3.

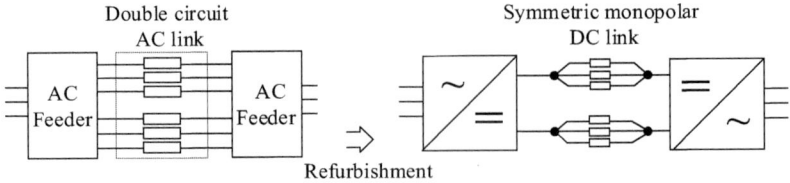

Figure 3.2   Refurbishing AC links for DC operation

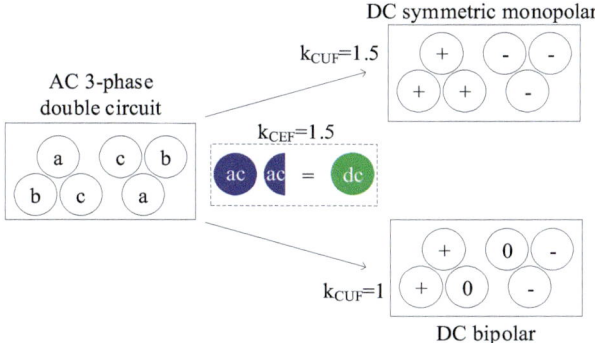

Figure 3.3   Power capacity enhancement by converting AC circuit for DC operation

The capacity enhancement factor k_{CEF} represents the amount of conductor area required to transfer AC power equivalent to that under DC operation. In the shown case, the value indicates that 1.5 times more AC conductor area is required to transfer the equivalent DC power. In general, the approximate k_{CEF} can be estimated from (3.1):

$$k_{CEF} = k_{vr} \underbrace{\left(\frac{v_{dc}}{v_{ac}} \right)}_{k_v} \underbrace{\left(\frac{i_{dc}}{i_{ac}} \right)}_{k_i} \underbrace{\left(\frac{1}{\cos \theta} \right)}_{k_{pf}} \tag{3.1}$$

Herein, the voltage enhancement factor k_v is defined based on the link insulation performance as a ratio of the pole to ground DC voltage v_{dc} and phase to ground AC voltage v_{ac}. While a conservative estimate of $k_v = \sqrt{2}$ can ensure similar or better insulation performance under DC as compared to AC [5,17], the possibility of selecting higher values of k_v to achieve significant capacity enhancements have been reported in the literature [18,19]. The factor k_i accounts for the current rating enhancement with DC operation. For short link lengths typical of medium voltage levels, k_i is in the range of 1.01–1.03 based on the skin affect associated with conductors of different cross-sectional area, while for systems with longer link lengths, the influence of capacitive currents can result in a higher value. Further, k_i can be higher if the AC to DC refurbishment impacts the derating associated with the thermal proximity of the link conductors. The factor k_{pf} incorporates the influence of reactive power support provided by the Voltage Source Converter (VSC) at the receiving end of the refurbished DC link. Assuming a minimum power factor $\cos \theta = 0.9$, k_{pf} is in the range of 1–1.1. Finally, the factor k_{vr} is associated with the voltage regulation and link length dependent inductive voltage drop and is in the range of 1.03–1.05. A discussion on the influence of varying system parameters and operating conditions on these factors is offered in [5], which forms the basis for the assumption that $k_{CEF} = 1.5$ is a reasonable conservative estimate for DC capacity enhancement.

The actual system level capacity enhancement is described by the conductor utilization factor k_{CUF} depending on the topology of the original AC link, the refurbished DC link and the selected substation converter rating. It is related to k_{CEF} by (3.2):

$$k_{CUF} = k_{cr} \left(\frac{N_{dc}}{N_{ac}} \right) k_{CEF} \tag{3.2}$$

N_{ac} and N_{dc} are the number of active conductors involved in high power transfer for original AC and refurbished DC link respectively. For example, if a three-phase AC link is converted to a 2-pole DC link, then $N_{ac} = 3$ and $N_{dc} = 2$ resulting in a reduced k_{CUF} relative to the k_{CEF}, leading to lower achievable capacity gains with the refurbishment strategy.

The converter rating factor k_{cr} is defined as the ratio of rated converter active power to the maximum achievable active power capacity of the DC link conductors. The value of k_{cr} influences the installation cost of the substation converters and can be selected based on the required capacity enhancement in the system. With maximum possible value of $k_{cr} = 1$ and a conservative $k_{CEF} = 1.5$, it can be seen from Figure 3.3 that while no capacity enhancement is achieved when a three-phase double circuit AC link is refurbished to a bipolar DC link with two neutral conductors, 1.5 times higher capacity is achieved when it is refurbished to a symmetric monopolar DC system. A review of capacity enhancement claims in the literature are listed in Table 3.1.

While all listed studies reported a k_{CEF} above 1.5, the contributing factors and assumptions based on the system parameters are different. For example, the field implementation in [4] selected the most conservative DC voltage enhancement k_v and still achieved a higher k_{CEF} because the operating line current was increased from 66 A AC to 150–300 A dc. On the other hand, the case-study in [18] suggests that the achieved DC capacity gains were because voltage drop associated with long link length limited the transmission capacity of the original AC system below the thermal limit. Both [4,18] suggest that a higher DC voltage enhancement can be selected with consideration to the consequences on insulation performance. Based on empirical evidence provided in [20], $k_v = \sqrt{2}$ can offer capacity gains that are independent of link length as suggested in [5,17].

A higher AC to DC voltage enhancement potential ($k_v > \sqrt{2}$) is an important consideration for medium voltage DNs because short links dominate such systems

Table 3.1   Capacity enhancement claims with refurbished DC link operation

References	k_{CEF}	k_v	$v_{ll,rms}$	Capacity	Length	Type
			(kV)	(MVA)	(km)	
[4]	1.5–6.0	0.5–1	35	3	40	Cable
[5]	≥1.5	$\sqrt{2}$	11	10–30	10–20	Cable
[17]	1.59	$\sqrt{2}$	10	10–20	–	Cable
[18]	1.63	1.18	66	70	18	Overhead
[19]	3.5	$2\sqrt{3}$	145–420	100–2,500	>100	Overhead

and therefore, the capacity gains associated with voltage regulation and capacitive current are limited. Ref. [19] is one of the early works supporting AC to DC refurbishment strategy for extremely high voltage overhead networks. While the study considers a much higher k_v, the suggested value is corresponding to the modification of supporting tower and insulation structures of the overhead transmission system. A similar principle is being explored for medium voltage overhead distribution lines in [15]. However, the appropriate choice of k_v must correlate similar insulation performance under AC and DC voltages, particularly for underground cable systems as discussed in Section 3.2.4.

3.2.2   Operational considerations

The defined factor k_{CUF} represents the capacity enhancement during normal system operation. During $(n-1)$ contingencies, introducing reconfigurability between the refurbished AC–DC link systems can be seen as a method to maximize the utilization of the remaining healthy infrastructure to reduce the need for redundancy in the system [21]. This is advantageous because the added cost of redundancy has limited use during normal conditions where the system spends the majority of its operational lifetime. Figure 3.4(a) shows the reconfigurable AC–DC link architecture where the feeder switches can regulate the physical AC and DC power flow path between the Sending end Sub-Station (SSS) and the Receiving end Sub-Station (RSS) via the existing link conductors. The AC–DC reconfiguration based service restoration and design of selective protection scheme for fault isolation is an important research topic and some discussion on the possible opportunities and challenges will be discussed in the subsequent sections.

Figure 3.4(b) depicts a parallel AC–DC link system between SSS and RSS. Such systems inevitably emerge with partial AC to DC refurbishment wherein only some of the multiple AC links (refer Figure 3.1) are converted for DC operation [22]. Further, the reconfigurability between AC and DC operation can offer flexibility in choosing the system architecture for a given active and reactive power (P, Q) demand at the RSS. The optimal active power sharing (P_{ac}, P_{dc}) between the AC and DC link systems is a function of dynamically varying operating efficiency. The control objective of the RSS side VSC is to steer the active power in the system at the maximum efficiency while supporting the reactive power needs in the system for a given AC–DC configuration.

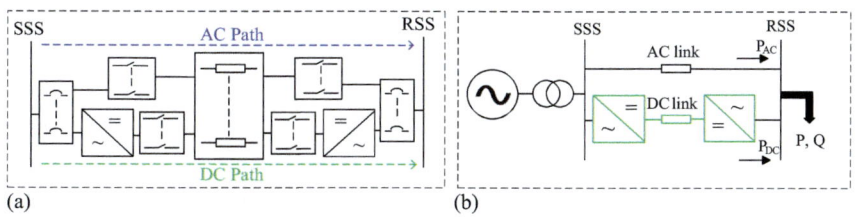

Figure 3.4   Operational considerations with (a) reconfigurability between AC and DC links and (b) parallel AC–DC operation

3.2.3   Converter station design considerations

The substation converters must be rated according to the maximum demand that they are expected to deliver during normal operation as well as $(n-1)$ contingencies. k_{cr} can be reduced while maintaining the same system capacity based on the operational mode selected such as parallel AC–DC operation and the reconfiguration strategy employed when a fault occurs [23]. For example, it is discussed that the required converter station capacity can be downsized by more than 50% to maintain the same system capacity during $(n-1)$ contingencies if a re-configurable architecture shown in Figure 3.4 is used. While the advantage of this derating opportunity is the significant reduction in installation costs, it can reduce the system efficiency under normal operating conditions because DC power transfer can be more efficient under specific situations. This trade-off must be investigated by the DNOs for proper sizing of the refurbished parallel AC–DC link converter stations depending on the annual load profile at the RSS.

Figure 3.5 shows two different converter station architectures that can used for DC link system.

In Figure 3.5(a), a common DC bus supplies power with each DC pole connected to $N_{dc}/2$ conductors. This architecture is similar to the one shown for refurbished DC system in Figure 2.2. Alternatively, $N_{dc}/2$ independently operating DC links can be formed as shown in Figure 2.5(b). There are three important research considerations that influence the choice between these architectures (i) The number of converters N_{conv} and rating of each can be optimized in case of common DC bus based on the functional and modularity requirements of the system while for independent DC links, at least $N_{dc}/2$ converters must be present, (ii) if AC circuit breakers are used to isolate the DC conductor to ground faults, modularity offered by independent DC link operation can limit the load shedding during fault isolation process, and (iii) if the VSC station architecture is employed within the reconfigurable AC–DC system shown in Figure 3.4(a), common DC bus approach can simplify the AC and DC feeder switching scheme. Therefore, the selected architecture can influence converter cost, capacity of the system during $(n-1)$ contingencies, protection, and the reconfigurable switching scheme. Some discussion on VSC configuration on capacity enhancement with DC refurbishment scheme is presented in [24].

Finally, the selection of topology and components of the operating converter relevant to this medium voltage and high power application is important. With current

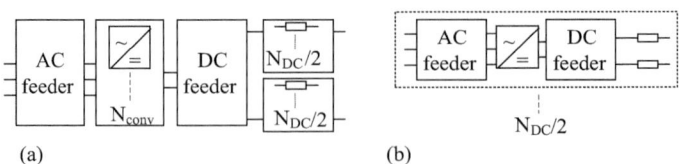

(a)   (b)

Figure 3.5   Converter station architecture (a) common DC bus and (b) independent dc links

power electronic technologies, use of VSCs has implicitly been assumed in this discussion and operation with more than two-levels can have advantages. Neutral point clamped (NPC-VSC) have acceptable performance up to three to five levels. Further, use of multilevel H-bridge converters have also shown promise [25,26]. The performance parameters of multilevel converters with optimum number of levels as a function of grid voltage is offered in [27]. An operating efficiency above 99% can be expected for Insulated Gate Bipolar Transistor (IGBT) switch based half-bridge Modular Multilevel Converter (MMC) as per the design considerations for optimum number of levels presented in [28,29]. The potential application of new semiconductor technologies can be an interesting research topic.

3.2.4   Link conductor considerations

The insulation performance under AC and DC medium voltages is important to determine the acceptable voltage enhancement factor k_v. The corresponding capacity enhancement estimates may be different for insulation types, overhead and cable infrastructure and joints and accessories of the link system. In general, it is expected that a much higher DC operating voltage can be imposed as compared to AC considering partial discharges as an insulation performance indicator [30]. While preliminary empirical studies comparing insulation performance under AC and DC voltages support this expectation [20], it is difficult to estimate the accurate DC voltage enhancement factor ensuring safe operation without detrimental effect to insulation lifetime over several years. Therefore, with field implementation of the suggested AC to DC refurbishment strategy, a gradual increase in k_v is recommended [4].

Further, it is known that space charge accumulation can result in local electric field enhancement and even inversion within the insulation under DC voltages. Based on qualitative understanding, the detrimental effect on insulation lifetime in case of frequent and sudden polarity reversal may encourage minimization of DC to AC reconfiguration during $(n-1)$ contingencies. However, many aspects need further investigation, such as: (i) magnitude of accumulated space charges at medium voltage levels (translated to imposed electric fields on the insulation) [31]. These are possibly low at the considered voltage levels and therefore, less significant. (ii) Time and temperature dependence of space charges when DC voltage is imposed or removed [32]; (iii) insulation performance under the expected space charge accumulation at different locations in cable insulation system over its operational lifetime (iv) understanding of impact on insulation with DC to AC reconfiguration based on (a) reconfiguration frequency and (b) time-lag after each such reconfiguration. Such performance indicators must be investigated with different insulation systems (example: XLPE, EPR) [33]. Keeping these considerations in mind, incremental gathering of on-field operational know-how as recommended in [4] is particularly important.

3.3   Parallel AC–DC link operation

Consider the original AC system operating with three-phase AC links as shown in Figure 3.6(a). It can be inferred that with a base power corresponding to a single

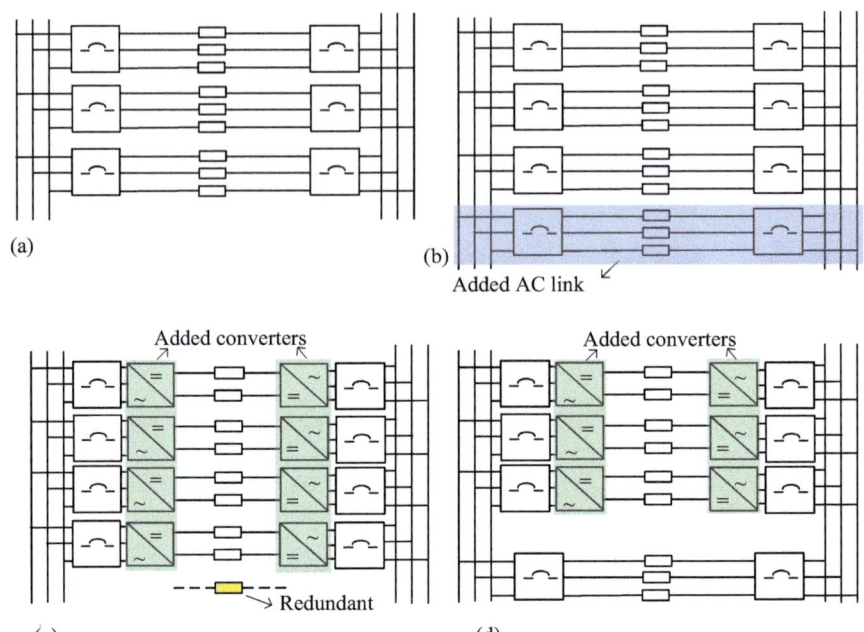

Figure 3.6   Different capacity enhancement strategies: (a) original system ($N_{ac} = 9$, $N_{dc} = 0$); (b) configuration C0 ($N_{ac} = 12$, $N_{dc} = 0$); (c) configuration C1 ($N_{ac} = 0$, $N_{dc} = 8$); (d) configuration C2 ($N_{ac} = 3$, $N_{dc} = 6$)

3-phase AC link, this system has a capacity of 3 p.u. during normal operation and 2 p.u during (n−1) contingencies because two healthy AC links remain after fault isolation when a single component failure occurs. Conventionally, an AC link is added to increase the capacity of this system to 3 p.u. during (n−1) contingency, shown as configuration C0 in Figure 3.6(b).

Figure 3.6(c) and (d) shows two different DC refurbishment strategies with $N_{ac} = 0$, $N_{dc} = 8$ (configuration C1) and $N_{ac} = 3$, $N_{dc} = 6$ (configuration C2) respectively. For $k_{CEF} = 1.5$, both systems are capable of delivering a power demand of 3 p.u. during (n−1) contingencies. While the post-fault reconfigurability between AC and DC link conductors is not shown here, both refurbishment schemes are incorporated within the reconfigurable architecture illustrated in Figure 3.4(a). Assuming that the configurations shown in Figure 3.6(b)–(d) can be imposed with a maximum system power demand of 3 p.u. during normal conditions and operate at maximum efficiency, the total converter rating must be 3 p.u. for C1 and 2.25 p.u. for C2 for substation at each end. Therefore, the converter costs are about 25% lower for parallel AC–DC (C2) as compared to complete DC (C1) refurbishment for maintaining the same system capacity during (n−1) contingencies.

Table 3.2   System parameters and assumptions

Parameter	Value
AC grid voltage (line to line)	10 kV
DC link voltage (pole to pole)	16.33 kV
Type, cross-sectional area and current rating of link conductor	Al, 400 mm 2,450 A
Base power	7.8 MVA
Required system capacity during (n−1) contingencies	3 p.u at power factor pf=0.9
Converter efficiency	99.34%
Cost[†] of installing a 3-phase AC link ($f_{c,cond}$)	100 per m
Cost[†] of converter station ($f_{c,conv}$)	50 per kVA
Cost of space requirements ($f_{c,s}$)	50,000 per station
Energy cost	0.1 per kWh

[†]A sensitivity analysis to the assumed cost parameters is offered in [22].

Table 3.3   Viability boundaries for different capacity enhancement strategies

	Zone 1	Zone 2	Zone 3	Zone 4	Zone 5
Most efficient	C0	C2	C2	C1	C1
Economically viable	C0	C0	C2	C2	C1

Consider the operating system parameters and assumptions listed in Table 3.2 with the objective to select the optimal grid reinforcement strategy (C0, C1, or C2) to enhance the capacity of the system shown in Figure 3.6(a) from 2 p.u. to 3 p.u.

The DC link pole-to-pole voltage (v_d) can be calculated from the r.m.s. line to line AC line voltage ($v_{ll,rms}$) from (3.3):

$$v_d = \frac{2k_v v_{ll,rms}}{\sqrt{3}} \tag{3.3}$$

The DC link of the parallel AC–DC configuration can be used to steer the active power in the system depending on the RSS demand (P_{RSS}). The approximate share (y_{approx}) of DC active power P_{dc} to operate the system at optimal efficiency is given by (3.4):

$$y_{approx} = \frac{P_{dc}}{P_{RSS}} = \frac{k_v N_{dc}}{N_{ac} + k_v N_{dc}} \tag{3.4}$$

For example, $y_{approx} \approx 0.74$ when $k_v = \sqrt{2}$ is selected for configuration C2 (refer Figure 3.6d). Here, y_{approx} is calculated by taking into account solely the power carrying capacity of the AC and DC link conductors and assuming that the substation converters are rated to deliver this power. The optimal efficiency point of DC power share y_{opt} varies as a function of the operating parameters such as RSS active and reactive power demand, converter efficiency, link length and conductor area. The dynamic deviation of y_{opt} from y_{approx}, which is further constrained by the substation converter size (k_{cr}) is explored in [34].

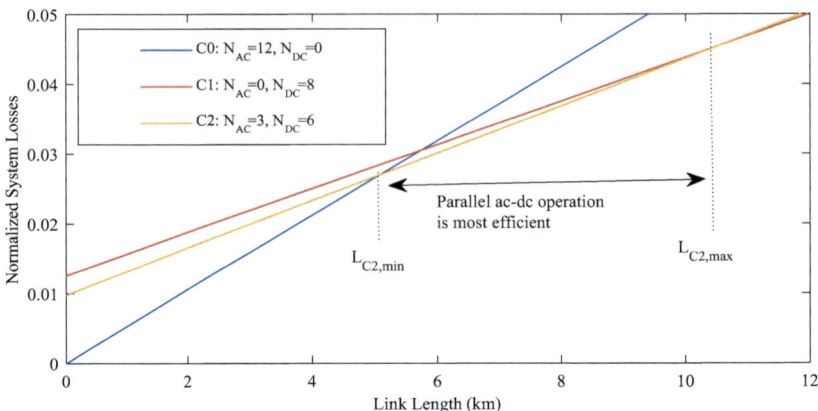

Figure 3.7   Normalized system losses for different grid reinforcement strategies with varying link length for RSS demand of 23.4 MVA at pf=0.9

Based on the AC and DC link conductor currents i_{ac} and i_{dc} respectively, the system losses ($P_{loss,sys}$) for the given link length l, conductor resistance per unit length (r_{ac}, r_{dc}) and the RSS power demand is given by (3.5):

$$P_{loss,sys} = N_{ac}i_{ac}^2 lr_{ac} + N_{dc}i_{dc}^2 lr_{dc} + P_{loss,conv} \qquad (3.5)$$

Here, $P_{loss,conv}$ is the sum of the total SSS and RSS converter loss in delivering the DC link active power while supporting the reactive power demand. It can be estimated that with RSS demand of 3 p.u. at pf=0.9, the normalised $P_{loss,conv}$ is 0.0125 p.u. for configuration C1 and 0.0097 p.u. for configuration C2 for the operating parameters mentioned in Table 2.2. Correspondingly, using (2.5), the variation of system losses for capacity enhancement strategy C0 (added AC link), C1 (complete DC refurbishment) and C2 (parallel AC–DC refurbishment) as a function of l is shown in Figure 3.7.

The observed slope is different for the three grid reinforcement strategies because AC link conductor losses per unit length are higher than DC for the given operating conditions. As a consequence, C0 has the highest loss variation with link length, followed by C2 and C1, respectively. For lower link lengths the converter losses dominate in DC operation, and these are lower for C2 as compared to C1 because the share of power transferred by the DC link is lower for the former. It can be observed that the indicated crossover points $L_{C2, min}$ and $L_{C2, max}$ describe the length range between which parallel AC–DC operation is the most efficient. For $l < L_{C2, min}$, AC reinforcement strategy C0 is most efficient, while complete DC refurbishment C1 is most efficient for $l < L_{C2, min}$. The variation in $L_{C2, min}$ and $L_{C2, max}$ depends on the AC grid voltage, DC link voltage, power demand at RSS, share of AC and DC power transfer, converter efficiency, number of conductors and their cross-sectional area. The associated mathematical relationship and sensitivity analysis is described in [22].

The economic choice between the reinforcement strategies depends on the 10 year payback associated with the operating losses for the given load demand and

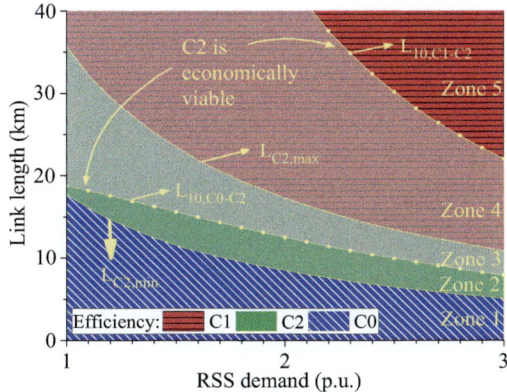

Figure 3.8   Viability boundaries for a 10 year payback period with varying link lengths and RSS demand at pf=0.9 (copyright: [22])

incurred investment costs. For the given configuration, the total cost of installation ($f_{c,ins}$) is given by (3.6):

$$f_{c,ins} = f_{c,cond}l + f_{c,conv}S_{conv} + f_{c,s} \qquad (3.6)$$

The installation costs increase with link length for AC reinforcement strategy C0 (1 M€ for $l = 10$ km and double this value for $l = 20$ km). The installation cost for DC refurbishment strategies are independent of link length and depend on the chosen converter rating. Since the maximum converter rating for C2 is approximately 75% of C1, its $f_{c,ins}$ is 1.85 M€ as against 2.44 M€ for the latter. These values are based on the cost assumptions described in Table 3.2. The economic and efficiency boundaries for selecting the appropriate capacity enhancement strategy with varying average RSS demand and link length is shown in Figure 3.8.

The four plot-lines, $L_{C2, min}$, $L_{10,C0-C2}$, $L_{10,C1-C2}$ and $L_{C2, max}$, divide the graph area into five zones. Herein, $L_{10,C0-C2}$ describes the link length above which the return on higher investment with C2 has a 10 year simple payback as compared to C0. Similarly, $L_{10,C1-C2}$ describes the link length above which the return on higher investment with C1 has a 10 year simple payback as compared to C2. The observed efficiency and economic viability boundaries for zones 1–5 are listed in Table 3.3. In zone 1, installing an additional AC link is the most efficient and economic solution. Similarly, in zones 3 and 5, DC refurbishment with C2 and C1 are the most viable capacity strategy respectively. The transitional zones 2 and 4 describe the region where an increase in DC capacity can be more efficient but not economically viable.

The conclusion is that the break-even distance between AC and DC power transfer is a smooth transition wherein parallel AC–DC operation can be the most economic choice for a band of link lengths described by Zone 3 ∪ Zone 4. The width and boundaries of this viability band can shift with different system parameters and cost assumptions than those listed in Table 3.2. Particularly at medium voltage levels of 10–66 kV, short link lengths up to a few tens of kilometers and capacities in the range of few MWs, power transfer with parallel AC–DC operation is preferred as compared to solely AC or DC transmission within this multi-dimensional viability band.

3.4   Reconfigurable AC–DC link architecture

Consider the refurbishment of a double circuit AC link to DC operation shown in Figure 3.2, wherein the entire system must be shut down in case a converter fault occurs. By providing modularity as suggested in Figure 3.5, some system capacity can be preserved during faults. Protection design aims to isolate the faulty component and restore the operation of the remaining healthy infrastructure as soon as possible to minimize the magnitude and duration of load shedding. Adequate redundant infrastructure is installed to ensure that the required capacity can be maintained during $(n-1)$ contingencies. While increased modularity can reduce the need for redundancy, this method has both practical and cost limitations. For example, increasing the number of converters N_{conv} in Figure 3.5(a) significantly increase the number of power electronic components, which is detrimental to reliability, space requirement and unit system cost. Similarly, modularity in number of point-to-point DC links suggested in Figure 3.5(b) is limited to $N_{dc}/2$ by the number of link conductors and the loss in capacity is equivalent to a single DC link, which translates to increased incurred cost of redundancy.

Another way of minimizing the cost of necessary redundancy is to introduce reconfigurability in the system architecture. This offers post-fault flexibility to maximize the functional utilization of the remaining healthy infrastructure. For example, if the same converter fault occurs in a reconfigurable system, the healthy DC link conductors can be re-utilized in AC operation to enhance capacity. The concept is depicted in Figure 3.9, where the six link conductors x_1-x_6 can operate either in AC or DC condition by choosing the state of feeder switches connecting the AC and DC bus at the substation.

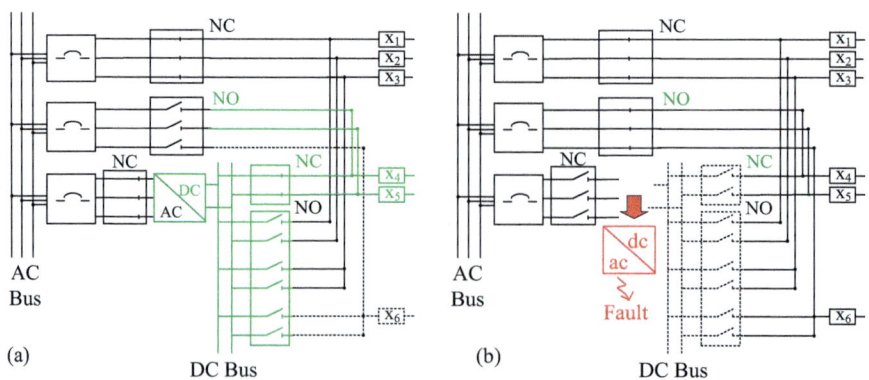

Figure 3.9   Re-configurable ac–dc architecture (a) normal operation and (b) AC bypass during converter faults

Figure 3.9(a) depicts the normal operation with x_1-x_3 forming a three-phase AC link, x_4-x_5 operating as DC link and x_6 is a redundant conductor. This is a possible refurbishment strategy with one of the double circuit AC link (refer Figure 3.2 for original AC system) converted to DC operation. It can be inferred that without reconfigurability, the k_{CUF} of this system during normal operation is equal to 1. Further, a single component failure anywhere in the system results in the loss of one line and therefore, no capacity enhancement over the original AC system could be achieved even with DC refurbishment. However, Figure 3.9(b) shows that when a converter fault occurs, conductors x_4-x_6 can be re-connected to the AC bus after isolating the DC bus to form a second three-phase AC link. Therefore, the capacity of this system is increased to twice due to AC–DC reconfigurability in operation.

Similarly, when a link conductor to ground fault occurs in the system initially operating in the state shown in Figure 3.9(a), reconfigurations can improve the maintainable capacity. Figure 3.10(a) shows that when a short circuit occurs in the AC conductor x_3, it is isolated and the feeder switches are reconfigured such that x_4-x_6 form the new AC link while x_1-x_2 form the DC link in the post-fault scenario. Figure 3.10(b) shows the scenario where x_6 replaces the faulty DC link conductor x_5. The capacity is improved in both considered $(n-1)$ contingencies because the AC and DC links can function at their design rating even with a single component failure.

Reconfiguration strategies with different types of fault scenarios, the related converter downsizing potential and trade-off with operating efficiency of the system is discussed in [35]. The study shows that with use of reconfigurability, the substation converters can be downsized by as much as 75% to achieve the same required capacity during $(n-1)$ contingencies. However, this reduction in converter size decreases the system efficiency during healthy operating conditions because the DC link is unable to operate at its optimal power share. Considering that the system is expected to operate in normal conditions for most of its lifetime, a 10-year simple

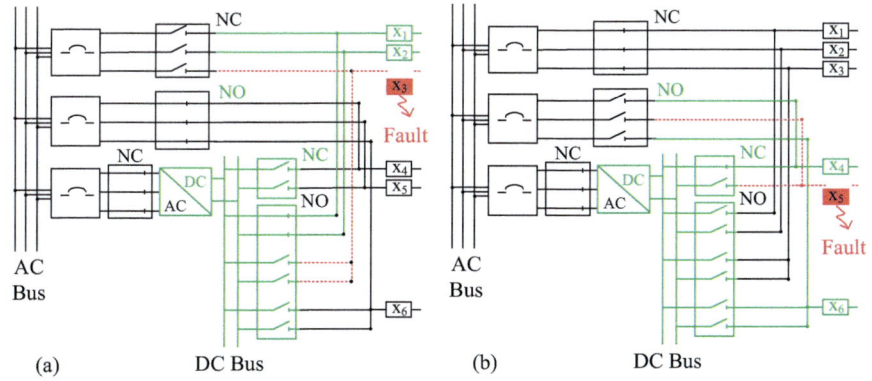

Figure 3.10   Reconfigurability during (a) AC conductor to ground fault and (b) DC conductor to ground fault

payback calculation suggests that the potential for converter de-rating increases with increasing grid voltage and decreasing link length of a refurbished parallel AC–DC system operating within a reconfigurable architecture proposed in Figure 3.4(a).

Consider a system originally operating with $N_{ac,ori}$ AC conductors. Using base power equivalent to a single three-phase AC link ($P_{base} = 3v_{ac}i_{ac}$), the per unit system capacity during normal operation is $N_{ac,ori}/3$ and that during (n−1) contingency is $(N_{ac,ori} - 3)/3$. For example, with $N_{ac,ori} = 6$, the system can deliver a maximum power of 2 p.u. during normal operation and 1 p.u. when a single AC link fails. Since the number of AC conductors is a multiple of three and number of DC conductors is even, the original system can be refurbished to DC in $N_{ac,ori}/3$ possible configurations while ensuring that the number of redundant conductors (N_{red}) is minimum. The resultant number of operating AC and DC conductors is given by (3.7) and (3.8), respectively:

$$N_{ac} = (n - 1) * \left(\frac{N_{ac,ori}}{3} \right) \tag{3.7}$$

$$N_{dc} = \begin{cases} N_{ac,ori} - N_{ac}, & \text{if } N_{ac,ori} - N_{ac} \text{ is even} \\ N_{ac,ori} - N_{ac} - 1, & \text{otherwise} \end{cases} \tag{3.8}$$

Here, n can take integer values between 1 to $N_{ac,ori}/3$ and the possible refurbished AC–DC configurations for different $N_{ac,ori}$ are listed in Table 3.4. k_{CUF} and $k_{CUF,n-1}$ indicate the capacity enhancement corresponding to the fixed refurbished DC configuration compared to the original AC system during normal and fault conditions respectively. For a $k_{CEF} = 1.5$, the k_{CUF} and $k_{CUF,n-1}$ can be derived from (3.2), given by (3.9) and (3.10) respectively:

$$k_{CUF} = \frac{\left(\frac{N_{ac}}{3}\right) + \left(\frac{N_{dc}}{2}\right)k_{cr}}{\left(\frac{N_{ac,ori}}{3}\right)} \tag{3.9}$$

$$k_{CUF,n-1} = minimum \left(\frac{\left(\frac{N_{ac,n-1}}{3}\right) + \left(\frac{N_{dc,n-1}}{2}\right)k_{cr,n-1}}{\left(\frac{N_{ac,ori,n-1}}{3}\right)} \right) \tag{3.10}$$

Table 3.4   *AC–DC reconfigurations, converter size and capacity enhancement with refurbished DC operation*

$N_{ac,ori}$	N_{ac}	N_{dc}	N_{red}	P_{conv}	k_{cr}	k_{CUF}	$k_{CUF,n-1}$	$k_{CUF,n-1,r}$
6	0	6	0	3 p.u.	1	1.5	2	2
				2 p.u.	0.67	1	1.33	2
	3	2	1	1 p.u.	1	1	1	2
9	0	8	1	4 p.u.	1	1.33	1.5	2
				3 p.u.	0.75	1	1.125	1.5
	3	6	0	3 p.u.	1	1.33	1.5	1.5
				2 p.u.	0.67	1	1.167	1.5
	6	2	1	1 p.u.	1	1	1	1.5

Here, the converter rating factor $k_{cr} \geq 1$ is determined by the installed substation converter power rating (P_{conv}) as the ratio of the total DC link conductor power for the given operating configuration. $N_{ac,n-1}$, $N_{dc,n-1}$ and $k_{cr,n-1}$ depend on the state of remaining conductors and converters after AC conductor, DC conductor or a converter fault occurs respectively. $k_{CUF,n-1}$, therefore, describes the capacity enhancement factor of a given configuration during (n−1) contingencies. $k_{CUF,n-1,r}$ is the achievable capacity enhancement factor if post-contingency reconfigurability between AC and DC operation is employed.

It can be observed from Table 3.4 that decreasing P_{conv} results in a decrease in k_{CUF} and $k_{CUF,n-1}$ for the given configuration. However, if reconfigurability is introduced, the converter can be significantly downsized even while maintaining the same system capacity enhancement ($k_{CUF,n-1,r}$). For example, a system with $N_{ac,ori} = 9$ has 2 p.u. capacity during (n−1) contingencies. Three DC refurbishment strategies are possible and as installed converter rating is varied between 1 and 4 p.u., a post-contingency capacity of 3 p.u. ($k_{cr,n-1,r} = 1.5$) can be maintained using re-configurations, leading to a potential of about 75% cost reduction. However as a result, the trade-off in lower operating efficiency during normal conditions (as described in Figure 3.8) is important to consider over the system lifetime.

3.5 Hybrid AC–DC distribution systems

3.5.1 DC interlinks in radial networks

The concept of DC based active power routing in the DN is shown in Figure 3.11. Here, a Back-To-Back Modular Multilevel Converter (BTB-MMC) based DC interlink can control the bidirectional power flow between nodes 6 and 8 of the network. Further, the power electronic converters at each node can offer ancillary services to the connected grid, such as fault ride through, reactive power and voltage support. In practice, this restructuring can be realized either by refurbishing existing tie-lines or by installing a new DC link interconnection between selected nodes in the system, depending on the economic viability in relation to the functionalities offered.

Conventionally, Normally Open (NO) tie lines in AC distribution systems are used for network reconfiguration to achieve load balancing, loss reduction and service restoration [37–39]. While these studies investigate the optimal branch switching configuration to realize the mentioned objectives for given power flows in the DN, a radial constraint is imposed on the selected operating topology. Weakly meshing such systems with limited number of loops can improve functionality and toward this concept, the DC interlink can provide an efficient and controllable capacity enhancement solution. For example, it is shown that the DN losses of a 33-bus weakly meshed system reduce with increasing DC power transfer as compared to the radial network with similar load powers at the nodes [36]. The power flow between bridged feeders in a radial network was regulated using controllable back-to-back VSC based DC links to show that the loadability improved by about 150% as compared to the base case [40]. The paper showed that assuming

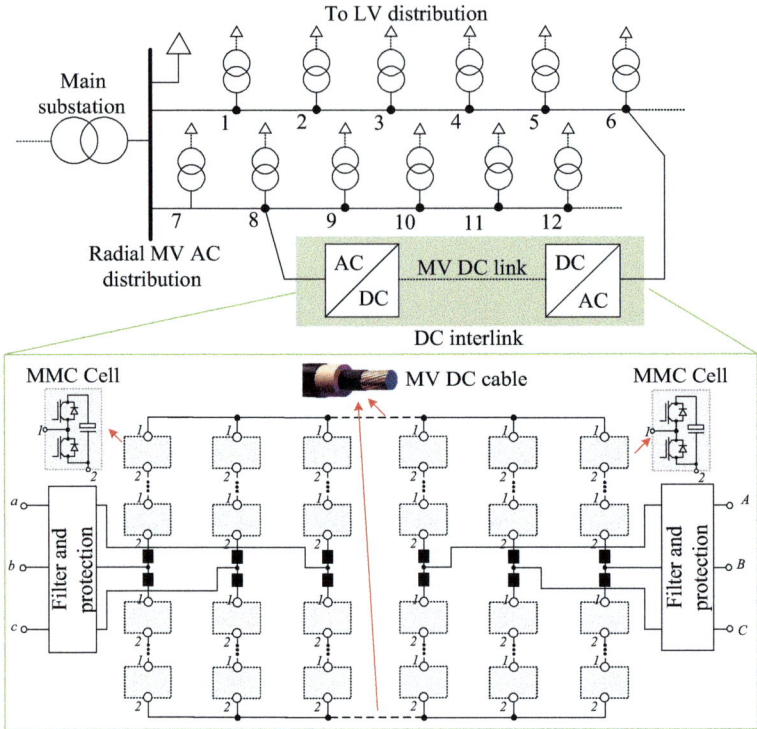

Figure 3.11   Power electronics assisted DC interlink based power redirection in radial AC distribution networks (adapted from [36])

a installed 6 MVA DC link technology cost of 250 €/kVA, a payback of 1.4 years is possible because of extra income due to 21% increase in peak PV power with improved DG penetration. Extending the concept, it was shown that storage integrated DC interlink can improve voltage at remote terminals, reduce feeder transformer loading by 50% and reduce system losses by 2–4% with increasing photovoltaic (PV) power generation [41].

The emerging hybrid AC–DC DN is fundamentally a composite version of a parallel AC–DC system shown in Figure 3.4 (b) and therefore, the underlying principles discussed in Section 3.3 are applicable. In this context, the sizing and placement of the DC interlink in a given AC DN is a multi-objective optimization considering the efficiency and capacity gains with DC power redirection in addition to the ancillary services offered by the grid connected VSCs.

3.5.2   Protection, restoration and availability

Large-scale deployment of DC interlinks in MV-DN will require a redesign of existing AC protection systems [42]. The paper explores different fault protection strategies in such hybrid AC–DC systems using AC Circuit breakers (ACCBs), DC

Circuit Breakers (DCCBs) and fault blocking converters. The fast fault propagation in the network associated with the low impedance of DC systems, exacerbated by the slow response of ACCBs (several hundred microseconds), makes selective isolation a challenge. Nevertheless, if the embedded DC interlinks are restricted to point-to-point topology within the AC MV-DN, ACCB based solution is possible. While DCCBs and full-bridged converter based topologies can offer rapid fault-clearance time, these methods involve higher capital costs and power losses that are challenging to justify at MV-level. A promising solution based on multi-line hybrid circuit breaker minimizes the installation costs by sharing the solid-state based current commutation path for protection of several DC lines [43].

At the system level, the two important network protection requirements include (i) proper selectivity and speed to minimize the downstream load shedding and (ii) service restoration with adequate capacity during (n−1) contingencies [21,44]. The former necessitates design of protection devices considering relay coordination associated with fault response time, current magnitude and direction. The latter involves the trade-offs in redundancy and reconfigurability in maintaining the required power capacity when the operation of a healthy grid section is restored post-fault isolation. The field tests conducted in [45] indicate that the maximum service restoration time including fault isolation, switch motion and system reconfiguration is 500–800 ms. It is discussed in [35] that the costs associated with the inevitable loss in load corresponding to this restoration time is negligible compared to the benefit in downsizing the redundancy requirements achieved with AC–DC reconfigurability described in Section 3.4.

Different (n−1) contingencies on AC lines, DC lines, generators and converters were studied to demonstrate that the number of network violations were reduced by more than 70% when embedded DC networks were used to support the AC network for increasing the security of supply and resilience during outages [46]. Figure 3.12 (a) shows the pre-fault DN with two radial laterals with several nodal power load taps protected with protection relays R11–R14 and R21–R24. Conventionally, these

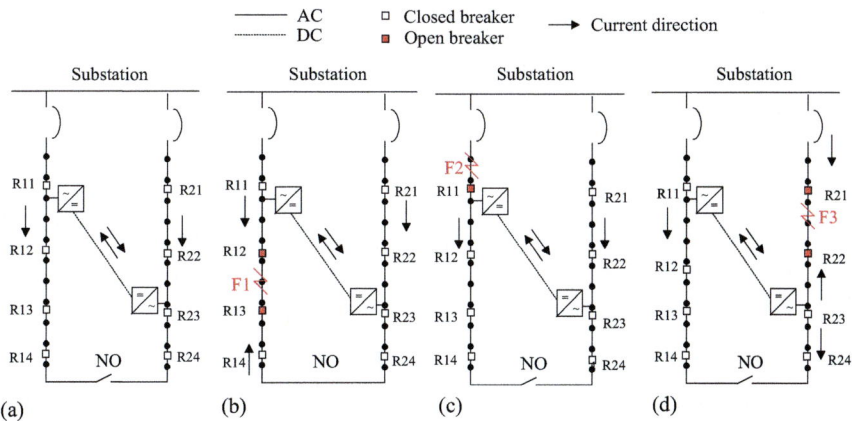

Figure 3.12   Improving availability of the radial AC DN using DC interlinks [14]

laterals are interconnected at end points using a NO tie-line. The tie-line is used to re-energize loads to healthy downstream section through another feeder after fault isolation. Therefore, outage management in this case needs intervention after fault clearance and the new network topology resulting with the closure of the tie switch can have high system losses, overloaded branches and poor voltage profile along the healthy feeder. A different strategy to improve availability is to weakly mesh the system by using a DC interlink with controlled bi-directional power capability, shown to be installed between relay location R11–R12 and R22–R23.

Figure 3.12(b) shows that when fault F1 is isolated using R12 and R13, the tie-switch re-energizes loads downline of R13. In this case, power redirection using the dc-interlink can prevent overloads in branches upstream of R23. Further, depending on the distance of R13 from the tie-line, the nodal voltage profile can be improved. For fault locations F2 and F3 in Figure 3.12(c) and (d), the tie-switch operation can be avoided while ensuring power availability to healthy sections. It can further be observed that the original current direction can be preserved in the active relays.

The main conclusion from this illustrative discussion is that the DC interlink can improve the system availability while minimizing the post-clearance tie-line switching operations, prevent branch overloads by introducing parallel pathways for controlled power flow and eliminate node voltage violations. However, these are competing objectives governing the optimal location and capacity of the installed dc-interlink. For example, if the NO tie-line itself is refurbished to operate as dc, switching operation can be avoided during fault F1 as well. On the other hand, the ability of the DC link to prevent branch overloads will be limited. Furthermore, it can be inferred from Section 3.3 that the system efficiency during normal operation can reduce with end-node DC link power redirection. Therefore, protection coordination and service restoration goal can add an interesting research dimension to the placement and sizing of DC power routers in radial AC distribution networks.

References

[1]   Sijm J, Gockel P, de Joode J, r van Westering W, and Musterd M. The demand for flexibility of the power system in the Netherlands, 2015–2050. In: *Report of Phase 1 of the FLEXNET Project*; 2017. pp. 1–140.

[2]   Luong NH, Grond MOW, La Poutré H, and Bosman PAN. Scalable and practical multi-objective distribution network expansion planning. In: *2015 IEEE Power Energy Society General Meeting*; 2015. pp. 1–5.

[3]   Ismael SM, Abdel Aleem SHE, Abdelaziz AY, and Zobaa AF. Practical considerations for optimal conductor reinforcement and hosting capacity enhancement in radial distribution systems. *IEEE Access*. 2018; 6: 27268–27277.

[4]   Liu Y, Cao X, and Fu M. The upgrading renovation of an existing XLPE cable circuit by conversion of AC line to DC operation. *IEEE Transactions on Power Delivery*. 2017; 32(3): 1321–1328.

[5]   Shekhar A, Kontos E, Ramírez-Elizondo L, Rodrigo-Mor A, Bauer P. Grid capacity and efficiency enhancement by operating medium voltage AC

cables as DC links with modular multilevel converters. *International Journal of Electrical Power & Energy Systems*. 2017; 93: 479–493.

[6] Deboever J, Peppanen J, Maitra N, Damato G, Taylor J, and Pate J. Energy storage as a non - wires alternative for deferring distribution capacity investments. In: *2018 IEEE/PES Transmission and Distribution Conference and Exposition (T D)*; 2018. pp. 1–5.

[7] Mateo C, Rodríguez Calvo A, Reneses Guillén J, Frías Marín P, and Sánchez Miralles A. Cost−benefit analysis of battery storage in medium−voltage distribution networks. *IET Generation, Transmission Distribution*. 2016; 10(3): 815–821.

[8] Ziari I, Ledwich G, Ghosh A, and Platt G. Integrated distribution systems planning to improve reliability under load growth. *IEEE Transactions on Power Delivery*. 2012; 27(2): 757–765.

[9] Assis TML, *et al.* Impact of multi - terminal HVDC grids on enhancing dynamic power transfer capability. *IEEE Transactions on Power Systems*. 2017; 32(4): 2652–2662.

[10] Piccolo A and Siano P. Evaluating the impact of network investment deferral on distributed generation expansion. *IEEE Transactions on Power Systems*. 2009; 24(3): 1559–1567.

[11] Zhao J, Wang J., Xu Z, Wang C, Wan C, and Chen C. Distribution network electric vehicle hosting capacity maximization: a chargeable region optimization model. *IEEE Transactions on Power Systems*. 2017; 32(5): 4119–4130.

[12] Palensky P and Dietrich D. Demand side management: demand response, intelligent energy systems, and smart loads. *IEEE Transactions on Industrial Informatics*. 2011; 7(3): 381–388.

[13] Dragičević T, Wheeler P, and Blaabjerg F. *DC Distribution Systems and Microgrids*. The Institution of Engineering and Technology (IET). 2018. pp. 1–469.

[14] Shekhar A, Ramírez-Elizondo L, Feng X, Kontos E, and Pavol B. Reconfigurable DC links for restructuring existing medium voltage AC distribution grids. *Electric Power Components and Systems*. 2017; 45(16): 1739–1746.

[15] Rentschler A, Kuhn G, Delzenne M, and Kuhn O. Medium voltage DC, challenges related to the building of long overhead lines. In: *2018 IEEE/PES Transmission and Distribution Conference and Exposition (T D)*; 2018. pp. 1–5.

[16] Shekhar A, Kontos E, Mor AR, Ramírez-Elizondo L, and Bauer P. Refurbishing existing MVAC distribution cables to operate under DC conditions. In: *2016 IEEE International Power Electronics and Motion Control Conference (PEMC)*; 2016. pp. 450–455.

[17] Zhang L, Liang J, Tang W, Li G, Cai Y, and Sheng W. Converting AC distribution lines to DC to increase transfer capacities and DG penetration. *IEEE Transactions on Smart Grid*. 2019; 10(2): 1477–1487.

[18] Larruskain DM, Zamora I, Abarrategui O, and Aginako Z. Conversion of AC distribution lines into DC lines to upgrade transmission capacity. *Electric Power Systems Research*. 2011; 81(7): 1341–1348.

[19] Clerici A, Paris L, and Danfors P. HVDC conversion of HVAC lines to provide substantial power upgrading. *IEEE Transactions on Power Delivery*. 1991; 6(1): 324–333.

[20]   Shekhar A, Feng X, Gattozzi A, *et al.* Impact of DC voltage enhancement on partial discharges in medium voltage cable—an empirical study with defects at semicon-dielectric Interface. *Energies.* 2017; 10(12): 1968.

[21]   Shekhar A, Kontos E, Ramírez-Elizondo LM, and Bauer P. Ac distribution grid reconfiguration using flexible DC link architecture for increasing power delivery capacity during $(n-1)$ contingency. In: *2017 IEEE Southern Power Electronics Conference (SPEC)*.IEEE; 2017. pp. 1–6.

[22]   Shekhar A, Ramírez-Elizondo LM, Batista T, and Bauer P. Boundaries of operation for refurbished parallel AC-DC reconfigurable links in distribution grids. *IEEE Transactions on Power Delivery.* 2020; 35(2): 549–559.

[23]   Shekhar A, Ramirez-Elizondo L. and Bauer P. Reliability, efficiency and cost trade-offs for medium voltage distribution network expansion using refurbished AC–DC reconfigurable links. In: *2018 International Symposium on Power Electronics, Electrical Drives, Automation and Motion (SPEEDAM)*. IEEE; 2018. pp. 242–247.

[24]   Larruskain DM, Zamora I, Abarrategui O, and Iturregi A. VSC-HVDC configurations for converting AC distribution lines into DC lines. *International Journal of Electrical Power & Energy Systems.* 2014; 54: 589–597.

[25]   Yazdani A and Iravani R. *Voltage-Sourced Converters in Power Systems.* John Wiley & Sons Inc; 2010. pp. 127–159.

[26]   Ma K and Blaabjerg F. Multilevel converters for 10 MW wind turbines. In: *Proceedings of the 2011 14th European Conference on Power Electronics and Applications*; 2011. pp. 1–10.

[27]   Huber JE and Kolar JW. Optimum number of cascaded cells for high-power medium-voltage AC—DC converters. *IEEE Journal of Emerging and Selected Topics in Power Electronics.* 2017; 5(1): 213–232.

[28]   Shekhar A, Soeiro TB, Qin Z, Ramirez Elizondo LM, and Bauer P. Suitable submodule switch rating for medium voltage modular multilevel converter design. In: *2018 IEEE Energy Conversion Congress and Exposition (ECCE)*; 2018. pp. 3980–3987.

[29]   Shekhar A, Larumbe LB, Soeiro TB, Wu Y, and Bauer P. Number of levels, arm inductance and modulation trade-offs for high power medium voltage grid-connected modular multilevel converters. In: *2019 10th International Conference on Power Electronics and ECCE Asia (ICPE 2019 – ECCE Asia)*; 2019. pp. 1–8.

[30]   Morshuis PHF and Smit JJ. Partial discharges at DC voltage: their mechanism, detection and analysis. *IEEE Transactions on Dielectrics and Electrical Insulation.* 2005; 12(2): 328–340.

[31]   Stancu C and Notingher P. Computation of the electric field in aged underground medium voltage cable insulation. *IEEE Transactions on Dielectrics and Electrical Insulation.* 2013; 20(5): 1530–1539.

[32]   Uehara H, Li Z, Cao Y, Chen Q, and Montanari GC. The effect of thermal gradient on space charge pattern in XLPE. In: *2015 IEEE Conference on Electrical Insulation and Dielectric Phenomena (CEIDP)*; 2015. pp. 138–141.

[33] Tefferi M, Li Z, Uehara H, Chen Q, and Cao Y. Characterization of space charge and DC field distribution in XLPE and EPR during voltage polarity reversal with thermal gradient. In: *2017 IEEE Conference on Electrical Insulation and Dielectric Phenomenon (CEIDP)*; 2017. pp. 617–620.

[34] Shekhar A, Soeiro TB, Wu Y, and Bauer P. Optimal power flow control in parallel operating AC and DC distribution links. *IEEE Transactions on Industrial Electronics*. 2021; 68(2): 1695–1706.

[35] Shekhar A, Soeiro TB, Ramírez-Elizondo LM, and Bauer P. Offline reconfigurability based substation converter sizing for hybrid AC–DC distribution links. *IEEE Transactions on Power Delivery*. 2020; 33(5): 2342–2352.

[36] Shekhar A, Soeiro TB, Ramírez-Elizondo L, and Bauer P. *Weakly Meshing the Radial Distribution Networks with Power Electronic Based Flexible DC Interlinks*. ICDCM. 2019.

[37] Baran ME and Wu FF. Network reconfiguration in distribution systems for loss reduction and load balancing. *IEEE Transactions on Power Delivery*. 1989; 4(2): 1401–1407.

[38] Jiang D and Baldick R. Optimal electric distribution system switch reconfiguration and capacitor control. *IEEE Transactions on Power Systems*. 1996; 11(2): 890–897.

[39] Shirmohammadi D and Hong HW. Reconfiguration of electric distribution networks for resistive line losses reduction. *IEEE Transactions on Power Delivery*. 1989; 4(2): 1492–1498.

[40] Romero-Ramos E, Gomez-Exposito A, Marano-Marcolini A and Maza-Ortega JM. Assessing the loadability of active distribution networks in the presence of DC controllable links. *IET Generation, Transmission Distribution*. 2011; 5(11): 1105–1113.

[41] Chaudhary SK, Guerrero J, and Teodorescu R. Enhancing the capacity of the AC distribution system using DC interlinks—a step toward future DC grid. *IEEE Transactions on Smart Grid*. 2015; 6(4): 1722–1729.

[42] Li G, Zhang L, Joseph T, Liang J and Yan G. Comparisons of MVAC and MVDC systems in dynamic operation, fault protection and post-fault restoration. In: *IECON 2019 - 45th Annual Conference of the IEEE Industrial Electronics Society*. Vol. 1; 2019. p. 5657–5662.

[43] Kontos E, Schultz T, Mackay L, Ramirez-Elizondo L, Franck C, and Bauer P. Multiline breaker for HVDC applications. *IEEE Transactions on Power Delivery*. 2018; 33(3): 1469–1478.

[44] Burstein AW, Cuk V, and de Jong ECW. Effect of network protection requirements on the design of a flexible AC/DC-link. *The Journal of Engineering*. 2018; 2018(15): 1291–1296.

[45] Yip T, Jing-huan Wang, Bingyin Xu, Kaijun Fan, and Tianyou Li. Fast self-healing control of faults in MV networks using distributed intelligence. *CIRED − Open Access Proceedings Journal*. 2017; 2017(1): 1131–1133.

[46] Teixeira Pinto R, Aragüés Peñalba M, Gomis Bellmunt O and Sumper A. Optimal operation of DC networks to support power system outage management. *IEEE Transactions on Smart Grid*. 2016; 7(6): 2953–2961.

Chapter 4

Bidirectional isolated DC–DC converters—enabling technology for MVDC networks with distributed generation

Rik W. De Doncker[1] and Jingxin Hu[1]

4.1 Introduction

The transition from a predominantly fossil fuel-based power generation towards renewable power sources, predominantly wind turbines and photo-voltaic systems, inevitably leads towards an energy supply system that greatly depends on power electronics to feed the energy in the electrical grid. At the same time, the increasing electrification of many sectors, in particular the heating and cooling (heat pumps) and the transportation sectors, which are all driven and controlled by power electronic converters, require considerably higher capacities of the distribution grid. As pointed out in the general introduction of this book, the distributed nature of modern power sources, storage systems, and prosumers requires a flexible, that is, interconnected distribution grid. As all power electronic driven systems are intrinsically DC sources or loads, a medium-voltage DC (MVDC) distribution system becomes evident, not only because it is more efficient and cost effective, but also increases the ampacity of cables.

Similar to the invention and development, 150 years ago, of "secondary generators," nowadays called AC transformers, the development and commercialization of medium-voltage (MV), multi-megawatt DC–DC converters, that is, so-called DC (electronic) transformers, is a key component to realize flexible, interconnected MVDC grids. These DC transformers can provide an interface between high- and MV, as well as medium- and low-voltage distribution grids and loads. In addition, they can control continuously the power flow between arbitrary grid segments. Nevertheless, in such environments, the DC transformer has to be able to fulfil all functions of a classical AC transformer. AC transformers not only transform the voltage and provide the required insulation (protection against lightning strikes) but also play an important role in the coordination of fault currents with protection gear. Indeed, the short circuit impedance of transformers are specified such that they limit short circuit currents at a level that can be handled by fuses and circuit breakers.

[1]E.ON ERC Institute for Power Generation and Storage Systems, FEN Research Campus, RWTH Aachen University, Aachen, Germany

Hence, compared to AC transformers, power electronic based DC transformers not only need to transform voltage and control power flow, but also need to offer similar efficiencies (up to 99%), provide the same insulation levels and limit fault currents, that is, offer fault-ride-through capabilities. Although the basic principle of the DC transformer may appear simple, it was a daunting task to comply with all these requirements and maintain high efficiency at high switching frequencies.

In the following sections, we describe and analyze galvanically isolated, bi-directional three-phase Dual Active Bridge (3ph-DAB) DC-to-DC converters that are suitable for multi-megawatt applications. As will be explained in the next section, a 3ph-DAB converter comprises two three phase inverters and a medium frequency transformer. Similar to three-phase AC transformers, the three-phase variant of the DAB offers a higher efficiency and has a more compact medium-frequency transformer, as compared to single-phase transformers. Furthermore, the DC link capacitors are substantially smaller in the 3ph-DAB as compared to a single-phase DAB. This feature reduces substantially the energy dump in case of a short circuit, which helps the coordination with DC protection gear. In addition, the 3ph-DAB can be constructed with three-phase MV source inverters that are standard components in large industrial drives and wind turbines. This allows a rapid product development of MVDC solid state DC transformers, which is an important factor in all engineering product development projects. Furthermore, control engineers who deal with three-phase systems are familiar with the Clarke three-phase to two-phase $\alpha\beta$-transformation, which allows them to apply to the 3ph-DAB all dynamic model-based control principles that have been developed over the past decades for three-phase power converters. Actually, the three-phase DAB came to mind of the author back in 1988 [1], based on his experience of field-oriented control of induction machines. In theory, rotating field machines can be considered as rotating transformers. Moreover, it was noted that the switching losses and the EMI of the inverter fed drive went down considerably when the drive operated in the so-called six-step mode (50% duty cycle). It was found that this mode of operation provides zero-voltage soft-switching (ZVS) operation of the power semiconductors and is therefore applied as the basic modulation principle of DAB converters.

In the following sections we will make extensive use of the Clarke transformation, which makes the analysis and control development a lot simpler. Advanced control modes for fast and robust current and power control, loss-optimized operational modes, such as soft-switching schemes, compensation for dc-offsets and dead time, and the fault ride-through (FRT) operation are introduced.

4.2   Dual-active bridge DC–DC converter

4.2.1   *Fundamental operation principle*

4.2.1.1   Introduction of three-phase dual-active bridge converter

The three-phase Dual Active Bridge (3ph-DAB) bidirectional, isolated DC–DC converter comprises two, three-phase converter bridges linked through a medium-frequency

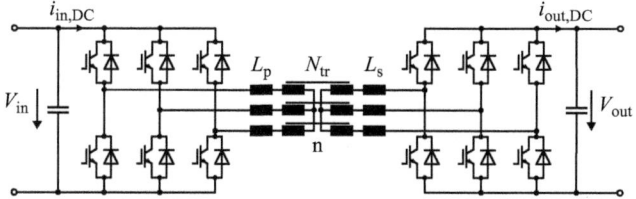

Figure 4.1   Schematic of the three-phase dual-active bridge converter

three-phase transformer as illustrated in Figure 4.1. Different from conventional PWM converters, the converters operate in the so-called square-wave mode (cf. six-step or block mode, 50% duty cycle). The power flow in the DAB can be controlled by phase-shifting the converter waveforms against each other. Hence, similar to three-phase synchronous generators that are connected to a fixed frequency grid, the power flow of the DAB can be controlled by setting a load angle between both converter bridges. This basic mode of operation is referred to as the single phase-shift (SPS) operation mode. SPS is widely used in DAB DC–DC converters as it enables ZVS operation of the power semiconductors and, in addition, provides buck-boost capability over a wide operating range. In this section, these basic operation principles are firstly introduced. The power transfer principle will be explained together with the current and voltage waveforms.

For many applications, that is, battery storage systems, electrolyzers, or photovoltaic systems, the 3ph-DAB converter is required to operate over a wide voltage and power range. To achieve a high performance covering the full operation range, especially at partial load, specific design considerations have to be taken to maintain ZVS operation of the semiconductor devices. In particular, the transferred power strongly depends on the size of the leakage inductance of the transformer. A proper selection of the transformer leakage inductance influences the maximum transferred power, the soft-switching boundaries and the system efficiency, as will be discussed later.

With the SPS modulation, the ZVS performance can be achieved. Therefore, the system efficiency can be increased especially for high-power applications. However, the ZVS feature cannot be guaranteed in light-load operations. Hence, it is necessary to analyze to ZVS boundaries, in order to keep the 3ph-DAB converter operating efficiently and reliably. These aspects will be presented in Section 4.2.1.3.

4.2.1.2   Basic operation principle

The traditional SPS modulation strategy of the 3ph-DAB converter was firstly published in 1988 for the on-board energy supply of the NASA space station program [1]. A transaction paper was published in 1991 [2]. Both the primary and secondary side bridges operate with a fixed duty cycle of 50%. As presented in Figure 4.2, each bridge produces a six-step voltage waveform at the AC terminals of the transformer that is, $\pm V_{in}/3$ and $\pm 2V_{in}/3$ for the primary bridge, $\pm V_{out}/3$ and $\pm 2V_{out}/3$ for the secondary bridge, where V_{in} and V_{out} denote the corresponding

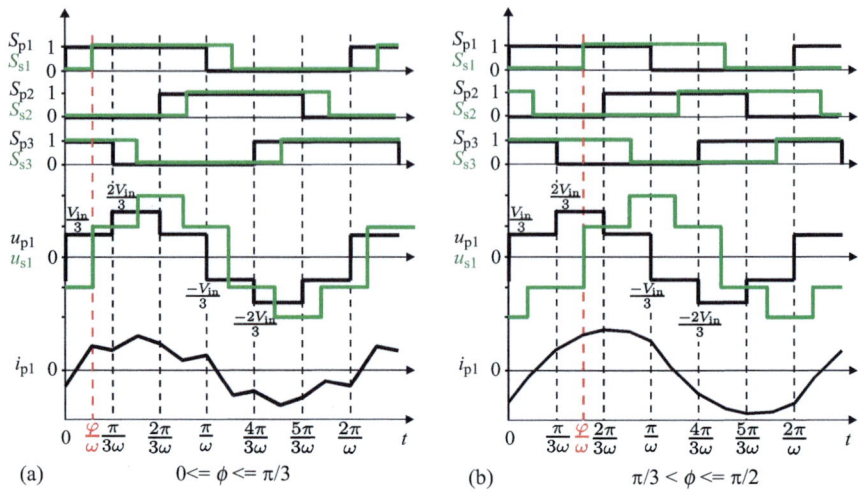

Figure 4.2   Steady-state operation of a 3ph-DAB converter, with $\omega = 2\pi f$

DC-link voltages. Applying a phase shift, that is, load angle φ, between the primary and secondary bridges, a voltage difference is applied across the transformer short circuit impedance. Note that the short circuit impedance of efficient transformers is predominantly determined by its leakage inductance L_σ. Assuming no saturation and neglecting the (large) magnetizing inductance, the circuit can be analyzed as a linear circuit, resulting in current waveforms that are piecewise linear. According to the value of the load angle φ, two different operating modes have to be considered, as depicted in Figure 4.2. The transferred power controlled by φ for both modes are given according to:

$$
P_{SPS} = \begin{cases} \dfrac{dV_{in}^2}{\omega L_\sigma} \cdot \left(\dfrac{2\varphi}{3} - \dfrac{\varphi^2}{2\pi} \right), & \text{for } 0 \le \varphi \le \dfrac{\pi}{3} \\[3mm] \dfrac{dV_{in}^2}{\omega L_\sigma} \cdot \left(\varphi - \dfrac{\varphi^2}{\pi} - \dfrac{\pi}{18} \right), & \text{for } \dfrac{\pi}{3} \le \varphi \le \dfrac{\pi}{2} \end{cases}
\tag{4.1}
$$

where $d = N_{tr} V_{out}/V_{in}$ denotes the voltage ratio, N_{tr} the turns ratio of the transformer, $\omega = 2\pi f$ with f the switching frequency, L_σ represents the total leakage inductance of the transformer referred to the primary side and f denotes the switching frequency. The reader is encouraged to compare Equation set (4.1) with the well-known power transfer equation of a synchronous generator connected to a fixed frequency voltage source, which is proportional to the product of induced stator voltage and the grid voltage, divided by the synchronous reactance and multiplied by sine of the load angle φ. Note that in Figure 4.2, the AC voltage of the secondary side bridge is lagging the primary side one. Hence, the power is transferred from the primary to secondary bridge. A reversed power flow direction can be achieved with a negative phase-shift angle, which means the AC voltage of the secondary bridge is leading the primary one. In the following the primary bridge is the one that leads the secondary bridge. In other words, a positive power flow is considered from the primary to the secondary bridge.

From (4.1), the maximum transferred power of the SPS modulation is achieved at $\varphi = \pi/2$. However, due to the large reactive power in the region $\pi/3 \leq \varphi \leq \pi/2$, the load angle φ is usually limited below $\pi/3$ for an efficient operation of the DAB [3]. For simplicity, only operation within the interval $0 \leq \varphi \leq \pi/3$ is considered in the following analysis.

4.2.1.3 Soft switching boundaries

With the SPS modulation, switches are turned on under zero-voltage conditions when the corresponding anti-parallel diodes are conducting. Take the commutation process of one leg in the primary bridge as an example as illustrated in Figure 4.3. The leg current is firstly conducted through the top switch S_{top} (left figure). Once S_{top} is turned off, the current is immediately commutated to the bottom diode D_{bot} (middle figure). If the bottom switch S_{bot} is turned on during the conduction of D_{bot}, ZVS is realized. Hereinafter, S_{bot} takes over the leg current once the current changes a direction. In DAB DC–DC converters, this current reversal happens due to switching of the converter behind the inductor. Actually, as shown in Figure 4.4, looking at a single-phase simplified circuit diagram, it can be seen that the single-phase DAB can be considered as a synchronous, bi-directional, isolated buck-boost circuit.

Compared with traditional hard-switching process in PWM converters, 3ph-DAB converters greatly benefit from the ZVS operating mode, which can be achieved with the SPS control, as the switching losses are significantly reduced. This improves the efficiency of the 3ph-DAB, increases reliability or reduces the costs of the thermal cooling system. However, the 3ph-DAB converter may lose the ZVS feature under light-load operating conditions especially when the voltage ratio d deviates from unity. To enable ZVS for both converters of a 3ph-DAB, the

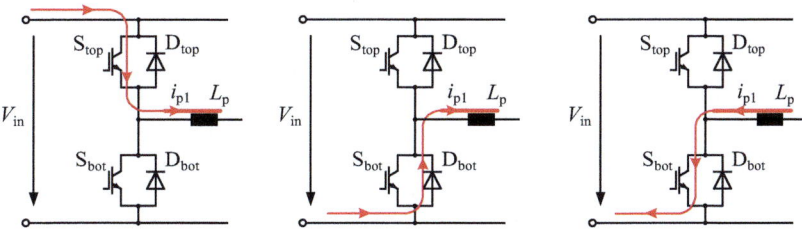

Figure 4.3   Commutation process under ZVS conditions

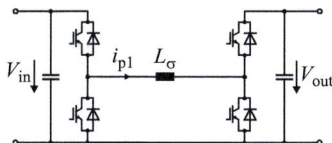

Figure 4.4   Single-phase simplified circuit diagram of a 3ph-DAB converter

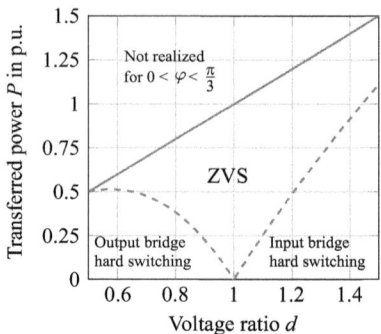

*Figure 4.5   Power transfer boundary and ZVS region of the SPS operation
(0 ≤ φ ≤ π/3)*

primary current i_{p1} has to lag the AC voltage v_{p1} of the primary bridge, while the secondary current referred to the primary side (equals i_{p1}) must be leading the AC voltage v_{s1} of the secondary bridge. This yields the ZVS conditions of a 3ph-DAB converter with SPS control, that the current zero crossing of i_{pi} (with $i = 1, 2, 3$) takes place between the switching events of the primary and secondary converter phase legs. In mathematical terms this means that $i_{pi}(0) \leq 0$ for the primary bridge and $i_{pi}(\varphi) \geq 0$ for the secondary bridge, respectively. Combining the primary current expression given in [2], the ZVS conditions as function of P and φ are:

$$P \geq \begin{cases} \dfrac{V_{in}^2}{\omega L_\sigma} \cdot \left(1 - \dfrac{3\varphi}{2\pi}\right)^{-1} \cdot \left(\dfrac{2\varphi}{3} - \dfrac{\varphi^2}{2\pi}\right), & \text{for primary bridge} \\[3mm] \dfrac{V_{in}^2}{\omega L_\sigma} \cdot \left(1 - \dfrac{3\varphi}{2\pi}\right) \cdot \left(\dfrac{2\varphi}{3} - \dfrac{\varphi^2}{2\pi}\right), & \text{for secondary bridge} \end{cases} \qquad (4.2)$$

The normalized power transfer boundary and ZVS region of the SPS modulation are shown in Figure 4.5. It is observed that the ZVS cannot be guaranteed under light-load conditions especially when the voltage ratio has a large deviation from unity. In order to extend the soft-switching range for the SPS modulation, different strategies are considered. Among them, advanced modulation schemes are promising solutions, which will be discussed in the following sections.

4.2.1.4   Design consideration, selection of leakage inductance

For many applications, 3ph-DAB converters are required to operate over a wide voltage or power range. It is reasonable to consider an operating range during the design procedure. A pre-defined operation range provides a reasonable boundary for the selection of semiconductor devices and passive components, which leads to an operation-oriented design procedure as summarized in Figure 4.6 [4].

Clearly, the leakage inductance L_σ is a key passive component for the power transfer in a 3ph-DAB converter, because it determines the maximum transferred power and the rms current in the three-phase transformer. A smaller leakage

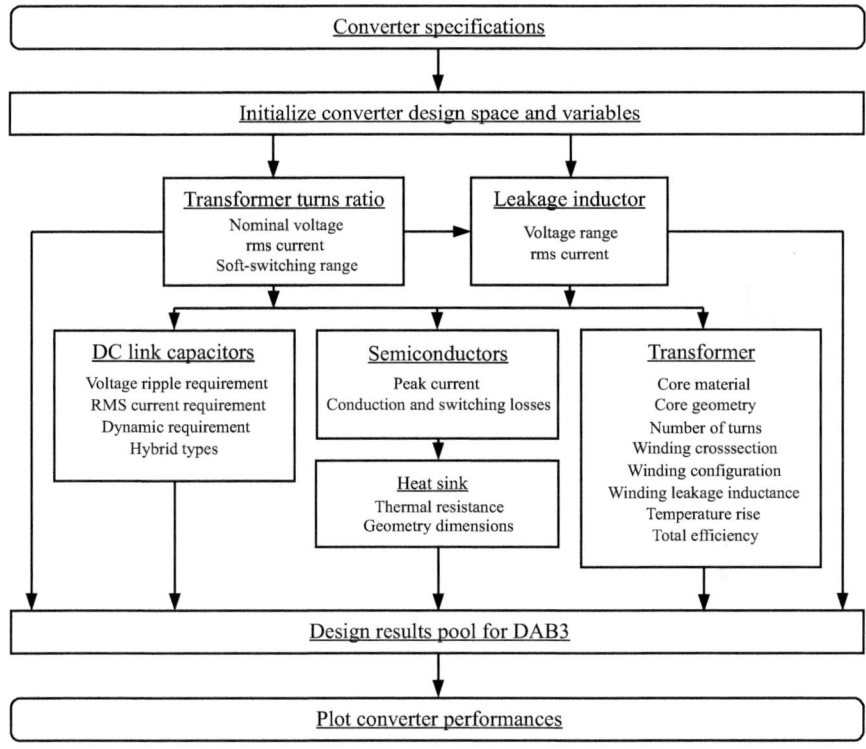

Figure 4.6   Operation-oriented design procedure

inductance yields a higher maximum transfer power, but leads to a larger hard-switching area and a higher rms current in the transformer. Therefore, it is necessary to optimize the size of the leakage inductance according to the application requirements.

Usually, the DC voltage range of both primary and secondary bridges are limited in a specific range. Hence the leakage inductance can be optimized, in order to achieve a minimum rms current over the whole operating range [4].

Power quality standards for the AC utility grid allow a voltage variation of typically +/−5%. Although power quality guidelines for MVDC grids are still being developed, the authors consistently proposed a practical voltage range for MVDC grids of +/−10% [2], which is easy to handle for DAB converters. In photovoltaic applications, the operation range is usually wider (up to 30%) and is determined by the maximum power point tracking (MPPT) control. Nevertheless, the leakage inductance can be optimized by achieving the largest overlap between the MPPT operation range and the soft-switching area, so that high efficiency operation over a wide range can be guaranteed [5]. The same principle can be also considered for a 3ph-DAB converter in battery applications if specific charging and discharging profiles are available, for instance in electric vehicles [6].

4.2.2   Advanced modulation schemes

As an interface converter to the renewable energy sources and battery storage systems, 3ph-DAB converters often need to operate in light-load conditions with a large variation of the DC-link voltage. According to the above analysis, the conventional SPS modulation presents a narrow soft-switching area when the primary and secondary voltage ratio d deviates from unity (all voltages referred to the primary side), so that the semiconductor devices may operate in the hard-switching mode under light-load conditions. This not only increases the switching losses of semiconductor devices, but also results in higher electromagnetic interference (EMI), that is, common mode and differential mode noise. Moreover, any deviation of d from unity also leads to an increased rms current, which increases the conduction losses of semiconductor devices and the winding losses of the medium-frequency transformer. Therefore, to increase the light-load efficiency for a wide voltage range operation, advanced modulation strategies are required to extend the soft-switching range and reduce the rms current.

In this section, the asymmetrical duty-cycle control (ADCC) method [7,8] is introduced for the 3ph-DAB converter, where the duty cycle of each bridge is variable. Note that the ADCC can be applied to a three-phase transformer without neutral wire as its star point is floating. By introducing the duty cycles of the primary and secondary bridges, that is, D_p and D_s, as additional degrees of freedoms, the transformer winding current can be shaped thereby realizing triangular or trapezoidal waveforms. Thus, three advanced operation modes are defined, that is, three-phase triangular-current modulation buck mode (TCM3-Buck), three-phase triangular-current modulation boost mode (TCM3-Boost) and three-phase trapezoidal-current modulation (TZM3), as shown in Figure 4.7. In these modulation modes, either ZVS or zero-current switching (ZCS) can be realized for all semiconductor devices during turn-on operation as listed in Table 4.1. Besides, part of semiconductor devices can realize ZCS turn-off. Thereby, the switching losses of the 3ph-DAB converter can be minimized in the three operation modes.

For each modulation mode of the ADCC method, the transformer winding current can be expressed by a piecewise-linear equation. As an example, the expressions of transformer phase voltages and winding currents for the TCM3-Buck mode are

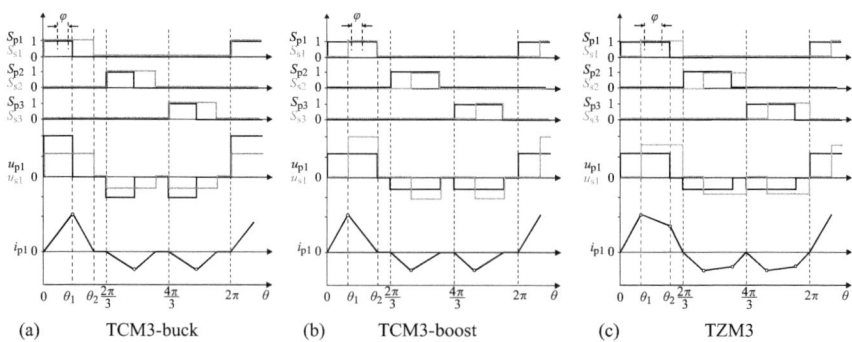

Figure 4.7   Key waveforms of three modulation modes in the ADCC method

Table 4.1 *Switching conditions of three modulation modes in the ADCC method*

Mode	Primary bridge		Secondary bridge	
	Turn-on	Turn-off	Turn-on	Turn-off
TCM3-buck	ZCS/ZVS	Hard/ZCS	ZCS	ZCS
TCM3-boost	ZCS	ZCS	ZVS/ZCS	ZCS/Hard
TZM3	ZCS/ZVS	Hard/ZCS	ZVS/ZCS	ZCS/Hard

Table 4.2 *Expressions of transformer voltages and winding current in the TCM3-Buck mode*

Interval	Primary voltage	Secondary voltage	Primary current
$\theta \in [0, \theta_1]$	$\frac{2V_{\text{in}}}{3}$	$\frac{2dV_{\text{in}}}{3}$	$\frac{(1-d)V_{\text{in}}}{3\pi fL_\sigma} \cdot \theta$
$\theta \in [\theta_1, \theta_2]$	0	$\frac{2dV_{\text{in}}}{3}$	$\frac{dV_{\text{in}}}{3\pi fL_\sigma} \cdot (\theta_2 - \theta)$
$\theta \in [\theta_2, \frac{2\pi}{3}]$	0	0	0
$\theta \in [\frac{2\pi}{3}, \frac{2\pi}{3} + \theta_1]$	$-\frac{V_{\text{in}}}{3}$	$-\frac{dV_{\text{in}}}{3}$	$\frac{(d-1)V_{\text{in}}}{6\pi fL_\sigma} \cdot \left(\theta - \frac{2\pi}{3}\right)$
$\theta \in [\frac{2\pi}{3} + \theta_1, \frac{2\pi}{3} + \theta_2]$	0	$-\frac{dV_{\text{in}}}{3}$	$\frac{-dV_{\text{in}}}{6\pi fL_\sigma} \cdot \left(\theta_2 - \theta + \frac{2\pi}{3}\right)$
$\theta \in [\frac{2\pi}{3} + \theta_2, \frac{4\pi}{3}]$	0	0	0
$\theta \in [\frac{4\pi}{3}, \frac{4\pi}{3} + \theta_1]$	$-\frac{V_{\text{in}}}{3}$	$-\frac{dV_{\text{in}}}{3}$	$\frac{(d-1)V_{\text{in}}}{6\pi fL_\sigma} \cdot \left(\theta - \frac{4\pi}{3}\right)$
$\theta \in [\frac{4\pi}{3} + \theta_1, \frac{4\pi}{3} + \theta_2]$	0	$-\frac{dV_{\text{in}}}{3}$	$\frac{-dV_{\text{in}}}{6\pi fL_\sigma} \cdot \left(\theta_2 - \theta + \frac{4\pi}{3}\right)$
$\theta \in [\frac{4\pi}{3} + \theta_2, 2\pi]$	0	0	0

Table 4.3 *Relationships of control parameters in three modulation modes of the ADCC method*

Mode	Duty cycle of the primary bridge	Duty cycle of the secondary bridge
TCM3-Buck	$D_{\text{p}}(\varphi) = \frac{d}{1-d} \cdot \frac{\varphi}{\pi}$	$D_{\text{s}}(\varphi) = \frac{1}{1-d} \cdot \frac{\varphi}{\pi}$
TCM3-Boost	$D_{\text{p}}(\varphi) = \frac{d}{d-1} \cdot \frac{\varphi}{\pi}$	$D_{\text{s}}(\varphi) = \frac{1}{d-1} \cdot \frac{\varphi}{\pi}$
TZM3	$D_{\text{p}}(\varphi) = \frac{d}{d+1} \cdot \frac{2\pi - 3\varphi}{3\pi}$	$D_{\text{s}}(\varphi) = \frac{1}{d+1} \cdot \frac{2\pi - 3\varphi}{3\pi}$

summarized in Table 4.2, while more details can be found in [7,8]. Besides, specific relationships between the control parameters φ, D_{p} and D_{s} are necessary to realize a triangular or trapezoidal current waveform, which are given in Table 4.3. Thereby, the transferred power of each modulation mode can be calculated with respect to the load angle φ as shown in Table 4.4. Thus, for a given operation point of the

Table 4.4   Transferred power of three modulation modes in the ADCC method

Mode	Transferred power
TCM3-Buck	$P_{TCM3-Buck} = \dfrac{d^2 V_{in}^2 \varphi^2}{\pi^2 f L_\sigma (1-d)}$
TCM3-Boost	$P_{TCM3-Boost} = \dfrac{d V_{in}^2 \varphi^2}{\pi^2 f L_\sigma (d-1)}$
TZM3	$P_{TZM3} = \dfrac{d V_{in}^2 \left(6\pi\varphi \left(1+d^2\right) - 9\varphi^2 \left(1+d+d^2\right) - \pi^2 (1-d)^2\right)}{9\pi^2 f L_\sigma (1+d)^2}$

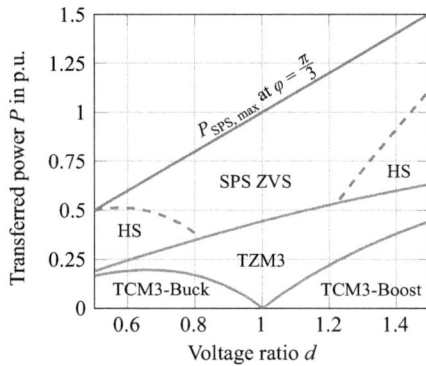

Figure 4.8   Power transfer boundary and soft-switching range of the ADCC method and SPS modulation

3ph-DAB converter, the control parameters φ, D_p and D_s can be calculated online in a feed-forward manner, which contributes to a fast transient response of the 3ph-DAB converter.

The power transfer boundary of each modulation mode with respect to the voltage conversion ratio is depicted in Figure 4.8. The boundary of the TZM3 mode is interconnected to the boundaries of the TCM3-Buck and TCM3-Boost modes. Thus, three modulation modes of the ADCC method can be seamlessly combined, which significantly extends the soft-switching range in light-load conditions over a wide voltage range compared to the conventional SPS modulation. Besides, due to the reduced duty cycles compared to the SPS modulation, the ADCC method also presents a lower rms current and lower peak flux density of the transformer in light-load conditions [7,8]. Therefore, the light-load efficiency of the 3ph-DAB converter can be substantially improved by the ADCC method, as indicated in Figure 4.9. Based on the above analysis, a hybrid modulation strategy that combines the ADCC method for light-load conditions and the SPS modulation for medium to heavy load conditions can realize a high-efficiency operation of the 3ph-DAB converter over a wide voltage and power range.

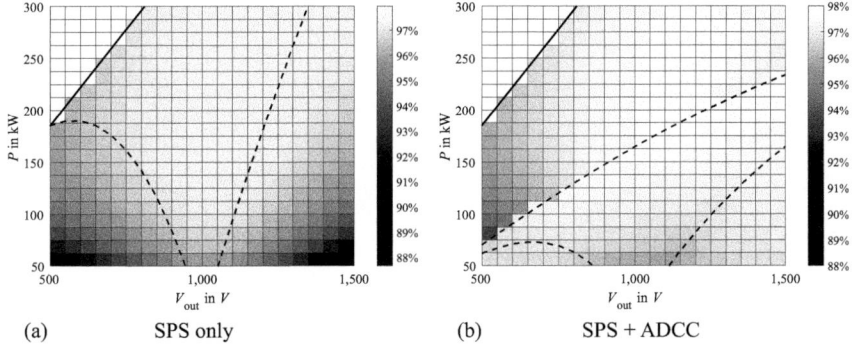

Figure 4.9   Simulated efficiency of a 300 kW 3ph-DAB converter with a fixed input voltage of $V_{in} = 1\ kV$, a variable output voltage of $V_{out} = 0.5 - 1.5\ kV$ and a switching frequency of 1 kHz [7]

4.2.3   Advanced control techniques

4.2.3.1   Instantaneous flux and current control

As a key component of controlling the power flow in DC grids, the 3ph-DAB converter needs to deal with various transient operating conditions including abrupt change of power flow in both magnitude and direction. However, any abrupt change of the load angle φ will lead to oscillations, because the DAB with the DC capacitors builds a higher order oscillatory system. Hence, the open loop duty cycle control that is optimized for efficiency does not guarantee stable operation of the DAB. During transients, volt-seconds imbalances lead to large winding current overshoots in the three-phase transformer windings [9]. Besides, an abrupt change of power, or instant start-up or shut-down can also induce DC components in the transformer magnetizing fluxes [8,10], which leads to magnetic saturation when the transformer is not substantially overdesigned. Therefore, eliminating DC components in both winding currents and magnetizing fluxes of the transformer in transient states is crucial to realize fast dynamic and robust control of the 3ph-DAB converter.

Due to the fact that a three-phase DC converter is considered here, the block-mode operation of the 3ph-DAB converter, that is, its three-phase winding currents and magnetizing fluxes can be conveniently modeled and analyzed using space vectors in the $\alpha\beta$ stationary frame using Clarke's Transformation. Using SPS control, each three-phase bridge sequentially switches between six switching states $s = \{1, 2, 3, 4, 5, 6\}$, as was shown in Figure 4.2. In the $\alpha\beta$-frame these states correspond to six space vectors (see Figure 4.10). Assuming a symmetric behavior and neglecting the resistive losses, the three-phase transformer of the DAB converter is simplified as a T-type equivalent circuit with the voltage vectors applied on the primary and secondary sides as shown in Figure 4.11. Thereby, the slopes of primary winding current $\frac{\mathrm{d}\vec{i}_{\mathrm{p,m/n}}}{\mathrm{d}t}$ and magnetizing flux $\frac{\mathrm{d}\vec{\psi}_{\mathrm{M,m/n}}}{\mathrm{d}t}$

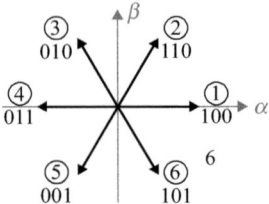

Figure 4.10   *Voltage vector of a three-phase bridge in αβ coordinates*

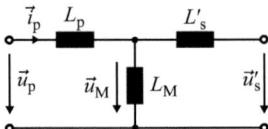

Figure 4.11   *Simplified equivalent circuit of a three-phase transformer in αβ coordinates*

for arbitrary switching states m for the primary side and n for the secondary side are obtained as:

$$\frac{d\vec{i}_{\text{p},m/n}}{dt} = \frac{\vec{u}_{\text{p},m}}{L_1} - \frac{\vec{u}'_{\text{s},n}}{L_2}$$

(4.3)

$$\text{with } L_1 = L_\text{p} + \frac{L_\text{M}L'_\text{s}}{L_\text{M} + L'_\text{s}} \text{ and } L_2 = L'_\text{s} + \frac{L_\text{M}L_\text{p}}{L_\text{M} + L_\text{p}},$$

$$\frac{d\vec{\psi}_{\text{M},m/n}}{dt} = \vec{u}_{\text{M},m/n} = \lambda_1\vec{u}_{\text{p},m} + \lambda_2\vec{u}'_{\text{s},n}$$

(4.4)

$$\text{with } \lambda_1 = \frac{L_\text{M}L'_\text{s}}{L_\text{M}L'_\text{s} + L_\text{p}(L_\text{M} + L'_\text{s})} \text{ and } L_2 = \frac{L_\text{M}L_\text{p}}{L_\text{M}L_\text{p} + L'_\text{s}(L_\text{M} + L_\text{p})}.$$

Applying the sequenced voltage vectors of the SPS modulation to (4.3) and (4.4), the trajectories of the transformer winding currents and magnetizing fluxes can be derived [10], as shown in Figures 4.12 and 4.13, respectively. In the steady state, the winding current trajectory with arbitrary load angle φ forms a regular hexagon centered at the origin of the αβ coordinates with the edge length determined by the load angle φ. Similarly, the steady-state magnetizing flux linkage trajectory with an arbitrary load angle φ is also a hexagon centered at the origin of the αβ coordinates, but it is always inscribed in the no-load magnetizing flux linkage trajectory (hexagon with dotted lines in Figure 4.12), which acts as a maximum volt-seconds boundary.

When the load angle φ changes abruptly the transient trajectories of the winding current and magnetizing flux linkage immediately drift from the initial trajectories with significant DC components respectively (see Figures 4.14 and 4.15). For the winding currents, the DC components are damped by the lumped

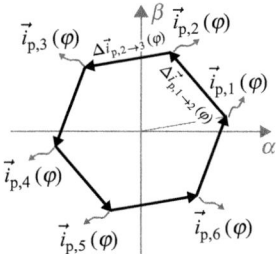

Figure 4.12   Steady-state αβ-frame trajectory of transformer winding currents for SPS modulation

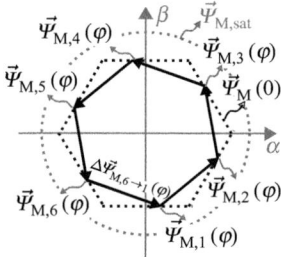

Figure 4.13   Steady-state αβ-frame trajectory of transformer magnetizing flux linkages for SPS modulation

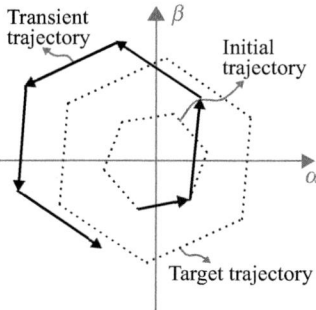

Figure 4.14   Trajectories of transformer winding current with a direct load angle change from φ_1 to φ_2

resistance R of the transformer windings and conducting semiconductor devices with a time constant of $\frac{L_\sigma}{R}$. For high-power transformers, the winding resistance is usually small which yields a large value of $\frac{L_\sigma}{R}$. This implies that the DAB converter will not reach the new reference steady state within several up to tens of switching cycles. The issue is even more severe with the magnetizing flux linkage. Since the

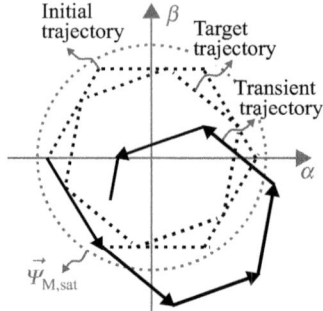

Figure 4.15   Trajectories of transformer magnetizing flux linkage with a direct load angle change from φ_1 to φ_2

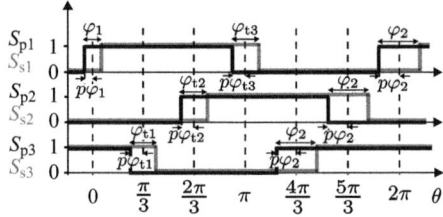

Figure 4.16   Switching pattern of the instantaneous flux linkage and current control (IFCC) method

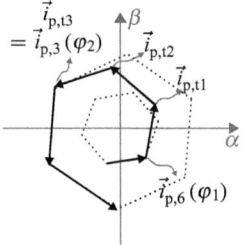

Figure 4.17   Trajectories of transformer winding current with a load angle change from φ_1 to φ_2 using the IFCC method

magnetizing inductance L_M is much larger than the leakage inductance L_σ of the transformer, the DC component in the magnetizing flux linkage has a significantly larger time constant of $\frac{L_M}{R}$, and the accumulative DC-bias flux linkage will ultimately drive the transformer in saturation, which negatively affects the operation of the 3ph-DAB converter.

To improve the dynamic performance, intermediate averaged load angles $\varphi_{t1} = \varphi_{t2} = \varphi_{t3} = \frac{\varphi_1 + \varphi_2}{2}$ are applied in the transient switching cycle (see Figure 4.16) so that the current trajectory will reach the new steady state within half a switching cycle without inducing overshoot currents (see Figure 4.17). This method is also known as Instantaneous Current Control (ICC) [9]. Furthermore, the load angle is split into two parts and distributed into both the primary and secondary bridges. When this double-side phase shift technique is applied as shown in Figure 4.16, the transient DC-bias magnetizing flux linkage can be completely eliminated [8,10] (see Figure 4.18). The reason is that the differential volt-seconds of the primary and secondary bridges instantaneously cancel each other out during transient states, when the distribution factor p of the double-side phase-shift modulation is set as:

$$p = \frac{L_p V_{out}}{L_p V_{out} + L_s N_{tr} V_{in}} \tag{4.5}$$

and $p\varphi$ is the corresponding load angle applied to the primary bridge.

Applying the double-side phase-shift modulation to the ICC method, the transformer current and magnetizing flux linkage can be instantaneously and simultaneously controlled within half a switching cycle in different transient conditions including power reversal (see Figure 4.19). This method is named as

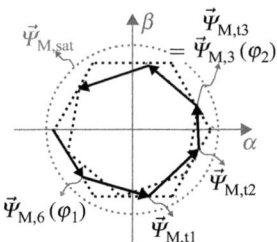

Figure 4.18   Trajectories of transformer magnetizing flux linkage with a load angle change from φ_1 to φ_2 using the IFCC method

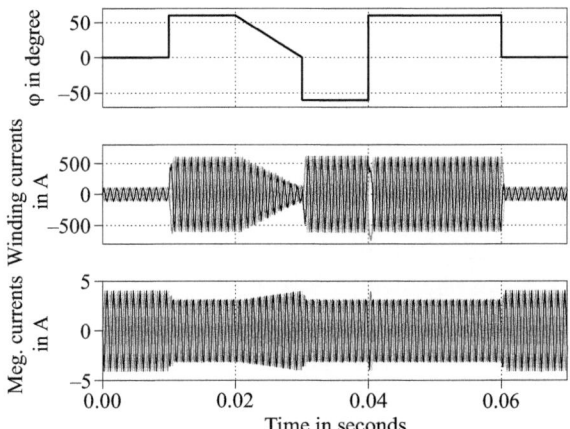

Figure 4.19   Transformer winding currents and magnetizing currents of the IFCC method with load changes and power reversal

instantaneous flux linkage and current control (IFCC). As an extension to the IFCC, the switching sequences for the start-up and shut-down cycle can be redesigned to allow soft-magnetization and de-magnetization of the transformer [10].

4.2.3.2   Anti-saturation control for medium-frequency transformer in the steady state

During commissioning of a 5 MW, 5 kV$_{DC}$ MV 3ph-DAB converter, it has been observed in steady state that one or more transformer phases operated with unexpected high harmonic currents. It was found that these undesired harmonic components were caused by transformer saturation effects. Detailed information can be found in [11,12]. The saturation is an undesired effect since it causes unequal voltage sharing of the windings, additional losses, and increased audible noise due to magnetostriction of the core. This harmonic content is caused by DC-components in the magnetizing current, which can occur due to various imperfections, such as minor differences in switching speeds and delay times of the semiconductors, asymmetries between the transformer phases and DC voltage ripple.

Many compensation techniques that are proposed in literature are not easily applicable, as they require either DC magnetic field measurements in the core or extremely high precision DC current sensors. Note that in well-designed transformers (without an airgap) the magnetizing AC current is orders of magnitude smaller than the load current. In the following sections a novel and simple feedback controller, which is based on an observer that estimates the DC-magnetizing currents using the star-point voltage, is introduced to avoid DC steady state currents in the three-phase transformer.

Transformers are designed such that the magnetizing currents do not saturate the transformer core when operating at maximum voltage. However, if the voltage applied to the transformer includes a small DC component, a fairly large magnetizing current, which is only limited by the DC resistance of the windings, can occur. These DC current components can force the transformer core into saturation, which leads to asymmetric voltage waveforms of the phase voltages and the start-point voltage (referred to the DC voltage mid-point), as shown in Figure 4.20(b).

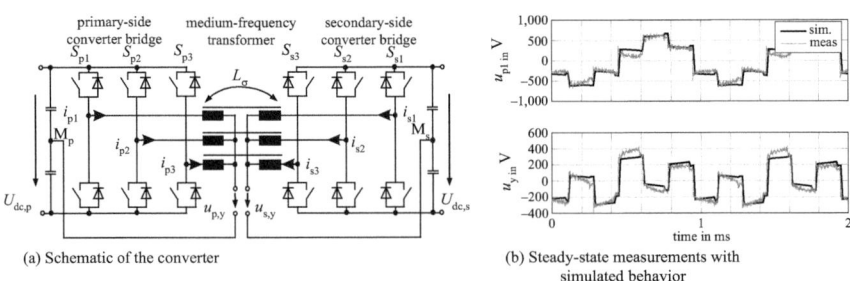

(a) Schematic of the converter

(b) Steady-state measurements with simulated behavior

Figure 4.20   Three-phase dual-active bridge with star-point voltage and experimental test results up to 1100 V at 500 kW

As mentioned above, in a 3ph-DAB converter, a DC voltage component may arise from offsets in the power semiconductor switching instants, different forward voltage drops of the semiconductor switches, or harmonic components in the DC-link voltage, as shown in a schematic manner in Figure 4.21. The deviation of the volt-seconds compared to the undisturbed operation is shaded gray.

Several approaches exist to simulate saturation effects in magnetic materials. In Figure 4.22(a), a simplified function is shown that characterizes the $L_m(i_m)$ characteristic of the magnetic core. A simplified *BH*-characteristic is also shown to illustrate the impact of a DC components on the saturation level of the core.

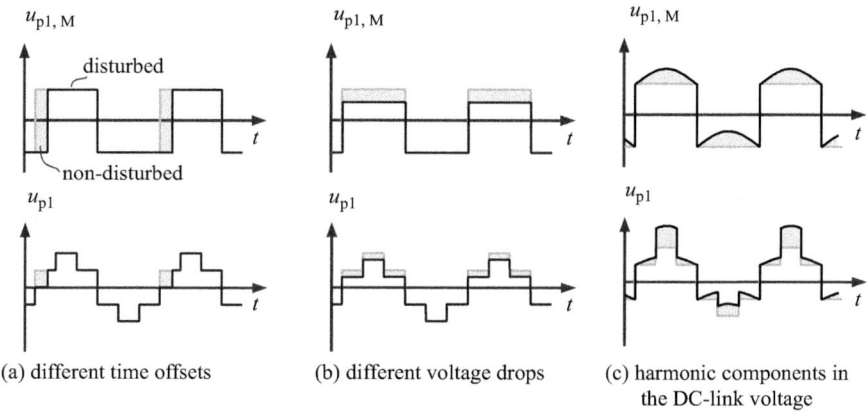

(a) different time offsets   (b) different voltage drops   (c) harmonic components in the DC-link voltage

Figure 4.21   Different root causes for DC voltages across the magnetizing inductance

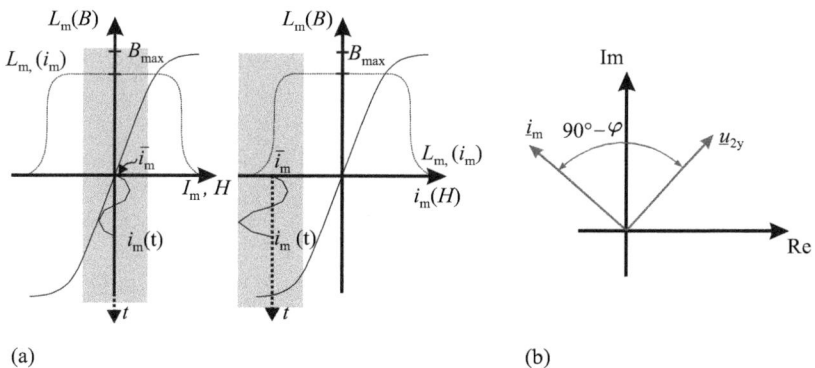

(a)   (b)

Figure 4.22   Simulation-based investigation to model saturable inductors relations between the complex star-point voltage and the three-phase magnetizing current space vector, that is, applying Clarke's transformation on the periodic averaged magnetizing currents

Embedding this saturation model in simulations, the relationships between magnetizing current and the star voltage as given in Figure 4.22(b) can be derived. A direct relation between the averaged magnetizing currents (after being transformed using Clarke's transformation) $i_{m,dc}$, the load angle φ and the vector of the second harmonic of the star-point voltage u_{2y} can be found:

$$\arg\left(\overrightarrow{i_{m,dc}}\right) \approx \arg\left(\overrightarrow{u_{2y}}\right) + 90° - \varphi$$

$$\left|\overrightarrow{i_{m,dc}}\right| \approx c_2 \left|\overrightarrow{u_{2y}}\right| \tag{4.6}$$

The factor c_2 describes the nearly linear relationship between the amplitude of the averaged magnetizing current and the second harmonic star-point voltage. This factor mainly depends on the characteristic of the used transformer. Hence, an observer of the DC magnetizing current can be built by sensing the second harmonic of the star-point voltage.

The block diagram of the closed loop controller, including the plant, is shown Figure 4.23. The components of the DC magnetizing current observer are subdivided in the star-point voltage preconditioner, the second harmonic extractor and the magnetizing current estimator. A simple PI-controller followed by the model-based duty cycle calculator controls the modulator of the 3ph-DAB.

The *Star-Point Voltage Preconditioner* filters the influences due to the modulation by the compensation control itself and other impacts such as dead-time effects or other delays. The *Second-Harmonic Extractor* calculates the real and imaginary part of the second harmonic u_{2y}. Subsequently the second harmonic is extracted in the *Estimator* which applies the equations given above, resulting in the Clarke transformed averaged magnetizing currents. The *PI controllers* manipulate the average voltage vector $\overrightarrow{u_{m,dc}}$. This reference voltage is fed to a *duty cycle calculator* and *modulator* which controls the converter switching instances such that the DC-offsets of the magnetizing currents are eliminated.

The compensated and uncompensated cases are shown experimentally in Figure 4.24. The anti-saturation controller is able to stabilize the system and only a small distortion in the star-point voltage remains, which indicates an effective suppression of the DC-offsets in the magnetizing currents. Beside the tremendous noise reduction, in the experimental test setup the no-load losses were reduced from 15.8 to 14.0 kW. Particularly, the experiment showed that the star-point voltage is a very good indicator to estimate the DC-magnetizing currents in transformers, and thus a perfect quantity to adjust the duty cycles of the converter switching states using a simple PI controller. Compared to other methods to measure DC-offsets in magnetizing currents, here only one voltage sensor to measure the star-point voltage is required, resulting in a very cost effective and reliable solution.

4.2.3.3 Fault current limiting control

Protection against short-circuit faults remains one of the main challenges in DC grids, since compared to AC systems, DC fault currents are developing much faster and have no natural zero crossing. This imposes significant challenges on the

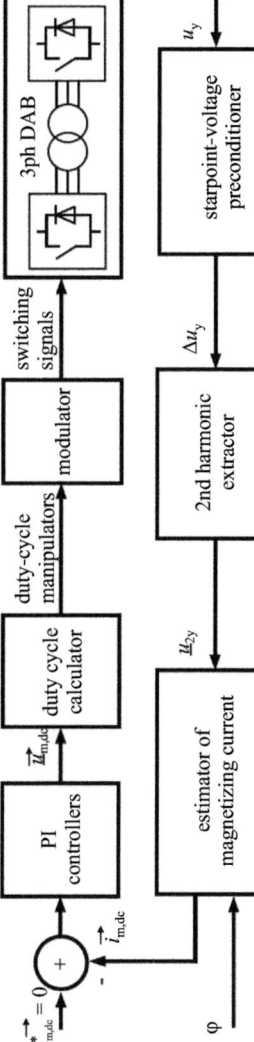

Figure 4.23   Diagram of the closed-loop anti-saturation controller

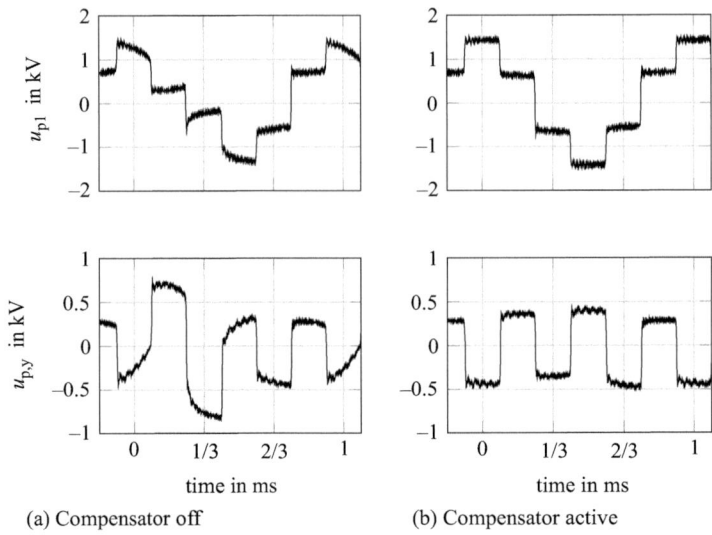

(a) Compensator off   (b) Compensator active

Figure 4.24   Control tested and proven at no-load operation at $U_{dc} = 2$ kV

development of DC circuit breakers, especially when various key performance factors such as interruption speed, cost, size and reliability have to be simultaneously considered.

In fact, as a pure power electronics-based system, DC grids can be alternatively protected by DC substations that use high-power DC–DC converters, for example, 3ph-DAB converters. The fast switching behavior and inherent controllability allows the DC–DC converter to control and limit the fault current instead of being shut down passively in short-circuit fault conditions. This further imposes two essential requirements on the DAB converter, resulting in the capability of FRT operation. First, the DAB converter shall be able to operate and maintain its current controllability when one of the DC-link voltages reduces to zero in the worst case, and, secondly, the DAB converter shall be able to provide sufficient fault current for fault detection and localization without causing electrical and thermal overstress on the semiconductor devices.

However, due to the very low or even zero DC-link voltage, high circulating currents occur when using the SPS control [13], which results in significant stress on the semiconductor devices. Moreover, under the zero-voltage DC fault condition, the controllability of DC output current is lost in the SPS modulation. Therefore, the SPS modulation is not suitable for the FRT operation. On the other hand, in the TCM3-Buck mode, the 3ph-DAB converter can handle a high voltage ratio by adjusting the duty cycle of the primary and secondary switches [7,13]. However, due to the triangular current waveform, the duty cycle of the unfaulty bridge needs to be minimized, which also significantly reduces the output DC current. Once the DC fault voltage reduces to zero, the DC output current of the TCM3-Buck mode decreases to zero accordingly. The main problem is that the significantly reduced volt-seconds in the faulty bridge disables the 3ph-DAB converter from controlling the transformer currents over a wide voltage range. This implies that the 3ph-DAB converter is not able to ride through DC fault using the TCM3-Buck mode.

To solve this issue, a modulation scheme named FRT-Buck was proposed in [8,13] which can be seamlessly combined with other modulation modes like SPS, TCM3, and TZM3. In the FRT-Buck mode (see Figure 4.25), the insufficient volt-seconds in the faulty bridge is compensated by using opposite voltage vectors in the other bridge. The transformer current can then be shaped similarly as in the TZM3 mode via manipulating the dwell time of the voltage vectors in both the primary and secondary bridge (see Figure 4.26) following (4.7):

$$V_{\text{in}} \cdot D_{\text{p},1} - V_{\text{in}} \cdot D_{\text{p},4} = V'_{\text{out}} \cdot D_{\text{s},1} \tag{4.7}$$

where $D_{\text{p},1} = \frac{\theta_1}{2\pi}$, $D_{\text{p},4} = \frac{1}{3} - \frac{\theta_2}{2\pi}$ and $D_{\text{s},1} = \frac{1}{3}$ correspond to the duty cycles of the voltage vectors $\overrightarrow{U_{\text{p},1}}$, $\overrightarrow{U_{\text{p},4}}$ and $\overrightarrow{U_{\text{s},1}}$. Due to the additional degree of freedom from

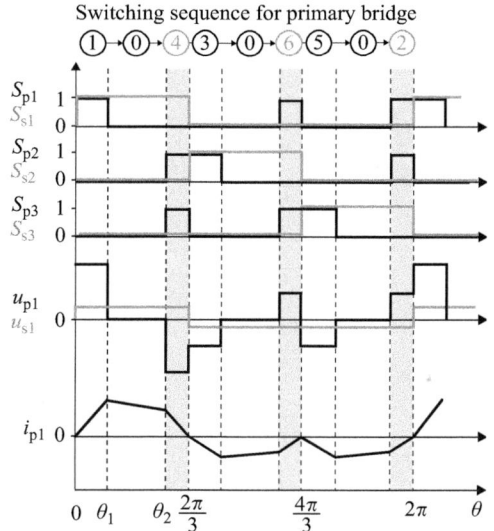

Figure 4.25   Switching sequence of the FRT-Buck mode

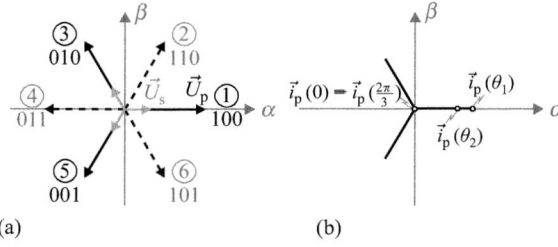

Figure 4.26   Voltage vectors and current trajectory in the FRT-Buck mode. (a) Voltage vectors. (b) Current trajectory

the opposite voltage vectors in the unfaulty bridge, the 3ph-DAB converter not only can operate and control the output current at zero fault voltage but also can realize soft switching with limited turned-off currents for all the semiconductor devices. The output DC current of the FRT-Buck mode is derived in (4.8), with its boundary shown in Figure 4.27:

$$\bar{i}_{\text{out, DC}} = \frac{2V_{\text{in}}}{fL_\sigma} \cdot \left[-\left(\frac{\theta_1}{2\pi} - \frac{1+d}{6}\right)^2 + \frac{1-d^2}{36} \right] \tag{4.8}$$

With the proposed FRT operation of the 3ph-DAB converter, it is possible to realize breakerless protection of the DC grid as shown in Figure 4.28, using the following steps:

- Step I: Fault detection and localization.
- Step II: De-energization of DC grid.

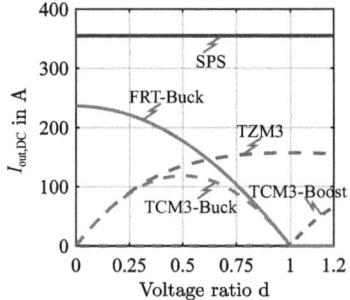

Figure 4.27   DC output current boundary of the FRT-Buck mode and other modulation schemes

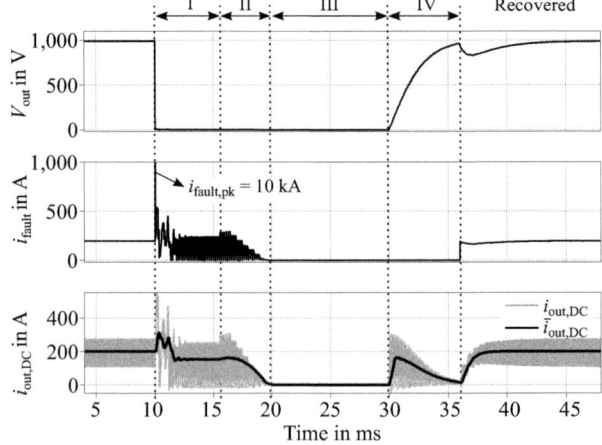

Figure 4.28   Simulation waveforms of breakerless protection scheme for DC grids

- Step III: Fault isolation by high-speed mechanical switch.
- Step IV: Re-energization of DC grid [8].

4.2.3.4 Practical considerations of dead time

It is a well-known fact that power semiconductor devices employed at MVDC levels are switching at significant lower speeds as compared to their low-voltage counterparts. Thus, a relatively large dead time between turn-off of one and turn-on of the corresponding device in the same phase leg is indispensable [14]. This large dead time can cause a significant error in the achieved power level, particularly at lower values of load angle φ. These errors can be compensated in the forward control loop itself if the dead time is incorporated in the model equations.

Considering the dead time phase equivalent δ ($\delta = \omega t_d$, with t_d being the dead time), different operating modes can be identified based on the voltage ratio d and the current behavior during the dead time. The definition of these modes is given in Table 4.5. The power transferred between the two bridges in different modes is given by Equation set (4.9). It is evident that in mode 2, the transferred power is independent of the phase shift between the two bridges. The parameter γ represents the phase equivalent of the time duration between the zero-crossing of the current and the turn-on of the next device. This can be analytically described by Equation set (4.10). It is important to mention that (4.10) shall be used only for $\delta \neq 0$ and is physically meaningless otherwise:

$$
P = \frac{V_{in}^2}{\omega} \frac{d}{L_\sigma} \frac{1}{12\pi}
\begin{cases}
\begin{pmatrix} 8\pi\varphi + 3\gamma^2 d + 3\gamma^2 - 6\varphi^2 - 6\delta\gamma - 4\pi\gamma \\ + 6\gamma\varphi - 6\delta\gamma d - 4\pi\gamma d + 6\varphi\varphi d \end{pmatrix}, \text{Mode 1a} \\
2(4\pi\varphi + 3\gamma^2 - 3\varphi^2 - 6\delta\gamma - 4\pi\gamma + 6\gamma\varphi), \text{Mode 1b} \\
3\gamma^2 d - 6\delta^2 + 3\gamma^2 + 8\pi\delta - 4\pi\gamma - 4\pi\gamma d, \text{Mode 2a} \\
\begin{pmatrix} 3\gamma^2 d + 6\delta^2 + 3\gamma^2 - 6\gamma d - 8\pi\delta + 4\pi\gamma \\ - 6\delta\gamma d + 4\pi\gamma d \end{pmatrix}, \text{Mode 2b} \\
\begin{pmatrix} 8\pi\varphi + 3\gamma^2 d + 6\delta^2 - 3\gamma^2 + 6\varphi^2 + 6\delta\gamma \\ -8\pi\delta - 12\delta\varphi - 4\pi\gamma - 6\delta\gamma d + 4\pi\gamma d \end{pmatrix}, \text{Mode 3} \\
(8\pi\varphi - 6\varphi^2), \text{Mode 0}
\end{cases}
$$

$$(4.9)$$

Table 4.5 *Definition of operating modes in DAB converter considering dead time* t_d ($\delta = \omega \cdot t_d, 0 \leq \varphi \leq \pi/3$) [15]

Modes	Conditions	
1a	$0 < \varphi > \delta$	$d < 1$
1b		$d \geq 1$
2a	$0 < \varphi < \delta$	$d < 1$
2b		$d \geq 1 \wedge \varphi < \varphi_{2b-3}$
3	$\varphi_{2b-3} \leq \varphi \leq \varphi_{3-1b}$	$d \geq 1$
0	$\delta = 0 \vee \varphi > \varphi_{1a-0}$	$d < 1$
	$\delta = 0 \vee \varphi > \varphi_{1b-0}$	$d \geq 1$

$$\gamma = \begin{cases} \dfrac{6\delta(1+d) - 4\pi(1-d) - 6\varphi d}{3(d+1)}, & \text{Mode 1a} \\[3mm] \dfrac{3\delta(1+d) - 2\pi(1-d) - 3\varphi d}{3d}, & \text{Mode 1b} \\[3mm] \dfrac{6\delta - 4\pi + 4\pi d}{3(d+1)}, & \text{Mode 2a} \\[3mm] \dfrac{4\pi + 6\delta d - 4\pi d}{3(d+1)}, & \text{Mode 2b} \\[2mm] 0, & \text{Mode 3} \\[1mm] 0, & \text{Mode 0} \end{cases} \qquad (4.10)$$

To apply the power Equation set (4.9) in the converter controller, the boundary conditions for the operating modes need to be identified. The analytical expression for these boundary points are given by Equations (4.11)–(4.13) and their definitions in Figure 4.29:

$$\varphi_{1a-0} = \varphi_{1b-0} = \frac{3\delta(1+d) + 2\pi(d-1)}{3d} \qquad (4.11)$$

$$\varphi_{3-1b} = \frac{3\delta + 2\pi(d-1)}{3d} \qquad (4.12)$$

$$\varphi_{2b-3} = \frac{2\pi}{3} + \delta - \sqrt{\frac{9\delta^2(d-1) + 4\pi^2(5d-3) - 24\pi\delta(d-1)}{9(d+1)}} \qquad (4.13)$$

The interested reader is referred to [16] for a detailed description of the entire calculation procedure, and the effectiveness of the proposed compensation method.

4.2.4 Three-phase, three-level dual-active bridge converters

A three-phase, three-level DAB converter consists of two, three-level bridges on both sides of the transformer as depicted in Figure 4.30. The MV side can be realized with the neutral-point-clamped (NPC) topology [17]. The advantage of the NPC topology is the lower blocking-voltage requirement (half as required by

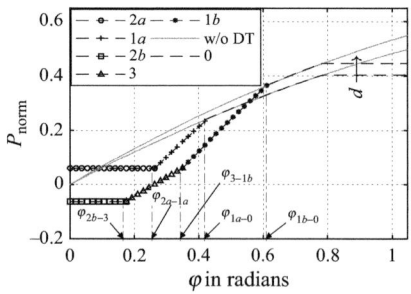

Figure 4.29   Normalized power curved with non-zero dead time and the definition of boundary points between operating modes

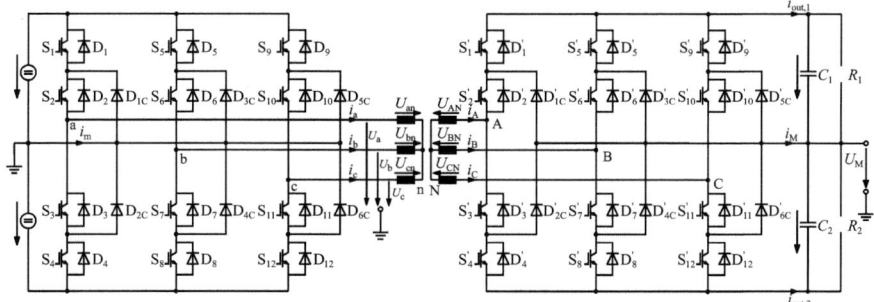

Figure 4.30   3ph-DAB in NPC configuration

Table 4.6   Boundary conditions of different operating cases

Case	Boundary conditions with $\varphi \in [0, \pi/3]$ and $k_1, k_2 \in [1/3, 2/3]$		
1	$k_1 \leq \left(\frac{1}{3} + \frac{\varphi}{2\pi}\right)$	$k_2 \leq \left(\frac{1}{2} - \frac{\varphi}{2\pi}\right)$	
2	$k_1 > \left(\frac{1}{3} + \frac{\varphi}{2\pi}\right)$	$k_2 < \left(\frac{1}{2} - \frac{\varphi}{2\pi}\right)$	$(k_1 - k_2) \leq \frac{\varphi}{2\pi}$
3	$k_1 > \left(\frac{1}{3} + \frac{\varphi}{2\pi}\right)$	$k_2 < \left(\frac{1}{2} - \frac{\varphi}{2\pi}\right)$	$(k_1 - k_2) > \frac{\varphi}{2\pi}$
4	$k_1 \geq \left(\frac{1}{3} + \frac{\varphi}{2\pi}\right)$	$k_2 \geq \left(\frac{1}{2} - \frac{\varphi}{2\pi}\right)$	
5	$k_1 < \left(\frac{1}{3} + \frac{\varphi}{2\pi}\right)$	$k_2 > \left(\frac{1}{2} - \frac{\varphi}{2\pi}\right)$	$\left(k_1 - k_2 + \frac{1}{6}\right) \geq \frac{\varphi}{2\pi}$
6	$k_1 < \left(\frac{1}{3} + \frac{\varphi}{2\pi}\right)$	$k_2 > \left(\frac{1}{2} - \frac{\varphi}{2\pi}\right)$	$\left(k_1 - k_2 + \frac{1}{6}\right) < \frac{\varphi}{2\pi}$

two-level DAB converter for the same DC-bus voltage) for the active devices on the MV side. Depending upon the target voltage levels, it is also possible to achieve the three-levels by using the so-called T-type NPC topology [18,19] to minimize conduction losses. The switching principle of the three-level 3ph-DAB converter can be summarized as follows:

- The top and bottom devices of the primary- and secondary-side bridge are switched with duty cycles k_1 and k_2, respectively. Consequently, the middle devices of the two bridges are switched on during two intervals $(1 - k_1) \cdot T_{sw}$ and $(1 - k_2) \cdot T_{sw}$, with T_{sw} being the switching cycle.
- The three phases within each three-phase converter are mutually phase-shifted by $2\pi/3$ radians.
- The respective phases in the two bridges are phase-shifted with a load angle φ.

Thus, there are three control or modulation variables, which can be used to control the power transferred between the two bridges. Considering the modulation space in which $\varphi \in [0, \pi/3]$ and $k_1, k_2 \in [1/3, 1/2]$, six different contiguous operational cases can be identified. The boundary conditions for these operational cases are given in Table 4.6. The interested readers can refer to [15] for typical waveforms of these operational cases. Using these boundary conditions, the switching states of converter bridges can be derived. With these switching states and the steady-state symmetry condition $i_x(\omega t) = -i_x(\omega t + \pi)$ with $x = a, b, c$, the equations for the power transferred from the primary to the secondary bridge for all cases can be derived and are given in Equation set (4.14). A similar piecewise linear approach can be used to derive all other important converter variables, among others the transformer currents and the DC-side current ripple, as given in [15]:

$$
P = \frac{V_{in}^2}{\omega L_\sigma}
\begin{cases}
\dfrac{d}{72\,\pi}\left(\begin{array}{l} 12k_1\pi^2 + 12k_2\pi^2 - 4\pi^2 - 27\varphi^2 \\ \quad - 108k_1^2\pi^2 + 108\pi\varphi k_1 + 72k_1k_2\pi^2 \end{array}\right), \text{case 1} \\[1.2em]
\dfrac{d}{12\pi}\left(\begin{array}{l} 2\pi\varphi - 2k_1\pi^2 + 2k_2\pi^2 - 3\varphi^2 - 12k_1^2\pi^2 \\ \quad + 12\pi\varphi k_1 + 12k_1k_2\pi^2 \end{array}\right), \text{case 2} \\[1.2em]
\dfrac{d}{6}(6k_2 + 1)(\varphi - \pi k_1 + \pi k_2), \text{case 3} \\[1.0em]
\dfrac{d}{12\pi}\left(\begin{array}{l} -6\varphi^2 + 12\pi\varphi k_1 - 12\pi\varphi k_2 - 12\pi^2 k_1^2 + 12\pi^2 k_1 k_2 \\ +8\pi\varphi - 2\pi^2 k_1 - 12\pi^2 k_2^2 + 14\pi^2 k_2 - 3\pi^2 \end{array}\right), \text{case 4} \\[1.2em]
\dfrac{d}{72\pi}\left(\begin{array}{l} -45\varphi^2 + 108\pi\varphi k_1 - 72\pi\varphi k_2 - 108\pi^2 k_1^2 + 72\pi^2 k_1 k_2 \\ +36\pi\varphi + 12\pi^2 k_1 - 72\pi^2 k_2^2 + 84\pi^2 k_2 - 22\pi^2 \end{array}\right), \text{case 5} \\[1.2em]
\dfrac{d}{72\pi}\left(\begin{array}{l} -54\varphi^2 + 144\pi\varphi k_1 - 108\pi\varphi k_2 + 42\pi\varphi - 144\pi^2 k_1^2 \\ +144\pi^2 k_1 k_2 - 108\pi^2 k_2^2 + 96\pi^2 k_2 - 23\pi^2 \end{array}\right), \text{case 6}
\end{cases}
$$

$$(4.14)$$

The three-level 3ph-DAB converter inherits the ZVS properties of the two-level converter. It also provides a possibility to turn-on the middle devices in the NPC bridge at zero-current, called ZCS. Figure 4.31 shows the turn-on and turn-off of one of the middle devices in of the NPC phase. The dotted and the solid lines show the current path before and after the turn-on/off transitions respectively.

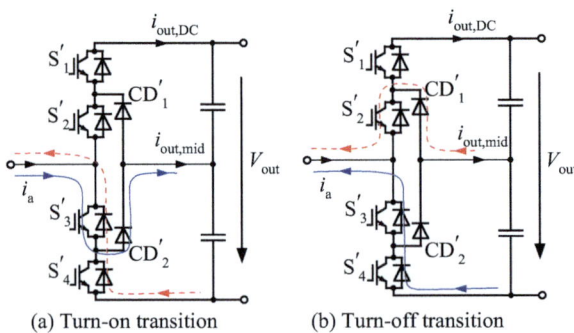

(a) Turn-on transition   (b) Turn-off transition

Figure 4.31   Zero-current turn-on and hard turn-off transition in the NPC phase leg; dashed red and solid blue paths indicate the current flow before and after the transition, respectively [15]

Table 4.7   Possibility of ZVS and ZCS in different operational cases [15]

Switching	Case 1	Case 2	Case 3	Case 4	Case 5	Case 6
Primary-side ZVS	Yes/No	Yes/No	Yes	Yes	Yes	Yes
Secondary-side ZVS	Yes	Yes	Yes	Yes	Yes	Yes
Secondary-side ZCS	No/Yes	No/Yes	No	Yes	Yes	Yes

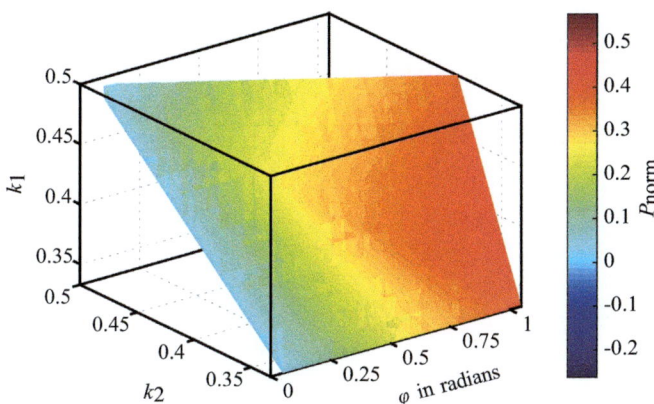

Figure 4.32   Locus of modulation parameter for maximum soft-switching in the three-level DAB converter for d = 1 [15]

Owing to symmetry, the conclusions apply to the other phases and the other middle devices as well. Table 4.7 shows the possibilities of ZVS as well as ZCS in different operational cases of the converter.

If the modulation parameters are carefully chosen so that the converter operates only in cases 4, 5, and 6, it is possible to achieve maximum soft-switching of the converter. Because of the relative positions of the gate signals, it turns out that in these cases, if only the secondary-side ZCS condition is satisfied, the ZVS turn-on of the top and bottom devices in both bridges is automatically ensured. Figure 4.32 shows the calculated modulation parameters for maximum soft-switching in the converter at different levels of the normalized power P_{norm} (given by dividing (4.14) by $V_{\text{in}}^2/\omega L_\sigma$). These optimal modulation parameters can be stored in the form of look-up tables (LUTs) since online calculation is computationally intensive. A detailed treatment of the soft-switching as well as the calculation of LUTs can be found in [15].

Similar to the two-level 3ph-DAB the 3L-NPC topology requires balanced DC-link voltages. Especially in three-wire bipolar MVDC grids, that is, with neutral wire connected to the DC midpoint voltage, non-symmetrical loads can occur. With the three-level space-vector modulation, small vectors in redundant switching stages could be used to control the neutral point current [20]. As this method would require a high ratio between the switching frequency and the fundamental output

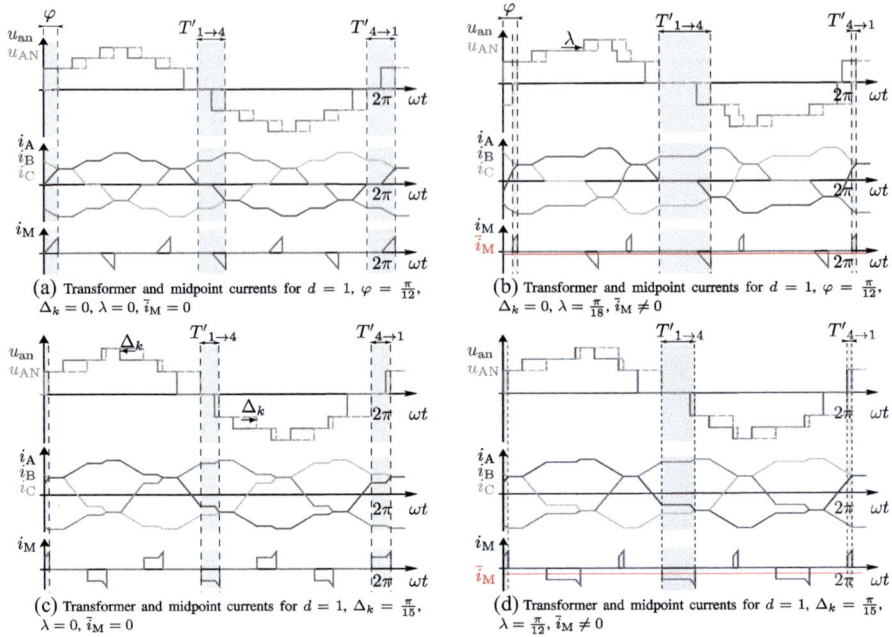

Figure 4.33   Impact of different values of λ and Δk on transformer and midpoint currents

frequency it is not applicable in the 3L-DAB3 converter. Thus, besides the methods for voltage balancing in a two-level DABs [21], a new strategy is introduced here.

To achieve balanced DC-link voltages under non-symmetrical loads, the average midpoint current i_M needs to be controlled. During symmetrical operation, positive and negative midpoint current pulses cancel out each other and thus no current is charging/discharging the DC-link capacitors on average. To influence the average midpoint current, a new modulation parameter λ is defined, which delays the turn-on transition of the lower outer switches [S_4, S_8, S_{12}], $\left[S_4', S_8', S_{12}'\right]$ and the inversely operated upper inner switches [S_2, S_6, S_{10}], $\left[S_2', S_6', S_{10}'\right]$. In addition to λ the parameters k_1 and $k_2 \in \left[\frac{1}{3}, \frac{1}{2}\right]$ as introduced in the previous paragraph are used. For $k_1 = k_2$ the influence on the midpoint current is relatively small, as shown in Figure 4.33(a) and (b). In order to increase the voltage-time per pulse, different values for k_1 and k_2 can be chosen, which increase the pulses of i_M. The relationship between k_1 and k_2 is defined as:

$$\Delta k = (k_2 - k_1)\pi \tag{4.15}$$

$$k_1 = \frac{5}{12} - \frac{\Delta k}{2\pi} \tag{4.16}$$

$$k_1 = \frac{5}{12} + \frac{\Delta k}{2\pi} \tag{4.17}$$

The influence of Δk is visualized in Figure 4.33(c) and (d). As the pre-calculation of the modulation parameters and the influence on the midpoint current for a given operating point is very complex and strongly non-linear, look-up tables are generated for each modulation parameter to achieve an optimal setting under the given condition. The interested reader is referred to [22] for a detailed description of the entire procedure, the implementation and the effectiveness of the proposed modulation scheme.

Acknowledgements

The authors would like to acknowledge the kind supports and valuable contributions from the following researchers and colleagues:

- Dr. -Ing. Peter Lürkens, Chief Engineer, E.ON ERC Institute for Power Generation and Storage Systems, RWTH Aachen University, for reviewing and commenting on this chapter.
- Dr. -Ing. Johannes Voss, Chief Engineer, E.ON ERC Institute for Power Generation and Storage Systems, RWTH Aachen University, for his valuable contribution to Section 4.2.3.2.
- Mr. Philipp Joebges, Chief Engineer, E.ON ERC Institute for Power Generation and Storage Systems, RWTH Aachen University, for his valuable contribution to Sections 4.2.3.4 and 4.2.4.
- Dr. -Ing. Hafiz Abu Bakar Siddique, Hardware Engineer, SEG Automotive, Germany, for his valuable contribution to Sections 4.2.3.4 and 4.2.4.
- Mr. Zhiqing Yang, Research Associate, E.ON ERC Institute for Power Generation and Storage Systems, RWTH Aachen University, for his valuable contribution to Section 4.2.1.

References

[1] R. W. De Doncker, D. M. Divan and M. H. Kheraluwala, "A three-phase soft-switched high power density DC/DC converter for high power applications," in Conference Record of the 1988 IEEE Industry Applications Society Annual Meeting, Pittsburgh, PA, 1988.

[2] R. W. De Doncker, D. M. Divan and M. H. Kheraluwala, "A three-phase soft-switched high-power-density DC/DC converter for high-power applications," *IEEE Transactions on Industry Applications*, vol. 27, no. 1, pp. 63–73, Jan–Feb. 1991, doi: 10.1109/28.67533.

[3] R. U. Lenke, "A contribution to the design of isolated dc-dc converters for utility applications," Dissertation, E. ON Energy Research Center, RWTH Aachen University, Aachen, 2012, ISBN: 978-3-942789-05-9.

[4] Z. Yang, J. Hu, G. Pasupuleti and R. W. De Doncker, "Operation-oriented design procedure of a three-phase dual-active bridge converter for a wide operation range," in *2018 IEEE Energy Conversion Congress and*

Exposition (ECCE), Portland, OR, 2018, pp. 2835–2842, doi: 10.1109/ECCE.2018.8557389.

[5]   J. Hu, P. Joebges, G. C. Pasupuleti, N. R. Averous and R. W. De Doncker, "A maximum-output-power-point-tracking-controlled dual-active bridge converter for photovoltaic energy integration into MVDC grids," *IEEE Transactions on Energy Conversion*, vol. 34, no. 1, pp. 170–180, March 2019, doi: 10.1109/TEC.2018.2874936.

[6]   F. Marra, G. Y. Yang, C. Træholt, E. Larsen, C. N. Rasmussen and S. You, "Demand profile study of battery electric vehicle under different charging options," in *2012 IEEE Power and Energy Society General Meeting*, San Diego, CA, 2012, pp. 1–7, doi: 10.1109/PE.

[7]   J. Hu, N. Soltau and R. W. De Doncker, "Asymmetrical duty-cycle control of three-phase dual-active bridge converter for soft-switching range extension," in *2016 IEEE Energy Conversion Congress and Exposition (ECCE)*, Milwaukee, WI, 2016, pp. 1–8, doi: 10.1109/ECCE.2016.7854888.

[8]   J. Hu, "Modulation and dynamic control of intelligent dual-active-bridge converter based substations for flexible dc grids," Dissertation, E.ON Energy Research Center, RWTH Aachen University, Aachen, 2019, ISBN 978-3-942789-68-4.

[9]   S. P. Engel, N. Soltau, H. Stagge and R. W. De Doncker, "Dynamic and balanced control of three-phase high-power dual-active bridge DC–DC converters in DC-grid applications," *IEEE Transactions on Power Electronics*, vol. 28, no. 4, pp. 1880–1889, April 2013, doi: 10.1109/TPEL.2012.2209461.

[10]   J. Hu, S. Cui, S. Wang and R. W. De Doncker, "Instantaneous flux and current control for a three-phase dual-active bridge DC–DC converter," *IEEE Transactions on Power Electronics*, vol. 35, no. 2, pp. 2184–2195, Feb. 2020, doi: 10.1109/TPEL.2019.2922299.

[11]   J. Voss, "Multi-megawatt three-phase dual-active bridge," Dissertation, E. ON Energy Research Center, RWTH Aachen University, Aachen, 2019, ISBN: 978-3-942789-73-8.

[12]   J. Voss, S. P. Engel and R. W. De Doncker, "Control method for avoiding transformer saturation in high-power three-phase dual-active bridge DC-DC converters," *IEEE Transactions on Power Electronics*, vol. 35, no. 4, pp. 4332-4341, April 2020, doi: 10.1109/TPEL.2019.2938585.

[13]   J. Hu, S. Cui and R. W. De Doncker, "DC fault ride-through of a three-phase dual-active bridge converter for DC grids," in *2018 International Power Electronics Conference (IPEC-Niigata 2018 -ECCE Asia)*, Niigata, 2018, pp. 2250–2256, doi: 10.23919/IPEC.2018.8507672.

[14]   N. Soltau, "High-power medium-voltage DC-DC converters: design, control and demonstration," Dissertation, E.ON Energy Research Center, RWTH Aachen University, Aachen, 2017, ISBN: 978-3-942789-42-4.

[15]   H. A. B. Siddique, "The three-phase dual-active bridge converter family: modeling, analysis, optimization and comparison of two-level and three-level converter variants," Dissertation, E.ON Energy Research Center, RWTH Aachen University, Aachen, 2019, ISBN: 978-3-942789-75-2.

[16] H. A. B. Siddique, P. Joebges and R. W. De Doncker, "Modelling and compensation of dead time in three-phase dual-active bridge DC-DC converter," *EPE Journal*, vol. 30, no. 3, pp. 122–138, 2020, doi: 10.1080/09398368.2020.1742518.

[17] A. Nabae, I. Takahashi and H. Akagi, "A new neutral-point-clamped PWM inverter," *IEEE Transactions on Industry Applications*, vol. 17, no. 5, pp. 518–523, 1981, doi: 10.1109/TIA.1981.4503992.

[18] M. Schweizer and J. W. Kolar, "Design and implementation of a highly efficient three-level t-type converter for low-voltage applications," *IEEE Transactions on Power Electronics*, vol. 28, no. 2, pp. 899–907, Feb 2013, doi: 10.1109/TPEL.2012.2203151.

[19] N. H. Baars, J. Everts, C. G. E. Wijnands and E. A. Lomonova, "Evaluation of a high-power three-phase dual active bridge DC-DC converter with three-level phase-legs," in *2016 18th European Conference on Power Electronics and Applications (EPE'16 ECCE Europe)*, Karlsruhe, 2016, pp. 1-10, doi: 10.1109/EPE.2016.7695616.

[20] D. H. Lee, S. R. Lee and F. C. Lee, "An analysis of midpoint balance for the neutral-point-clamped three-level VSI," in *PESC 98 Record. 29th Annual IEEE Power Electronics Specialists Conference (Cat. No.98CH36196)*, Fukuoka, Japan, 1998, pp. 193–199, vol.1, doi: 10.1109/PESC.1998.701899.

[21] A. Filbà-Martínez, S. Busquets-Monge and J. Bordonau, "Modulation and capacitor voltage balancing control of a three-level NPC dual-active-bridge DC-DC converter," in *IECON 2013–39th Annual Conference of the IEEE Industrial Electronics Society*, Vienna, Austria, 2013, pp. 6251–6256, doi: 10.1109/IECON.2013.6700163.

[22] P. Joebges, A. Gorodnichev and R. W. De Doncker, "Modulation and active midpoint control of a three-level three-phase dual-active bridge DC-DC converter under non-symmetrical load," in *2018 International Power Electronics Conference (IPEC-Niigata 2018 -ECCE Asia)*, Niigata, Japan, 2018, pp. 375–382, doi: 10.23919/IPEC.2018.8507930.

Chapter 5

Multiport DC power converters for MVDC applications

Gregory J. Kish[1]

5.1 Introduction

The power system landscape is rapidly evolving. The global push to reduce greenhouse gas emissions is driving change into the way electric power is generated, stored and delivered. Perhaps one of the biggest technological shifts being witnessed by the classical AC power grid is the utilization of DC power, across the full spectrum of generation, transmission and distribution levels. High-voltage DC (HVDC) transmission first emerged in the 1950s owing to its potential benefits such as low-loss delivery of bulk power over long distances and enhanced control of system power flows. The number of HVDC installations has steadily risen over the years, accompanied by advancements in semiconductor and converter technology, with the last decade or so seeing explosive levels of global growth. With the promise of realizing similar benefits for lower voltage and power applications, medium-voltage DC (MVDC) systems are now starting to emerge at the distribution level. As the grid continues to evolve it is anticipated that MVDC systems will play a key role in the reinforcement of legacy AC distribution networks, as well as to increase system efficiency and reliability.

Akin to traditional medium-voltage AC (MVAC) distribution systems, there will arise a need to interconnect or mesh together MVDC systems. This will require use of power electronic converters. Multiport DC power converters that can interconnect multiple different DC systems are an appealing option for MVDC interconnects. This chapter introduces the multiport DC converter concept and explores its application in distribution level MVDC systems. The focus is on using modular multilevel converter (MMC) technology that can accommodate demands for high modularity, scalability and reliability while simultaneously being compatible with MVDC voltage and power levels on the order of tens of kilovolts and several tens of megawatts. An overview of different multiport DC converter technologies is provided, including both isolated and non-isolated topologies. The chapter concludes with case study simulations of a multiport DC converter for routing power between three different MVDC systems.

[1]Department of Electrical and Computer Engineering, University of Alberta, Alberta, Canada

5.2   Overview of MVDC systems

The interest in MVDC for power systems applications is accelerating. Potential applications include, but are not limited to, microgrids, grid connection of battery energy storage systems, collector networks for renewable-based power generation such as solar photovoltaic (PV) and offshore wind, and utility distribution grids. MVDC voltage levels will vary on the application (see Section 5.2.1), but are generally on the order of a few kilovolts to several tens of kilovolts. This is far lower than the hundreds of kilovolts encountered in HVDC transmission systems. MVDC emerged in part to satisfy the need for an alternative to HVDC where lower voltage (and power) levels are sufficient and shorter distances are involved. The main applications fueling interest in MVDC grids include [1] (1) grid connection of renewable energy resources such as offshore wind power (via MVDC collector networks), (2) reinforcement of high density urban distribution networks, and (3) DC load connections.

Figure 5.1 shows prominent examples of MVDC in power systems. The point-to-point MVDC link in Figure 5.1(a) is the distribution level counterpart to conventional HVDC transmission, where the voltage and power levels and distance between converter stations are commensurately reduced. Siemens MVDC Plus [2] is an example of an industry led initiative on this front. The recently commissioned DC Angle Project [3,4] in Europe is one of the few point-to-point distribution level MVDC links in the world. In Figure 5.1(b), a multiterminal MVDC grid is shown that mirrors the classical meshed AC grid. Once again, this may be viewed as a distribution level counterpart to the well known multiterminal HVDC grid architecture. Multiterminal DC grids are more complex than point-to-point DC links but provide added flexibility in system power flows and control, and therefore have been the focus of much research. However, to date, no real world MVDC

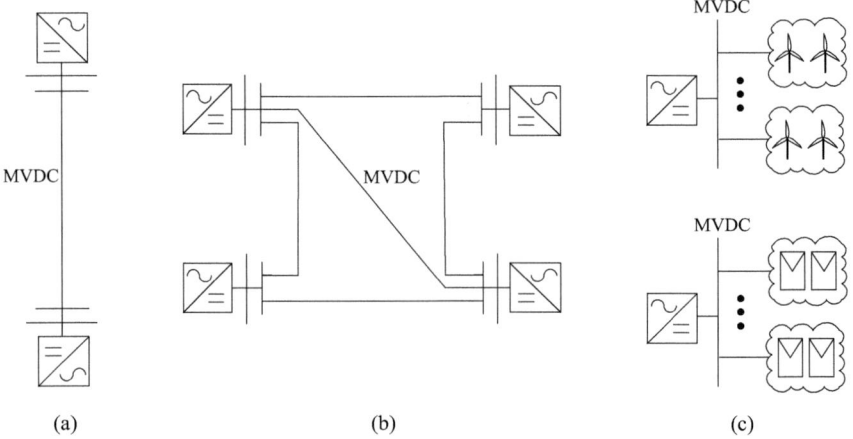

Figure 5.1   Example MVDC systems

multiterminal DC grid has been commissioned.* Lastly, Figure 5.1(c) illustrates MVDC collector networks for AC grid integration of offshore wind and onshore solar PV power plants. This scheme is garnering significant attention as a potentially more economical and efficient alternative to using standard AC collector networks [1,7,8]. Not shown in this figure are the DC–DC and/or AC–DC converters that would exist inside the wind and solar PV plants to create the main MVDC buses for connection to the grid interfacing DC–AC converter.[†]

5.2.1   MVDC voltage levels

MVDC voltages generally fall within the kilovolts to tens of kilovolts range. However, there is no general consensus on upper and lower voltage thresholds within this range. This is due in part to a wide range of emerging applications (i.e., from data centers and residential/industrial loads to offshore wind power plants and multiterminal DC grids) and a lack of accepted standards. Voltage levels play a large role in selection and rating of power system components such as converters, semiconductor devices, cables, etc. Moreover, voltage ratings (in conjunction with the nominal current ratings) dictate the overall system power levels. It is clear that voltage ratings are an important aspect of MVDC systems.

Research into MVDC is still at a relatively early stage when compared to other technologies, such as MVAC and HVDC. Consequently, the development of MVDC standards, guidelines and practices are lagging. Guidelines have started to emerge for MVDC shipboard power systems that recommend standardized voltage levels [9], while a recently published MVDC grid feasibility study [10] provides recommendations on voltage levels for MVDC distribution systems.

The following are notable examples of MVDC voltage levels either from a research perspective or real world installation.

- Considered in [7] as voltages ranging from 10 to 70 kV.
- A survey of voltage levels for MVDC systems provided in [11] indicate a range from about 5 to 40 kV, but can be as high as 80 kV.
- The working group in [10] targets MVDC distribution systems and applications with voltages ranging from ±0.75 to ±50 kV. However, depending on the power supply capacity of the corresponding AC distribution network, several different MVDC voltage level sequences are recommended, e.g., ±100 kV / ±35 kV/ ±0.4 kV and ±60 kV/ ±20 kV/ ±0.4 kV.
- Siemens MVDC Plus [2] portfolio proposes three design variants: (1) ±24 kV, 30–70 MW, (2) ±30 kV, up to 90 MW, and (3) ±50 kV, up to 150 MW.
- Angle DC project in the UK utilizes ±27 kV (bipolar) [3,4]. This voltage level was chosen based on conversion of existing 33 kV AC cables and lines to DC.
- University campus MVDC grid design in [12] selected ±5 kV (bipolar).

---

*The recently constructed Zhangbei-1 project represents the first global demonstrator of a multiterminal HVDC grid [5,6].

[†]The onshore dc/ac converter can be replaced with high-step ratio dc/dc converter in cases where direct connection to an onshore HVDC network is desired.

This chapter focuses on MVDC in distribution systems, i.e., emphasis on applications shown in Figure 5.1. MVDC voltages are considered to be in the range of 5–80 kV (pole voltages), which covers the majority of pertinent distribution level applications in the above listed examples while not encroaching on LVDC and HVDC systems.

5.3   Overview of DC–DC conversion for MVDC systems

Power electronic converters will be the workhorse of future MVDC systems. They enable high-efficiency DC–DC power conversion and regulation of power flows as needed for network level applications. Of course, additional features beyond simple DC voltage level conversion can be realized courtesy of the investment in semiconductor technology, such as fast reversal of DC line currents (and hence tight power regulation) and fault blocking or fault current limiting capability. This is in stark contrast to classical AC systems where iron-cored static transformers are the dominant device typically providing only basic voltage matching.[‡]

This section gives an overview of traditional two-port DC–DC converters suitable for interconnecting MVDC systems and regulating their power exchanges. The multiport DC converter concept is introduced as an extension of two-port DC–DC converters. MVDC converters must be capable of handling voltage and power levels on the order of tens of kilovolts and several tens of megawatts. MMC technology is well suited to accommodate these requirements while maintaining high efficiency and quality of power. Therefore, the focus is on MMC-based DC–DC conversion.

5.3.1   *Modular multilevel converter technology*

The MMC is a dc–ac topology first proposed in the early 2000s [13] and has since gained widespread acceptance for high voltage and power applications, primarily HVDC and FACTS [14,15]. More recently, the MMC has been applied to a variety of MVDC applications such as electric drives [16], shipboard power systems [17], and bipolar DC distribution grids [18].

Figure 5.2 shows the conventional three-phase MMC. Each phase leg comprises upper and lower phase arms, which in turn are made up of N submodules (SMs) stacked in series. Each SM is a half-bridge or full-bridge switching cell that includes a capacitor and switches. The basic half-bridge SM type is adequate for dc–ac conversion and is the most common (and most cost-effective) type used, but full-bridge type can be used to gain extra functionality such as increased AC voltage modulation and DC fault blocking [19]. The nominal capacitor voltage in each SM is relatively small (e.g., 2 kV) so that standard off-the-shelf IGBTs can be used. In practice, tens to hundreds of SMs are installed per phase arm depending on whether the MMC is being used for MVDC or HVDC applications. The arm chokes

---

[‡]Some devices like phase-shifting transformers can provide additional functionality.

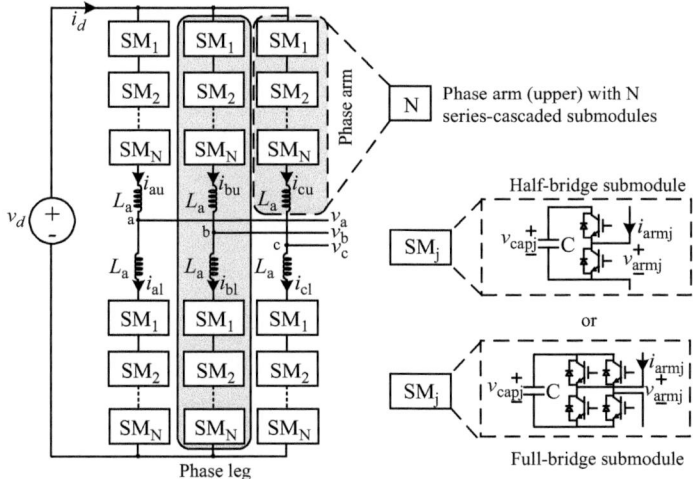

Figure 5.2   General structure for three-phase MMC

denoted by L_a are required to take up any voltage mismatch between the sum of inserted capacitors in each phase leg and the DC bus voltage v_d.

In comparison to the conventional two-level voltage-sourced converter (VSC), the MMC offers several advantages:

- Highly modular and scalable structure, well suited for MVDC and HVDC.
- $N + 1$ level AC phase voltage is obtained; large N circumvents need for conventional AC filtering requirements (low dv/dt).
- Achieve very low semiconductor switching losses and therefore high overall conversion efficiencies.
- Avoid challenges associated with direct series connection of many IGBTs; also, arm currents are continuous and therefore low di/dt.
- Inherent redundancy (and hence increased reliability) due to modular nature of switching cells.

These benefits do not come for free. Relative to the two-level VSC, the MMC employs more total number of switches and has a much higher installed total capacitive energy storage requirement (as capacitors must support ripple power). Also, there is increased control complexity due to large number of capacitors that must all be actively regulated, and also active control of internal circulating currents is typically required. However, these drawbacks are outweighed by the noted benefits for HVDC, MVDC and FACTS applications. Similar advantages are also obtained in comparison to traditional multilevel converters such as the three-level neutral-point-clamped (NPC) converter.

In MVDC systems, the MMC uses a lower number of SMs to accommodate lower voltages. For example, the dc–ac MMC in [18] uses $N = 20$ SMs per phase arm to support ± 10 kV DC link (capacitors rated at 1 kV). This is in contrast to

HVDC level converters that use hundreds of SMs per phase arm, e.g., $N = 400$ is used in [20] to support ±320 kV DC link (capacitors rated at 1.6 kV). A similar motivation is behind Siemens MVDC Plus DC–AC converters which are based on their established MMC HVDC technology but with a reduced SM count. However, MMC technology is not the only solution for MVDC systems. For example, the Angle DC project uses six series-connected 4.5 kVdc three-level NPCs per pole (thus 12 three-level NPCs per converter station) for ±27 kV bipolar MVDC link. But increasing to larger voltage and power levels would mean even a higher number of converters connected in series, leading to increased design and control complexity.

5.3.2   Conventional two-port DC–DC converters

Figure 5.3 depicts a traditional two-port DC–DC converter where subscripts 0 and 1 denote the DC terminals, i.e., the DC ports. All the power entering one port must exit the second port, minus any losses (assuming the converter does not have any internal long-term energy storage such as batteries). Generally, two-port converters can be classified into isolated and non-isolated topologies. The former is most commonly some variant of the classical dual-active bridge (DAB) structure [24] that uses two dc–ac converters with their respective AC links interconnected together through an intermediate AC transformer. Therefore, the two DC ports are galvanically separated. The latter is any topology that does not provide galvanic separation between the two DC ports, which typically means avoiding use of an intermediate AC transformer. A myriad of two-port DC–DC converter topologies suitable for MVDC and HVDC have been proposed, e.g., [25–28], many of which utilize the MMC concept.

Figure 5.4 shows representative examples of MMC-based isolated and non-isolated two-port DC–DC converters suitable for MVDC systems. Figure 5.4(a) is the isolated DAB based on MMC, hereinafter denoted as DAB-MMC, which is the focus of much ongoing research, e.g., [21,22]. The MMC arms comprise N_1 and N_2 series-cascaded SMs, see Figure 5.2. The two MMCs are separated via an AC transformer that must be rated to handle 100% of the power being transferred between ports. The AC frequency is a free design parameter (given the lack of AC grid connection). The utilization of two separate MMCs brings benefits such as DC fault blocking using only half-bridge SMs, and the ability to achieve large DC step ratios by suitable selection of transformer turns ratio. This comes at the expense of a two-stage conversion process, i.e., DC/AC/AC–DC. As an alternative to the DAB-MMC, the MMCs can be series-stacked which leads to the non-isolated

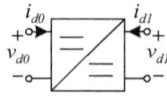

Figure 5.3   Generalized two port DC-DC converter

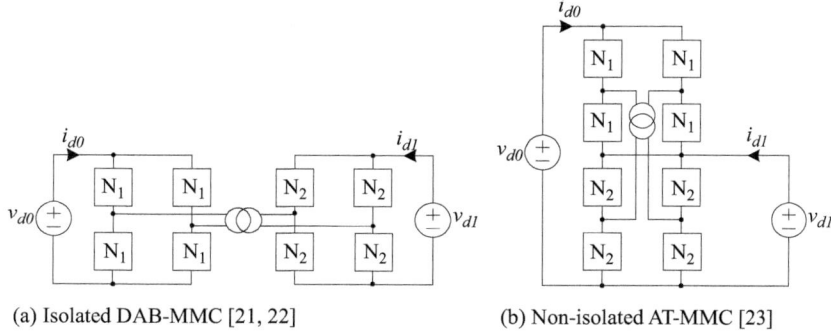

(a) Isolated DAB-MMC [21, 22]   (b) Non-isolated AT-MMC [23]

Figure 5.4   Exemplar isolated and non-isolated two-port DC–DC converters

autotransformer (AT) [23], shown in Figure 5.4(b), hereinafter denoted as AT-MMC. Galvanic separation between DC ports is lost, however, the transformer now processes only $(1 - v_{d1}/v_{d0})\%$ of the DC power transferred between ports. This partial power processing DC–DC topology offers significant potential savings in semiconductors and capacitive energy storage relative to the DAB-MMC [29]. Dc fault blocking for the AT-MMC requires a sufficient number of full-bridges SMs installed in the upper MMC in Figure 5.4(b).

The basic building block in Figure 5.4(a) and (b) is the single-phase dc–ac MMC. Three-phase (or a higher number of phase legs) MMCs can be used for cases where increased power rating is required. Other modular multilevel two-port DC–DC topologies can be created by re-arranging the MMC phase arms in different ways, e.g., [30–33].

5.3.3   Concept of multiport DC converters

Future MVDC systems comprising multiple different (or similar) voltage levels are anticipated to emerge, as motivated in Sections 5.1 and 5.2. Figure 5.5 illustrates how multiple MVDC buses could potentially be interconnected using conventional two-port DC–DC converters. Each two-port converter is designed for a certain DC step ratio, and therefore can only interconnect two buses. It is not practical to install a separate DC–DC converter to accommodate every possible DC power flow between buses, due to the high capital costs required, and therefore certain power transfers will encounter multiple conversion stages, i.e., between buses 1–3, 1–4, and 3–4 in Figure 5.5.

An intriguing line of research is the development of multiport DC converters that can interconnect more than two DC buses of similar and/or different voltage levels, using one component/structure [34]. These multiport converters are also referred to as *dc hubs* or *dc substations*. Figure 5.6 illustrates the multiport DC converter concept, generalized to $J + 1$ DC ports. Power flows can be routed

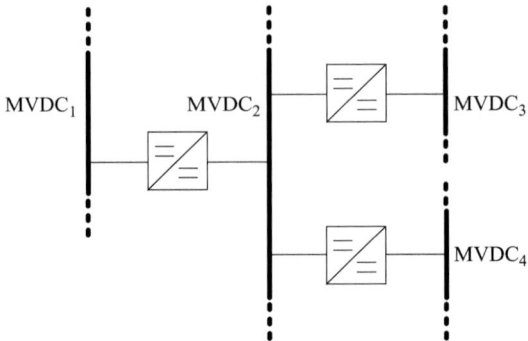

Figure 5.5   Motivation for multiport DC converters

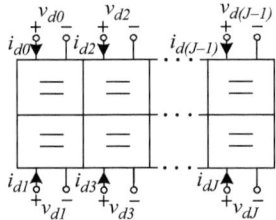

Figure 5.6   Generalized multiport DC converter

between some or all of the ports, depending on the design. Potential benefits of a multiport DC converter include reduced energy conversion stages and hence improved efficiency, lower total semiconductor requirement, smaller converter station foot-print (relative to a solution deploying separate DC–DC converters), and reduced cost. Multiport DC converters also offer operational simplicity with the ability to route power emanating from a common DC bus between multiple different DC ports. The caveat is that all DC lines must converge at a single physical location, similar to classical AC substations.

The majority of early literature on multiport DC converters focuses on low-voltage applications, e.g., [35,36], where the main target is renewable energy integration. It is typically not cost effective to directly scale up these architectures to handle the voltage and power levels encountered in MVDC grid applications. However, recent research has been directed toward higher voltage and power applications, e.g., [37]. Figure 5.7 gives example applications of multiport DC converters in the context of MVDC systems. These examples show a three-port structure for interconnecting MVDC lines, and a four-port structure for grid integration of renewable energy including provisions for battery energy storage. Figure 5.8 shows using the prospect of multiport DC converter to facilitate future MVDC expansion.

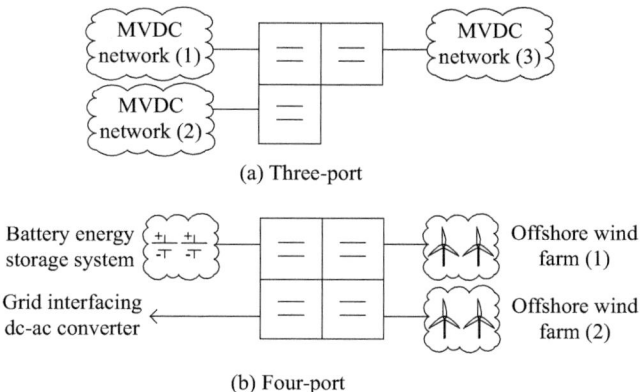

(a) Three-port

(b) Four-port

Figure 5.7   Multiport DC converter example applications

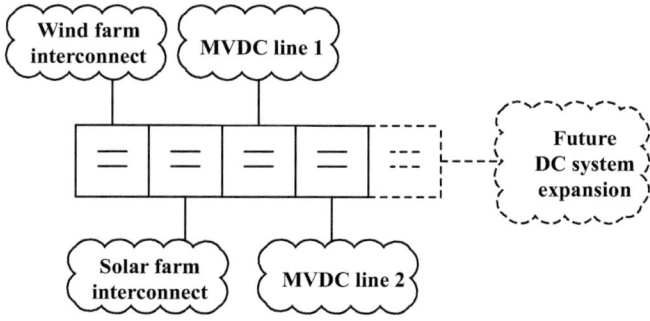

Figure 5.8   Potential expansion of MVDC systems using multiport DC converter

5.4   Multiport converter technology for MVDC interconnects

This section discusses multiport DC converter technology suitable for MVDC voltage and power levels on the order of tens of kilovolts and tens to hundreds of megawatts, respectively. Therefore, as elaborated in Section 5.3, the topologies under discussion focus on MMC technology.

5.4.1   Desired features and functionality

Obvious desirable attributes for multiport DC converters include high efficiency, compact footprint, reasonable cost, modularity and ease of scalability, and high reliability. Looking beyond these typical attributes, this section gives an overview of key features and functionality of multiport DC converters for use as DC hubs in MVDC systems. Some topologies may provide only a few, or potentially all of these, depending on considerations such as the capabilities of the underlying converter technology, the intended application, and capital investment constraints.

5.4.1.1   Multiple DC voltage levels

Ability to interconnect multiple different or similar DC voltage levels. For example, interconnecting 25, 30, 50, and 50 kV MVDC lines requires four DC ports characterized by three different DC voltage levels (two ports have the same nominal DC voltage). This is a distinguishing feature from two-port DC–DC converters that by design provide a (typically fixed) DC voltage ratio between input and output.

5.4.1.2   Flexible port power flows

Allow power flow between any combination of DC ports. For highest flexibility, each port would have bidirectional power flow capability and can arbitrarily split the power in-feed (or out-feed) to (or from) any other combination of DC ports. Of course, such high degree of flexibility may not be required for every application. For example, a central DC hub to regulate power transfers between multiple MVDC grids may need such flexibility, while a multiport DC converter for coalescing offshore wind MVDC collector buses for connection to an onshore MVDC link requires only uni-directional power flow capability.

5.4.1.3   High degree of modularity

Exploit topology modularity to ensure a highly reliable operation and to accommodate easy reconfiguration or expansion of the converter power circuit. The ideal multiport DC converter structure would have a high degree of modularity that readily permits (i) avoiding loss of one (or more) DC ports when experiencing failure of single components; (ii) adding new DC ports and/or increasing the power rating of existing DC ports. The latter makes the multiport DC converter nimble to evolving MVDC infrastructure, such as the addition of new MVDC lines or the upgrading of power transfer capability of existing MVDC lines.

5.4.1.4   Accommodate different DC system architectures

Ability to interconnect DC systems with different configurations, e.g., bipolar, symmetrical monopole, asymmetrical monopole. For example, a need may exist to connect two bipolar MVDC grids with a symmetrical monopole MVDC line. This places special demands on the underlying multiport converter structure. Accommodating different DC line configurations is typically easier for isolated topologies to handle due to the intermediate AC transformer that essentially decouples the DC ports from each other, while it may be more challenging to achieve for non-isolated topologies.

5.4.1.5   DC fault blocking capability

Ability to block or suppress the flow of fault currents through the converter, in response to a fault on the DC ports. This may be viewed as analogous functionality to a DC circuit breaker. Converter integrated DC fault blocking is an alternative to the use of external DC circuit breakers. Fault blocking capability could be provided on some or all of the DC ports, depending on the application, to prevent faults from propagating between MVDC systems. MMC topologies can exploit the internal SM capacitors to rapidly absorb fault energy by inserting them in the current conduction path. Some multiport DC converters have inherent DC fault blocking, while others require extra SMs of the appropriate type to block DC faults. It is also possible for some multiport DC converters to have fault-ride-through capability, whereby the converter can ride-through a DC fault (typically for short-term disturbances) while limiting fault currents below rated values to avoid semiconductor damage.

5.4.1.6   Continued power transfer with faulted DC port(s)

DC fault blocking as described in Section 5.4.1.5 incapacitates the converter normal operation, i.e., it functions strictly like a DC breaker or fault current limiter. However, it may be possible for a multiport DC converter to cease the flow of fault currents on the faulted port(s) while maintaining control of the healthy port(s). In this case, operation could continue albeit with reduced power flow capability. This advanced functionality comes at a cost, typically higher semiconductor investment to maintain control of the port power flows under low voltage conditions.

5.4.2   *Multiport dual active bridge*

The simplest multiport DC converter to conceptualize is the DAB-MMC topology in Figure 5.4(a) with more than two MMCs. This section introduces both isolated and non-isolated multiport variants of the DAB-MMC.

5.4.2.1   Isolated

The multiport extension of the conventional two-port DAB-MMC is shown in Figure 5.9, hereinafter referred to as MP DAB-MMC. $J + 1$ DC ports requires $J + 1$ MMCs interfaced with a central AC link. The power transfer capability of each port depends on the corresponding MMC design, i.e., not all MMCs need have the same power rating. The MP DAB-MMC builds on the well understood two-port DAB-MMC, retaining features such as inherent DC fault blocking, galvanic

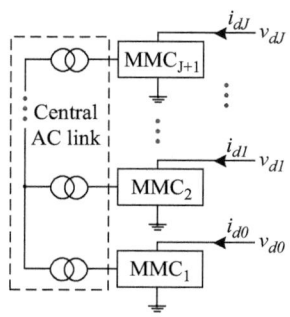

Figure 5.9   MP DAB-MMC

separation between DC ports, and good switching device utilization of the MMCs (courtesy of appropriate transformer turns ratios selection) [38]. The transformer isolation property also permits relative ease in accommodating different external DC line configurations, e.g., bipolar and monopolar. Of course, design of the inner AC circuit becomes increasingly more complicated as the number of MMCs increases. Also, relying on a central transformer creates a single potential point of failure, as its loss results in the loss of all $J + 1$ DC ports. Using multiple transformers to form the inner AC circuit may alleviate reliability concerns.

5.4.2.2 Non-isolated

The central AC link in Figure 5.9 processes the entire DC power throughput between ports. Ac transformer(s) rated for tens of kilovolts and several tens of megawatts can be large and bulky, and introduce extra losses. It is not mandatory that the AC link be realized with magnetics. In [39], an LCL circuit is proposed as a substitute for the AC transformer in a multiport DAB.[§] Galvanic separation between DC ports is consequently lost. An advantage of the LCL solution is that additional DC ports can be added without requiring AC link redesign. Extra care must be taken though when selecting the AC operating frequency, to ensure it is not close to global and local resonance frequencies introduced by the inductors and capacitors in the LCL circuit [38]. Also, the LCL circuit cannot block DC voltage offsets and therefore the external DC systems must be symmetrical monopole. The LCL-based MP DAB-MMC is being studied by researchers, for example [40].

5.4.3 *MP autotransformer*

The two-port AT-MMC in Figure 5.4(b) can also be expanded to a multiport realization. The multiport autotransformer (MP-AT) is shown in Figure 5.10. This topology was proposed as a non-isolated alternative to the MP-DAB that exploits a partial power processing architecture to reduce semiconductor investment, losses and converter station footprint [37]. A 50%–80% reduction in capital cost relative to a comparable DAB-MMC solution is claimed for HVDC applications [37] (a similar

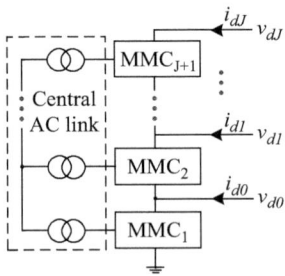

Figure 5.10   Multiport AT-MMC [37]

[§]Two-level VSCs are employed in this paper, but conceptually the LCL circuit can be used with MMCs.

savings would be anticipated for MVDC applications). However these savings do not take DC fault blocking capability into consideration. Other works have also studied multiport DC converters that use in part the AT-MMC concept [41].**

The MP-AT still uses transformers in the central AC link, however, they process less than 100% of the DC power throughput between DC ports. This implies smaller and less costly magnetics are required. However, the non-isolated MP-AT is not as flexible as the MP DAB-MMC with respect to accommodating different DC circuit configurations for the port connections. This is because of the lack of galvanic separation between DC ports. Also, unlike the MP DAB-MMC, the MP-AT does not have full DC fault blocking capability when using only half-bridge SMs in the MMCs. Self-blocking SMs such as full-bridge type would be required in some of the arms, depending on the application scenario.

5.4.4 MP DC-MMC

The MP DAB-MMC and MP-AT in Figures 5.9 and 5.10 are the two prominent MMC-based DC hub solutions that were firstly proposed by researchers. Both consist of MMCs interconnected by a centralized ac-link, and both require AC power transfer between the MMCs to enable their respective DC–DC conversion processes. Consequently, the reliance on a centralized AC link bears some consequences:

- Limited modularity and scalability, as each MMC is designed differently (to accommodate different DC port ratings) and requires re-design if the connected DC port rating changes;
- Failure of a single MMC would result in loss of an entire dc port;
- Failure of the central ac-link would render the entire multiport DC converter off-line, i.e., loss of all dc ports;
- Transformers in MP-AT must be designed to tolerate relatively large DC voltage stresses between windings, which complicates core design and manufacture.

The multiport DC–DC MMC (MP DC-MMC) proposed in [42,43] side steps these design and reliability issues by eliminating the centralized AC link. The MP DC-MMC, shown in Figure 5.11, is a completely modular solution for multiport DC–DC conversion. The topology displays some resemblance to the switched-mode architectures in [44,45], which are not suited for MVDC applications as defined in Section 5.2.1.

The general structure of the MP DC-MMC in Figure 5.11 uses J rows of subconverters (SCs) to establish $J + 1$ DC ports. The number of SCs in row j is denoted by K_j, and SC_{jk} denotes the SC located in row j, column k. Each SC_{jk} supports a DC voltage of V_{rj} and a DC current i_{jk} between its positive (+) terminals. The DC port voltages and currents are denoted by v_{dj} and i_{dj}, respectively. Power transfer between ports only involves the rows of SCs in-between those ports.

**The three-port dc converter proposed in [41] requires connection to an external ac grid.

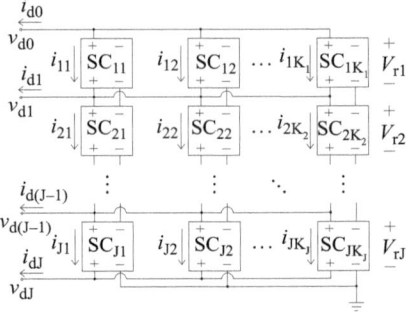

Figure 5.11   General structure for MP DC-MMC [42]

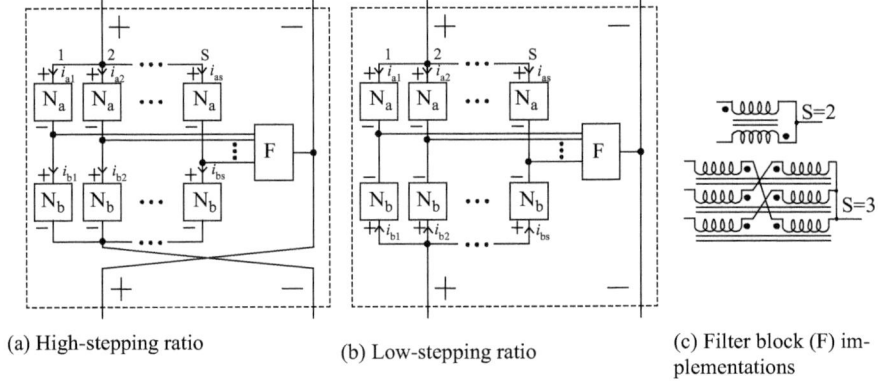

(a) High-stepping ratio   (b) Low-stepping ratio   (c) Filter block (F) implementations

Figure 5.12   Example subconverters (SCs) for the MP DC-MMC [42,43]

Each SC in Figure 5.11 is a DC–DC converter. Only DC connections exist between neighbouring SCs, and therefore the MP DC-MMC does not require a central AC link. Two different SC topologies are shown in Figure 5.12, which utilize two-port non-isolated DC–DC MMCs [42,43].[††] The SCs are shown generalized to s interleaved strings of SMs, with phase arms comprising N_a and N_b series-cascaded SMs. The high-stepping ratio (and low-stepping ratio) SC is best suited for *large* (and *small*) DC voltage steps between ports. Choosing the optimal SC for each row is application dependent. The filter (F) block in Figure 5.12(a) and (b) ideally passes only common-mode DC currents. The coupled inductor and zig-zag transformer implementations in Figure 5.12(c) leverage magnetics to suppress the flow of differential mode AC currents.

[††]These dc–dc MMCs belong to a class of converters that use internal ac currents for charge balancing of the SM capacitors [30–32].

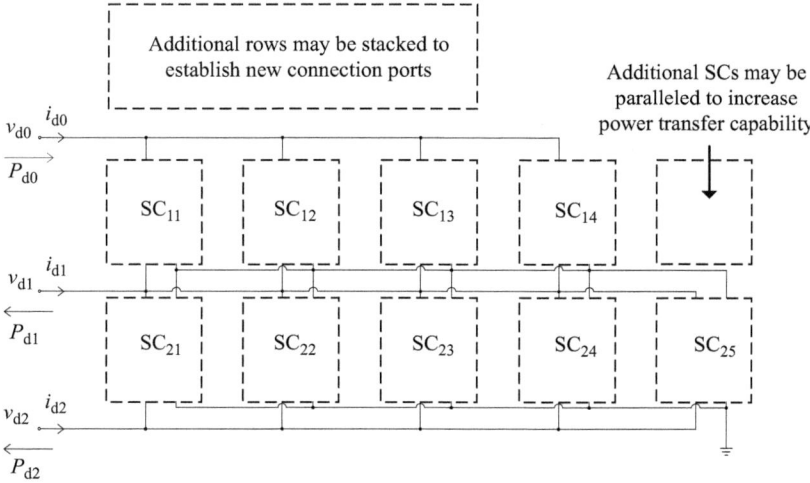

Figure 5.13   Expansion capabilities of MP DC-MMC by altering series/parallel connection of SCs

The operation of the MP DC-MMC in Figure 5.11 relies on two mechanisms: (i) voltage stacking and (ii) current splitting. The lowest port voltage v_{dJ} is established by the SCs in row J and subsequent higher DC port voltages are formed by adding new rows of SCs, wherein each series-stacked row of SCs introduces the requisite V_{rj}. This voltage stacking mechanism allows easy expansion of new ports by adding new rows of SCs. The DC port current i_{dj} can be split arbitrarily between SCs within a row by adding the requisite number of paralleled SCs, K_j. Assuming the SCs are identical then the current splits evenly between these paralleled SCs. This current splitting mechanism allows easy accommodation of future ports power flow requirements by adding new SCs in parallel. Figure 5.13 shows the high level implications of these two mechanisms for MP DC-MMC design and expansion.[‡‡]

5.4.5   Comparison of prominent multiport DC topologies for MVDC

The MP DAB-MMC, MP-AT, and MP DC-MMC have been introduced and briefly reviewed as prominent solutions for multiport DC converters in MVDC systems. However, these three topologies have different operating features and characteristics. The MP DAB-MMC is the only isolated topology while the other two relinquish galvanic separation between DC ports to realize savings in converter components (hence reduced cost) and size. Also, the MP DAB-MMC is the only topology with full DC fault blocking capability using half-bridge SMs, while the MP-AT and MP DC-MMC can obtain the same DC circuit breaker functionality as the MP DAB-MMC but with extra investment in number (and type) of SMs. These

---

[‡‡]Here, i_{d1} and i_{d2} would be negative valued to satisfy $P_{d0} + P_{d1} + P_{d2} = 0$.

Table 5.1   Comparison of modularity of MP DAB-MMC, MP-AT, MP DC-MMC

	MP DAB-MMC	MP-AT	MP DC-MMC
Add new dc port	Add new MMC, Redesign transformer*	Add new MMC, Redesign transformer	Add new row of SCs, i.e., increase J
Increase dc port power rating	Redesign MMC, and/or parallel arms	Redesign MMC, and/or parallel arms	Increase SCs in a row, i.e., increase K_j
Failure of one SC	Loss of a dc port	Loss of a dc port, reduced power flow between ports	Reduced power flow between ports
De-centralized control of SCs	No	No	Yes

* The non-isolated LCL-based MP DAB-MMC [39] does not require redesign of ac link.

differences stem from the fact the MP DAB-MMC uses two-stage dc/ac–ac/dc conversion between ports, while the MP-AT and MP DC-MMC are both partial power processing topologies. The choice of which topology to use for a certain MVDC application depends on a detailed cost-benefit analysis.

Table 5.1 [42] contrasts modularity features of the MP DAB-MMC, MP-AT and MP DC-MMC. The centralized AC link inherent to the MP DAB-MMC and MP-AT has repercussions, as adding new DC ports or changing the power transfer capability of existing DC ports requires redesign of MMCs and AC transformer (s).[§§] In comparison, owing to its high degree of modularity, the MP DC-MMC can accommodate changes to the DC ports by adding more SCs in series and/or parallel as required; the existing SCs are not impacted. Often the SCs can be of identical design. Also, the MP DC-MMC does not lose a DC port when a SC fails,[***] unlike the MP DAB-MMC and MP-AT when they experience failure of a single SC (i.e., MMC). The MP DC-MMC can also realize de-centralized controls of SCs, with the addition of new (or removal of existing) SCs not impacting the control/operation of the rest of the SCs.

Table 5.2 [42] provides a high level relative comparison of other key multiport features for the MP DAB-MMC, MP-AT and MP DC-MMC. The MP-AT has the lowest overall capital cost investment and highest efficiency, owing to its combination of partial power processing architecture and usage of a centralized AC link. The MP DC-MMC falls between the MP-AT and MP DAB-MMC. The MP DC-MMC has the highest degree of converter modularity and reliability as it avoids reliance on a centralized AC link. Lastly, the two-stage isolated architecture of the MP DAB-MMC gives it the best overall inherent DC fault blocking capability and flexibility to interconnect different DC line configurations.

[§§]Assuming multi-winding transformers are used.
[***]Assuming a sufficient number of SCs are installed in each row.

Table 5.2 *Relative comparison of key features of MP DAB-MMC, MP-AT,*
MP DC-MMC

	MP DAB-MMC	MP-AT	MP DC-MMC
Capital cost	$+++$	$+$ (✓)	$++$
Efficiency	$+$	$+++$ (✓)	$++$
Modularity	$++$	$++$	$+++$ (✓)
Converter reliability	$++$	$++$	$+++$ (✓)
DC fault blocking	$+++$ (✓)	$+$	$+$
Inherent flexibility for interfacing different DC line configurations	$+++$ (✓)	$+$	$+$

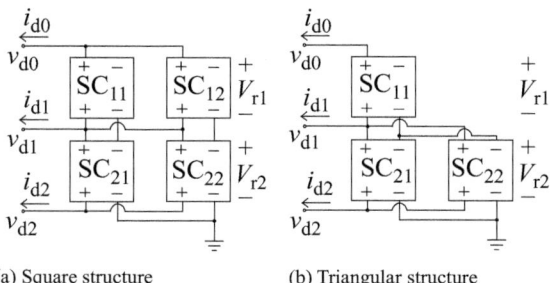

(a) Square structure   (b) Triangular structure

Figure 5.14 *Three-port MP DC-MMC designs*

5.5 Multiport DC converter case studies

This section examines three example applications of a multiport DC converter system for interconnecting MVDC lines of varying voltages. Two different three-port MP DC-MMC designs are used.[†††] The MP DC-MMC case studies under consideration are as follows:

1. Three-port 20/40/60 kV design, with square structure (ref. Figure 5.14(a)).
2. Three-port 25/30/50 kV design, with square structure (ref. Figure 5.14(a)).
3. Three-port 20/40/60 kV design, with triangular structure (ref. Figure 5.14(b)).

Three asymmetrical monopole MVDC systems are assumed as shown in Figure 5.14. Time-domain simulations are carried out in PLECS. The system parameters for the three different MP DC-MMC case study designs are given in Table 5.3. Case studies 1 and 3 use the same high stepping ratio SC in both rows of the MP DC-MMC, while case study 2 uses a combination of high stepping ratio (in row 1) and low stepping ratio (in row 2) SCs. This is to accommodate the different MVDC system voltage levels, i.e., 20/40/60 kV versus 25/30/50 kV. In all simulations, P_{d0}, P_{d1}, P_{d2}

[†††]The MP DC-MMC is chosen here, however, the MP-AT and MP DAB-MMC may also be used.

Table 5.3   Main MP DC-MMC parameters for case study simulations

	Row 1 (top row) SCs	Row 2 (bottom row) SCs
Case Study 1: 20/40/60 kV Square design (Figure 5.14(a)) ($J = 2$, $K_1 = K_2 = 2$)	High stepping (Figure 5.12(a)) $N_a = N_b = 20$, $s = 2$ $V_{cap} = 2$ kV, $f = 150$ Hz	High stepping (Figure 5.12(a)) $N_a = N_b = 20$, $s = 2$ $V_{cap} = 2$ kV, $f = 150$ Hz
	Row 1 (top row) SCs	Row 2 (bottom row) SCs
Case Study 2: 25/30/50 kV Square design (Figure 5.14(a)) ($J = 2$, $K_1 = K_2 = 2$)	High stepping (Figure 5.12(a)) $N_a = 20$, $N_b = 30$, $s = 2$ $V_{cap} = 2$ kV, $f = 150$ Hz	Low stepping (Figure 5.12(b)) $N_a = 30$, $N_b = 25$, $s = 2$ $V_{cap} = 2$ kV, $f = 150$ Hz
	Row 1 (top row) SCs	Row 2 (bottom row) SCs
Case Study 3: 20/40/60 kV Triangular design (Figure 5.14(b)) ($J = 2$, $K_1 = 1$, $K_2 = 2$)	High stepping (Figure 5.12(a)) $N_a = N_b = 20$, $s = 2$ $V_{cap} = 2$ kV, $f = 150$ Hz	High stepping (Figure 5.12(a)) $N_a = N_b = 20$, $s = 2$ $V_{cap} = 2$ kV, $f = 150$ Hz

denote the DC powers absorbed by ports $d0$, $d1$, $d2$ of the converter, respectively, and port $d0$ is assigned as the power flow slack bus (and hence $P_{d0} = -(P_{d1} + P_{d2})$). The lowest (and highest) MVDC line voltage corresponds to port $d2$ (port $d0$).

5.5.1   Case study 1: 20/40/60 kV, square design

This case study demonstrates a multiport DC converter interconnecting MVDC systems with different voltage levels. Figure 5.15 shows simulation results for withdrawing 60 MW and 30 MW from ports $d1$ (40 kV line) and $d2$ (20 kV line) of the MP DC-MMC, respectively, starting at $t = 1.2$ sec. The total 90 MW comes from the slack bus at port $d0$ which is connected to the 60 kV MVDC line. Thus, 90 MW is being transferred from the 60 kV line to the 40 kV line (67% split) and to the 20 kV line (33% split). An outer layer power controller is used to regulate the i_{d1} and i_{d2} needed to satisfy the commanded power flows. There are a total of four SCs in the MP DC-MMC; Figure 5.15 plots the arm currents and (total sum) capacitor voltages for SC_{11} and SC_{21}.

The simulations in Figure 5.16 demonstrate a polarity reversal in the power exchange between MVDC lines. Prior to $t = 1.2$ sec, 90 MW is being transferred from the 60 kV line to the 40 kV line (67% split) and to the 20 kV line (33% split). This power flow schedule is reversed at $t = 1.2$ sec.

5.5.2   Case study 2: 25/30/50 kV, square design

This case study demonstrates a multiport DC converter interconnecting MVDC systems with both different and similar voltage levels, i.e., 25 kV & 30 kV (similar nominal values) versus 50 kV. Figure 5.17 shows simulation results for withdrawing 45 MW and 37.5 MW from ports $d1$ (30 kV line) and $d2$ (25 kV line) of the MP DC-MMC, respectively, starting at $t = 1.2$ sec. The total 82.5 MW comes from the 50 kV MVDC line. Thus, 82.5 MW is sent from the 50 kV line to the 30 kV line (55% split) and to the 25 kV line (45% split). The arm currents and capacitor voltages for SC_{11} and SC_{21} are regulated with decentralized inner layer current

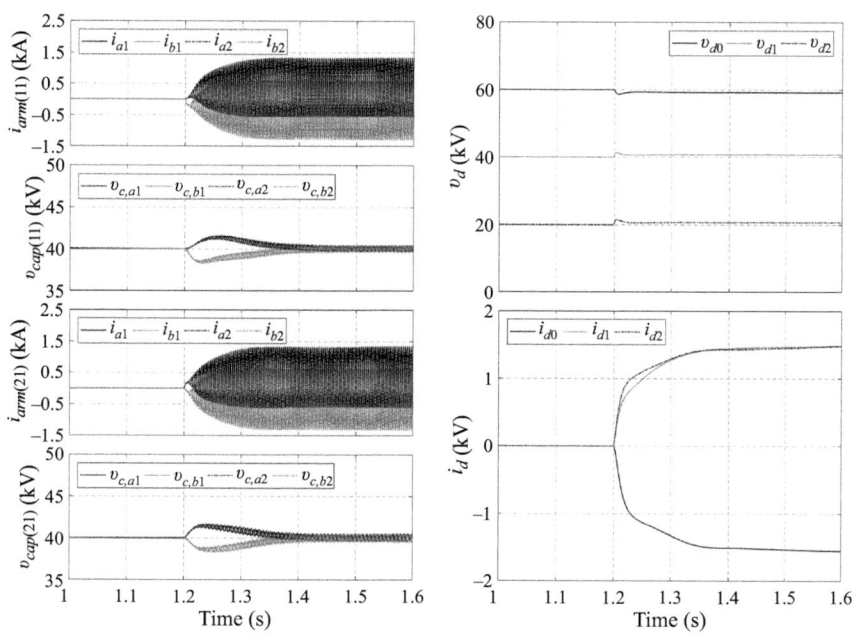

Figure 5.15   *Case study 1 (20/40/60 kV with square MP DC-MMC): $P_{d1} = 0$ MW, $P_{d2} = 0$ MW to $P_{d1} = -60$ MW, $P_{d2} = -30$ MW at $t = 1.2$ sec*

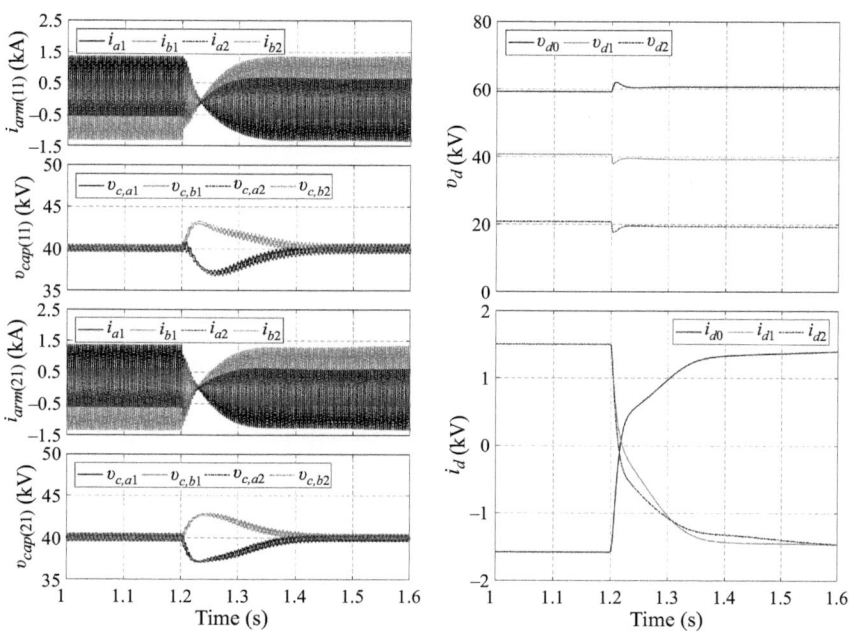

Figure 5.16   *Case study 1 (20/40/60 kV with square MP DC-MMC): $P_{d1} = -60$ MW, $P_{d2} = -30$ MW to $P_{d1} = 60$ MW, $P_{d2} = 30$ MW at $t = 1.2$ sec*

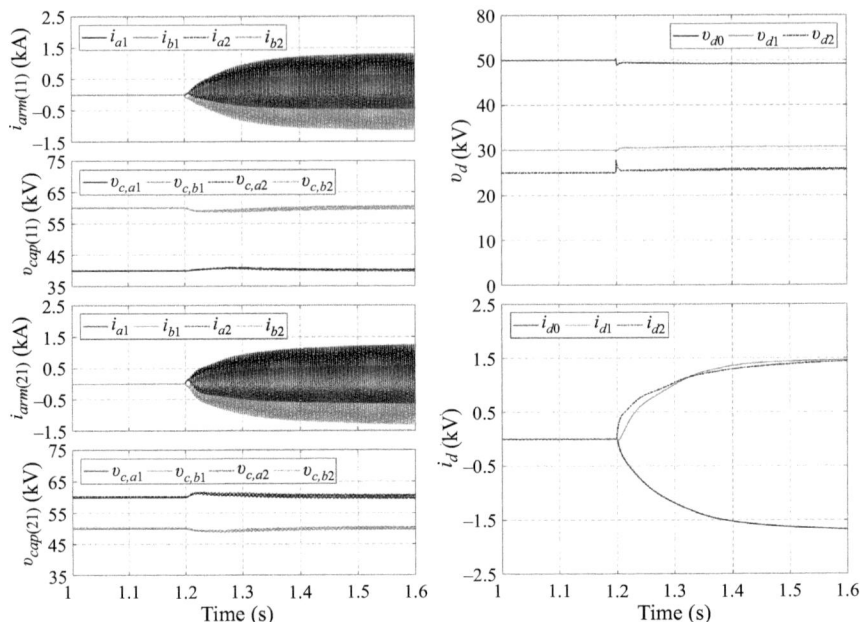

Figure 5.17   Case study 2 (25/30/50 kV with square MP DC-MMC): $P_{d1} = 0$ MW, $P_{d2} = 0$ MW to $P_{d1} = -45$ MW, $P_{d2} = -38$ MW at $t = 1.2$ sec

and capacitor voltage balancing controls. The total sum of capacitor voltages for the SCs correspond to the data in Table 5.3. That is, arms a and b in SC_{11} have nominal (total sum) DC capacitor voltages 20×2 kV $= 40$ kV and 30×2 kV $= 60$ kV, respectively, and arms a and b in SC_{21} have nominal (total sum) DC capacitor voltages 30×2 kV $= 60$ kV and 25×2 kV $= 50$ kV, respectively.

Figure 5.18 imposes a change in the direction of power flows between the MVDC lines. Prior to $t = 1.2$ sec, 83 MW is being transferred from the 50 kV line to the 30 kV line (55% split) and to the 25 kV line (45% split). This power flow schedule is reversed at $t = 1.2$ sec.

5.5.3   Case study 3: 20/40/60 kV, triangular design

The previous two case studies used the same square MP DC-MMC design in Figure 5.14(a). In this case study, a triangular MP DC-MMC structure is used as shown in Figure 5.14(b). A triangular design is best suited for the scenario where power is exchanged between only the highest and lowest voltage DC lines, i.e., between the top and bottom rows. Here this implies routing power between the 60 kV and 20 kV networks. In such a scenario, the lowest current is seen at the highest voltage level while the highest current is seen at the lowest voltage level. The bottom row should therefore have additional SCs in parallel to handle the higher current stress. This pattern holds as the number of rows increases (more SCs needed on the lower rows), hence the resulting triangular structure.

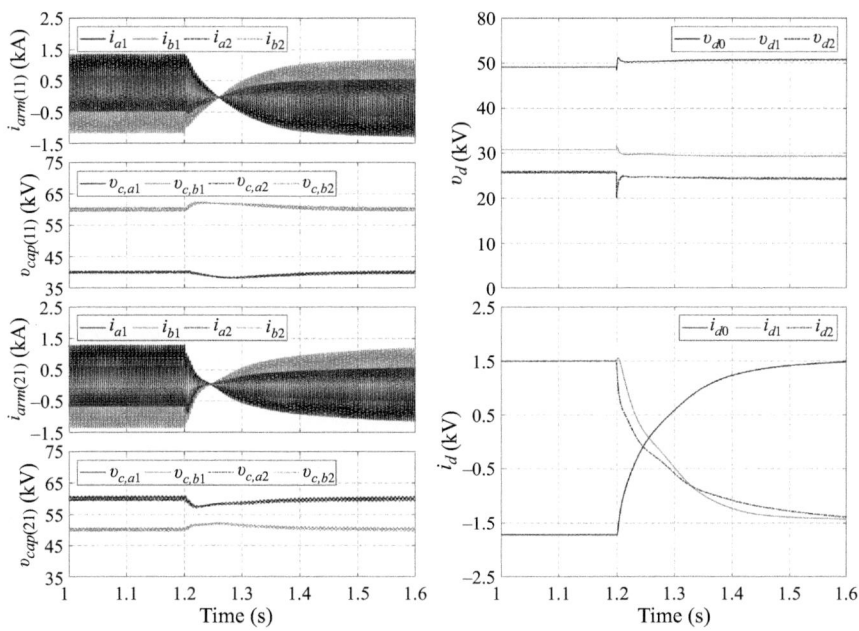

*Figure 5.18   Case study 2 (25/30/50 kV with square MP DC-MMC): $P_{d1} = -45$
MW, $P_{d2} = -38$ MW to $P_{d1} = 45$ MW, $P_{d2} = 38$ MW at $t = 1.2$ sec*

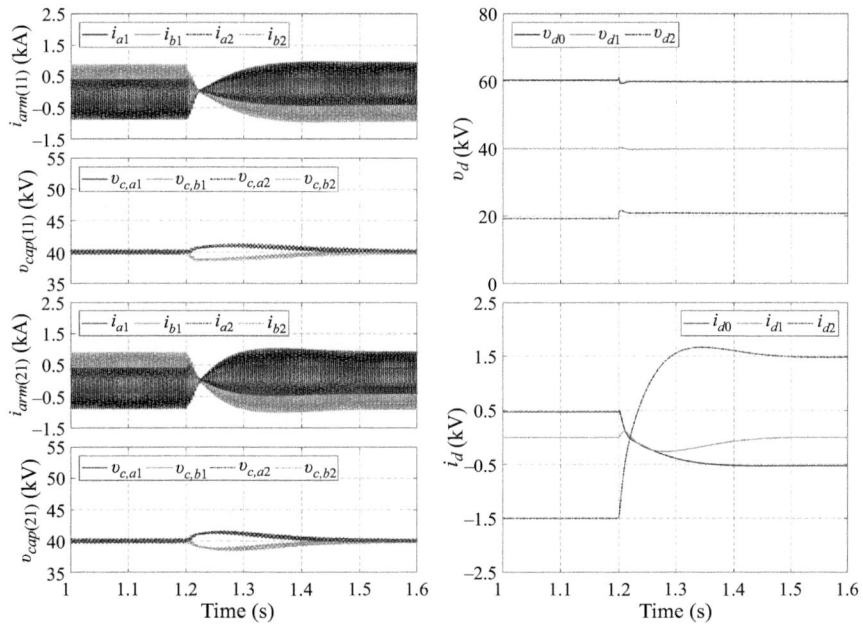

*Figure 5.19   Case study 3 (20/40/60 kV with triangular MP DC-MMC): $P_{d1} = 0$
MW, $P_{d2} = 30$ MW to $P_{d1} = 0$ MW, $P_{d2} = -30$ MW at $t = 1.2$ sec*

Figure 5.19 shows simulation results for routing 30 MW between ports $d0$ (60 kV line) and $d2$ (20 kV line) of the MP DC-MMC. Prior to $t = 1.2$ sec, 30 MW is being sent from the 20 kV line to the 60 kV line. This power flow schedule is reversed at $t = 1.2$ sec.

5.6 Summary

Multiport DC converters are an emerging option for interconnecting multiple MVDC systems of similar or different voltage levels. They offer potential benefits such as eliminating redundant energy conversion stages, improved efficiencies, and reduced converter station footprint when compared to conventional two port DC–DC converters. This chapter introduces the multiport DC converter concept and outlines possible applications in MVDC systems, such as power flow routers for meshed DC grids and as DC hubs for collection and distribution of power from renewable resources. An emphasis is placed on MMC type topologies that are suited to accommodate the voltage levels encountered in MVDC systems, i.e., tens of kilovolts. An overview of ideal features and functionalities of multiport DC converters in the context of MVDC systems is given. Examples include high degrees of modularity and reliability, flexibility of DC port power flows, and DC fault blocking capability. Three prominent multiport DC converter solutions are introduced, discussed and compared. These include both isolated and non-isolated topologies. Different case study simulations are carried out for one of the non-isolated topologies to illustrate its application as a three-port MVDC hub.

References

[1]   Z. Ma, W. Sheng, R. Li, *et al.* Study on the feasibility of MVDC. In: *Proc. CIGRE 2018 Paris Session*. Working Group Sub Committee C6.31; 2018. p. 1−12.

[2]   Siemens. *MVDC Plus: Medium Voltage Direct Current Managing the Future Grid (white paper)*. Siemens; 2017. Available online http://www.siemens.com/mvdc.

[3]   J. Yu, A. T. Moon, K. Smith, *et al.* Developments in the Angle-DC project; conversion of a medium voltage AC cable and overhead line circuit to DC. In: *Proc. CIGRE 2018 Paris Session*. Working Group B4–202; 2018. p. 1−10.

[4]   C. Long, J. Wu, K. Smith, *et al.* MVDC link in a 33 kV distribution network. *CIRED − Open Access Proceedings Journal*. 2017; 2017(1): 1308–1312.

[5]   T. An, G. Tang, W. Wang. Research and application on multi-terminal and DC grids based on VSC-HVDC technology in China. *High Voltage*. 2017; 2 (1): 1−10.

[6]   X. Zheng, R. Jia, L. Gong, *et al.* An optimized coordination strategy between line main protection and hybrid DC breakers for VSC-based DC grids using overhead transmission lines. *Energies*. 2019; 12(8).

[7] G. Bathurst, G. Hwang, L. Tejwani. MVDC − The new technology for distribution networks. In: *IET Int. Conf. on AC and DC Power Transmission*; 2015.

[8] W. Chen, A. Q. Huang, C. Li, *et al.* Analysis and comparison of medium voltage high power DC/DC converters for offshore wind energy systems. *IEEE Trans Power Electron.* 2013; 28(4): 2014−2023.

[9] IEEE Recommended Practice for 1 kV to 35 kV Medium-Voltage DC Power Systems on Ships. IEEE Standards Association; 2018. IEEE Std. 1709−2018.

[10] Medium voltage direct current (MVDC) grid feasibility study. CIGRE WG C6.31; 2020. 793.

[11] A. Giannakis, D. Peftitsis. MVDC distribution grids and potential applications: future trends and protection challenges. In: *20th European Conf. on Power Electronics and Applications (EPE ECCE Europe)*; 2018.

[12] F. Mura, R. W. De Doncker. Design aspects of a medium-voltage direct current (MVDC) grid for a university campus. In: *8th International Conference on Power Electronics − ECCE Asia*; 2011. p. 2359−2366.

[13] A. Lesnicar, R. Marquardt. An innovative modular multilevel converter topology suitable for a wide power range. In: *IEEE Bologna Power Tech Conference Proceedings.* vol. 3; 2003. p. 1−6.

[14] R. Marquardt. Modular multilevel converter: An universal concept for HVDC-networks and extended DC−bus-applications. In: *International Power Electronics Conference*; 2010. p. 502−507.

[15] H.-J. Knaak. Modular multilevel converters and HVDC/FACTS: a success story. In: *Proceedings of the 14th European Conference on Power Electronics and Applications*; 2011. p. 1−6.

[16] K. Shen, S. Wang, D. Zhao, *et al.* A discrete-time Low-frequency-ratio nearest level modulation strategy for modular multilevel converters with small number of power modules. *IEEE Access.* 2019; 7: 25792−25803.

[17] Y. Chen, Z. Li, S. Zhao, *et al.* Design and implementation of a modular multilevel converter with hierarchical redundancy ability for electric ship MVDC system. *IEEE Journal of Emerging and Selected Topics in Power Electronics.* 2017; 5(1): 189−202.

[18] S. Cui, J. H. Lee, J. Hu, *et al.* A modular multilevel converter with a zigzag transformer for bipolar MVDC distribution systems. *IEEE Trans Power Electron.* 2019; 34(2): 1038−1043.

[19] R. Marquardt. Modular multilevel converter topologies with DC-short circuit current limitation. In: *IEEE 8th International Conference on Power Electronics and ECCE Asia*; 2011. p. 1425−1431.

[20] J. Peralta, H. Saad, S. Dennetiere, *et al.* Detailed and averaged models for a 401-Level MMC−HVDC system. *IEEE Trans Power Del.* 2012; 27(3): 1501−1508.

[21] S. Kenzelmann, A. Rufer, M. Vasiladiotis, D. Dujic, F. Canales, and Y. R. de Novaes. A versatile DC−DC converter for energy collection and distribution using the modular multilevel converter. In: *Proceedings of the 14th European Conference on Power Electronics and Applications*; 2011. p. 1−10.

[22]   T. Luth, M. M. C. Merlin, T. C. Green, F. Hassan, and C. D. Barker. High - frequency operation of a DC/AC/DC system for HVDC applications. *IEEE Trans Power Electron*. 2014; 29(8): 4107−4115.

[23]   A. Schön and M.-M. Bakran. A new HVDC−DC converter for the efficient connection of HVDC networks. In: PCIM Europe Conf.; 2013. p. 525−532.

[24]   R. W. A. A. De Doncker, D. M. Divan, and M. H. Kheraluwala. A three-phase soft-switched high-power-density DC/DC converter for high-power applications. *IEEE Trans Ind Appl*. 1991;27(1): 63−73.

[25]   J. D. Pez, D. Frey, J. Maneiro, *et al*. Overview of DC−DC converters dedicated to HVdc Grids. *IEEE Trans Power Del*. 2019; 34(1): 119−128.

[26]   G. P. Adam, I. A. Gowaid, S. J. Finney, D. Holliday and B. W. Williams. Review of DC–DC converters for multi - terminal HVDC transmission networks. *IET Power Electronics*. 2016; 9(2): 281−296.

[27]   X. Zhang, X. Xiang, T. C. Green, *et al*. A Push−Pull modular-multilevel-converter-based low step−up ratio DC transformer. *IEEE Transactions on Industrial Electronics*. 2019; 66(3): 2247−2256.

[28]   S. Milovanovic, D. Dujic. High-power DC−DC converter utilising Scott transformer connection. IET Electric Power Applications. 2019.

[29]   A. Schön and M.-M. Bakran. Comparison of the most efficient DC−DC converters for power conversion in HVDC grids. In: *PCIM Europe Conference*; 2015. p. 518−526.

[30]   G. J. Kish, M. Ranjram and P. W. Lehn. A modular multilevel DC/DC converter with fault blocking capability for HVDC interconnects. *IEEE Trans Power Electron*. 2015; 30(1): 148−162.

[31]   S. Norrga, L. Angquist, A. Antonopoulos. The polyphase cascaded-cell DC/DC converter. In: *IEEE Energy Conversion Congress and Exposition*; 2013. p. 4082−4088.

[32]   J. A. Ferreira. The multilevel modular DC converter. *IEEE Trans Power Electron*. 2013; 28(10): 4460−4465.

[33]   G. J. Kish. On the emerging class of non-isolated modular multilevel DC−DC converters for DC and hybrid AC−DC systems. *IEEE Transactions on Smart Grid*. 2019; 10(2): 1762−1771.

[34]   Y. Tran, D. Dujic. A multiport medium voltage isolated DC−DC converter. In: *42nd Annual Conference of the IEEE Industrial Electronics Society (IECON)*; 2016. p. 6983−6988.

[35]   H. Tao, A. Kotsopoulos, J. L. Duarte, *et al*. Family of multiport bidirectional DC−DC converters. *IEE Proceedings − Electric Power Applications*. 2006; 153(3): 451−458.

[36]   Q.Mei, X. Zhen-lin, W. Wu. A novel multi-port DC−DC converter for hybrid renewable energy distributed generation systems connected to power grid. In: *IEEE International Conference on Industrial Technology*; 2008. p. 1−5.

[37]   W. Lin, J. Wen, S. Cheng. Multiport DC–DC autotransformer for inter-connecting multiple high voltage DC systems at low cost. *IEEE Trans Power Electron*. 2015; 30(12): 6648−6660.

[38]   E. Kontos, H. Papadakis, M. Poikilidis, *et al.* MMC-based multi-port DC hub for multiterminal HVDC grids. In: *PCIM Europe; International Exhibition and Conference for Power Electronics, Intelligent Motion, Renewable Energy and Energy Management*; 2017. p. 1−8.

[39]   D. Jovcic, W. Lin. Multiport high-power LCL DC hub for use in DC transmission grids. *IEEE Trans Power Del.* 2014; 29(2): 760−768.

[40]   L. Liu. *Dynamic Modelling and Simulation of a Multiport DC Hub with Closed-Loop Control.* University of Manitoba. Winnipeg, Canada; 2019.

[41]   F. Alsokhiry, Y. Al-Turki, I. Abdelsalam, *et al.* Multi - port converter for medium and high voltage applications. In: *7th International Conference on Renewable Energy Research and Applications (ICRERA)*; 2018. p. 150−155.

[42]   S. H. Kung, G. J. Kish. Multiport modular multilevel converter for DC systems. IEEE Trans Power Del. 2019; 34(1): 73−83.

[43]   S. Kung. *Multiport DC–DC Modular Multilevel Converter for MVDC and HVDC Networks.* University of Alberta, Edmonton, Canada; 2018.

[44]   K. Filsoof, P. W. Lehn. A Bidirectional modular multilevel DC−DC converter of triangular structure. *IEEE Trans Power Electron.* 2015; 30(1): 54−64.

[45]   K. Filsoof, P. W. Lehn. A Bidirectional multiple-input multiple-output modular multilevel DC–DC converter and its control design. *IEEE Trans Power Electron.* 2016; 31(4): 2767−2779.

Chapter 6

Modern control and mode visualization of bidirectional DC/DC converters

Brandon Grainger[1] and Zachary Smith[1]

6.1 Model reference controller design for bidirectional DC/DC converters

Constant power loads (CPLs) exhibit negative incremental impedance character-istics contributing to destabilizing effects in power systems [1]. Power electronic converters and motor drives, when tightly regulated, behave as CPLs. For the case of a motor drive, the inverter drives the motor and tightly regulates the speed to achieve constant speed. Assuming a linear relationship between torque and speed, motor torque will remain constant resulting in constant power consumption by the motor. For CPLs, the instantaneous value of the impedance is positive ($V/I > 0$), but the incremental impedance is always negative ($dV/dI < 0$). In the literature, the latter is referred to as negative incremental impedance instability [1–3].

As pointed out in [2], CPL induced instability can be resolved by modifying the DC system's hardware structure, by adding resistors, filters, or energy storage elements, but approaches based on feedback control can offer more efficient solutions. In addition to linear controllers featured by simple architectures and designs, many research teams have chosen nonlinear based control approaches to stabilize CPL scenarios to avoid limitations of linearization and to ensure that large-signal stability is guaranteed [3]. The primary nonlinear approaches include the use of Lyapunov-based design, hysteresis control, nonlinear passivity-based techniques, and boundary control. However, some of the disadvantages of these techniques include the use of proportional-derivative (PD) controllers which can be sensitive to noise, current and voltage transient overshoots, and have inherent dif-ficulty with implementation.

In this section, a model reference controller is developed for stabilizing CPLs that can be found in medium voltage (MV) DC systems. Model reference control (MRC) has the advantages of allowing the designer to select the pole and zero placements to adequately tune system behavior. The controlled closed-loop system,

[1]Department of Electrical and Computer Engineering, University of Pittsburgh Swanson School of Engineering, Pittsburgh, PA, USA

with a naturally unstable plant, replicates the dynamic system behavior of the designed reference model to ensure stable output. Moreover, this control architecture can be readily upgraded to include adaptation capabilities following model reference adaptive control design. The bidirectional DC/DC converter will serve as our testbed under evaluation and procedures are outlined for designing the model reference controller for applicability towards other dynamic systems.

6.1.1   System description

The current traces and operating waveforms of the dual active bridge DC/DC converter are provided in Figures 6.1 and 6.2, respectively. Plots correspond to the primary side transformer voltage, V_p, secondary side transformer voltage, V_s, inductor current, I_L, and input current, I_g. The most notable characteristic is the phase delay, ϕ, between both bridges, which controls the allowable power flow in

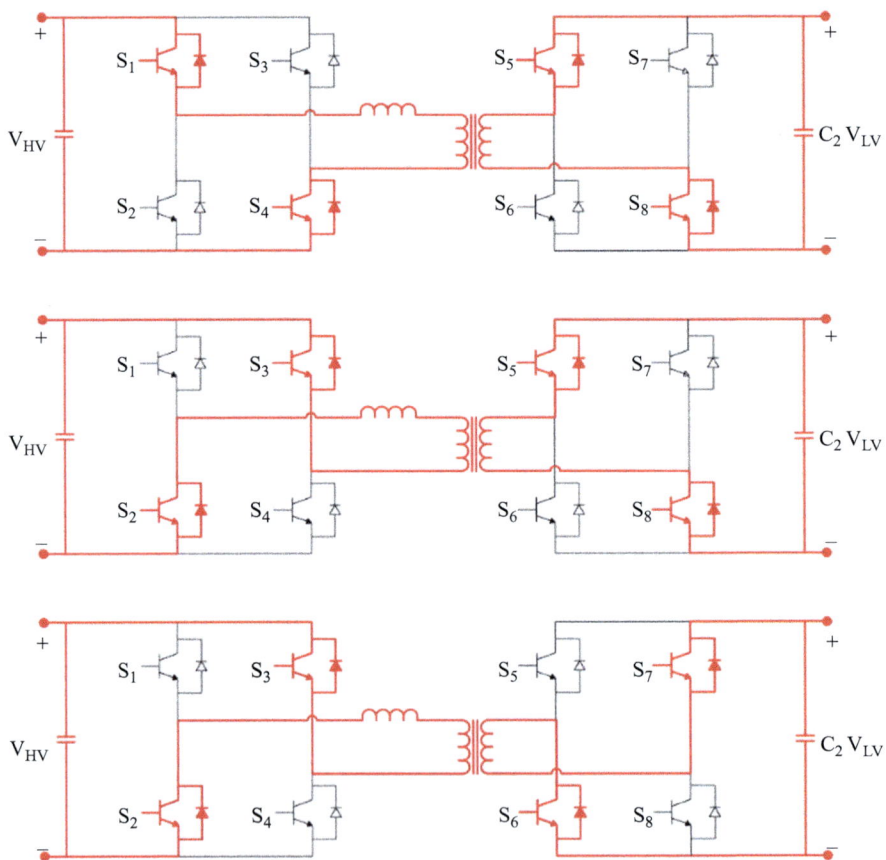

Figure 6.1   Current traces of dual active bridge DC/DC converter in various states

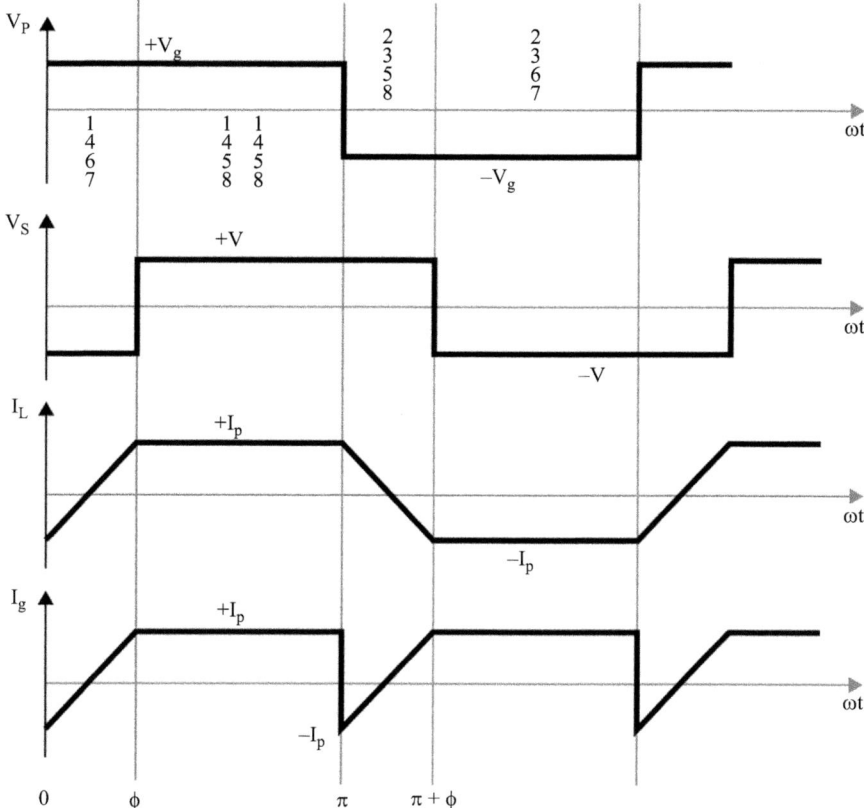

Figure 6.2   Operating waveforms of the dual active bridge DC/DC converter

the circuit. The relationship between the phase delay and duty cycle, d_h, is described by (6.1). Note that T_s is the switching period:

$$\phi = \frac{d_h T_s}{2} \tag{6.1}$$

The average inductor current can be described by (6.2) [4]. The output voltage
of the converter, which depends upon the converter duty cycle and output impe-
dance, Z_{out}, of the converter is described by (6.3). If the output voltage, $\widehat{V}$, and duty
cycle, $\widehat{d}_h$, are perturbed slightly and the system is then linearized, the small signal
transfer function between the output voltage and duty cycle can be shown to be
(6.4) where K_{DC} is a constant associated with the parameters of the DC/DC converter. Equation (6.4) is the plant that is unstable:

$$I_{avg} = \frac{n V_{DC} T_s}{2 L_{Tx}} d_h (1 - d_h) \tag{6.2}$$

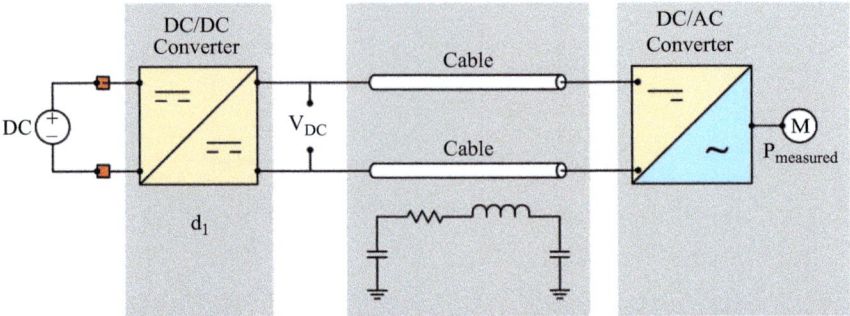

Figure 6.3   Motor drive system for studying the regulation of a constant power load

$$V = I_{avg}Z_{out} = \frac{nV_{DC}T_s}{2L_{Tx}}d_h(1 - d_h)Z_{out} \tag{6.3}$$

$$\frac{\widehat{V}}{\widehat{d}_h} = \frac{nV_{DC}T_s}{2L_{Tx}}(1 - 2d_h)Z_{out} \equiv K_{DC}Z_{out} \tag{6.4}$$

Figure 6.3 is an illustration that could be found in a future DC based grid system and will be the basis of our investigation. The MV bus is represented by an ideal voltage source. This is appropriate as we assume that the DC bus is well regulated. A bidirectional DC/DC converter interfaces the MV bus to the motor inverter. A single core, XLPE insulated, PVC sheathed, unarmoured cable bridges the power converters and is modeled as a coupled pi circuit as shown [5].

As described in [6], a common CPL is a DC/AC inverter that drives an electric motor. The motor and inverter reflect a CPL because the motor controller tightly regulates the speed of the machine. While operating at constant speed, the torque will be held constant and, therefore, the machine's output power. Thus, the behavior of the inverter and the motor is seen as a CPL to the bidirectional DC/DC converter. Noting that the DC/DC converter can be modeled as an average current source described by (6.2), Figure 6.3 can be simplified to Figure 6.4. Observe that the inverter and motor have been replaced by a CPL with negative resistance, Y. It is also assumed that the time constant of the inverter and motor is smaller than that of the DC/DC converter.

The output impedance associated with Figure 6.4 is described by (6.5) where all circuit parameters are defined in Figure 6.4. The poles and zeros of (6.5) are visually shown in Figure 6.5, and hence, the plant is naturally unstable. As power levels increase, the poles and zeros shift further into the unstable region of the complex plane. Line parameters used for this observation are provided in Table 6.1:

$$Z_{out} = \frac{LC_2Ys^2 + (RC_2Y - L)s + (Y - R)}{(LC_1C_2Y)s^3 + (RC_1C_2Y - LC_1)s^2 + (C_1Y - RC_1 + C_2Y)s - 1} \tag{6.5}$$

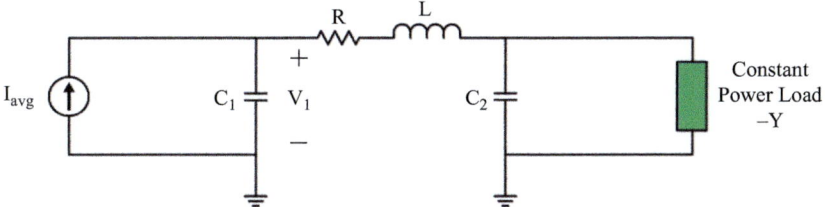

Figure 6.4   Simplified motor drive system model for studying the regulation of a constant power load

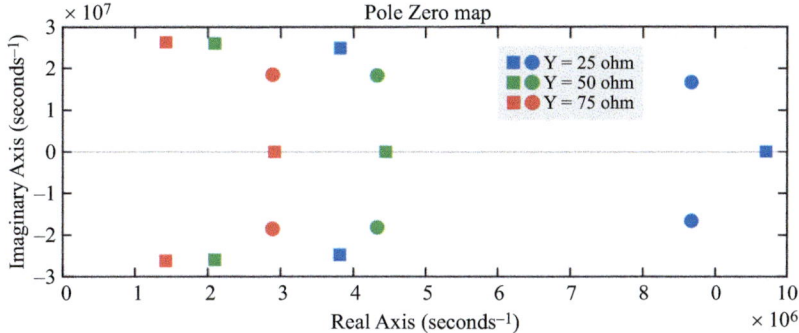

Figure 6.5   Pole-zero plot of naturally unstable transfer function at varying power levels

Table 6.1   Cable parameters

Parameter	Variable	Value
Line resistance	R	0.0176 Ω/km
Line inductance	L	0.248 mH/km
Line capacitance	$2C_{1,2}$	0.923 μF/km
Cable distance		5 m (16 feet)

6.1.2   Stability assessments using model reference control theory

The task at hand is to design a compensator to account for the instabilities found with CPLs. Power engineers are accustomed to using some form of PID controller. The integral term, I, is used to eliminate steady-state error, while the derivative term, D, is used to improve stability and system damping. In this portion of the chapter, it will be shown that a PD controller will not achieve satisfactory steady-state response performance nor will the system be stable. Thus, we consult the principles of MRC to aid in the stabilization of the CPL scenario described.

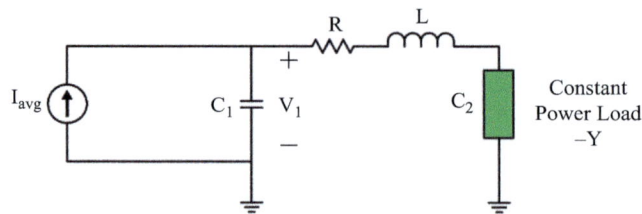

Figure 6.6   Second-order system model

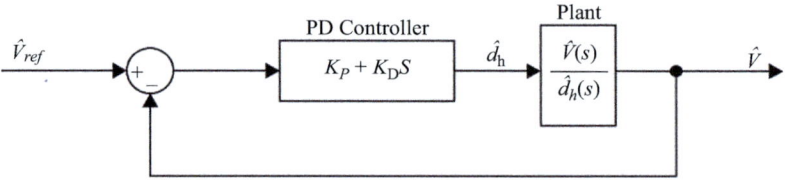

Figure 6.7   Closed loop architecture with PD controller

To appreciate the damping control effect of a PD controller, we first consider a simplified second-order model of the plant. The distance between the DC/DC converter and motor inverter is short (to minimize line inductance), the cable can be approximated with a line resistance and inductance in series (Figure 6.6) [7]. With the output capacitor of the dual active bridge DC/DC converter, the system model can be approximated as a second-order system with output impedance transfer function described by (6.6), where $Z_{2,out}$ represents the simplified second-order output impedance, K_{DC} was defined in the previous section, and the other circuit parameters are defined in Figure 6.6. The line resistance is assumed to be small:

$$\frac{\widehat{V}}{\widehat{d}_h} = K_{DC}Z_{2,out}(s) = K_{DC}\frac{Ls - Y}{LCs^2 - YCs + 1} \qquad (6.6)$$

The PD controller would be a natural choice in stabilizing a system response because this type of compensation introduces a phase lead, which creates a stabilizing effect. Generally, for a second-order plant, the derivative component will act to dampen the closed-loop system and thus stabilize the plant. Figure 6.7 shows a diagram with PD control in the closed loop.

The closed-loop transfer function of the system in Figure 6.7 can be written as (6.7) where a standard PD compensator of the form $K_P + K_D s$ is used to control the plant in (6.6), and K_P and K_D are the proportional and derivative gains, respectively. A couple of observations can be made on the performance of the PD control. First, under the premise of stability, from (6.8), as K_P increases the damping ratio, ς, increases. As K_D increases, the natural frequency, ω_n, and damping ratio decrease. This control behavior is outside the norm from the conventional impact of a PD controller on a stable and minimum-phase plant. Typically, a larger K_D and K_P usually achieve a higher damping ratio and larger natural frequency, respectively. Observing (6.9), the natural frequency extracted from (6.7), as K_P increases the natural frequency decreases.

The DC gain of the system is listed as (6.10). For the closed-loop system to be stable, both (6.8) and (6.9) need to be greater than 0. More formerly, $K_{DC}K_PL$ - $K_{DC}K_PY$ - $YC > 0$ and $K_{DC}K_PY < 1$. However, as $K_{DC}K_PY$ is always less than one, the system's DC gain will always be negative. Therefore, the PD controller here cannot achieve satisfactory steady-state performance while maintaining system stability:

$$\frac{\widehat{V}(s)}{\widehat{V}_{ref}(s)} = \frac{K_{DC}(K_P + K_D s)(Ls - Y)}{(K_{DC}K_DL + LC)s^2 + (K_{DC}K_pL - K_{DC}K_DY - YC)s + (1 - K_{DC}K_PY)} \tag{6.7}$$

$$2\varsigma\omega_n = \frac{K_{DC}K_PL - K_{DC}K_DY - YC}{K_{DC}K_DL + LC} \tag{6.8}$$

$$\omega_n^2 = \frac{1 - K_{DC}K_pY}{K_{DC}K_DL + LC} \tag{6.9}$$

$$DC_{Gain} = \frac{-K_{DC}K_pY}{1 - K_{DC}K_PY} \tag{6.10}$$

Under the premise of stability (here a necessary condition for stability is $K_{DC}K_P (Y - R) > 1$), the system's DC gain is always greater than 1. Therefore, the PD controller still cannot achieve satisfactory steady-state performance even with a capacitor added in parallel to the CPL.

6.1.3 Model reference control definitions and assumptions [8]

Thus far, it has been shown, analytically, that the plant cannot be controlled satisfactorily using a traditional PD controller. Our objective is to design a controller such that the behavior of the controlled plant remains close to the behavior of a desirable reference model, $M(s)$, despite uncertainties or variations in the plant parameters. Before establishing the mathematical framework, a few assumptions need to be described based upon the plant and reference model. First, the plant is a single-input, single-output, linear time-invariant system described by (6.11). Note that $\widehat{n}_p$ and $\widehat{d}_p$ are monic polynomials of order m and n, respectively. The plant is strictly proper, minimum phase, and not assumed to be stable. The gain of the plant, $k_p > 0$. Second, the reference model is described by (6.12) with numerator and denominator polynomials that are monic, coprime and of the same order as the plant. The reference model is stable, minimum phase with $k_m > 0$. For all notation,

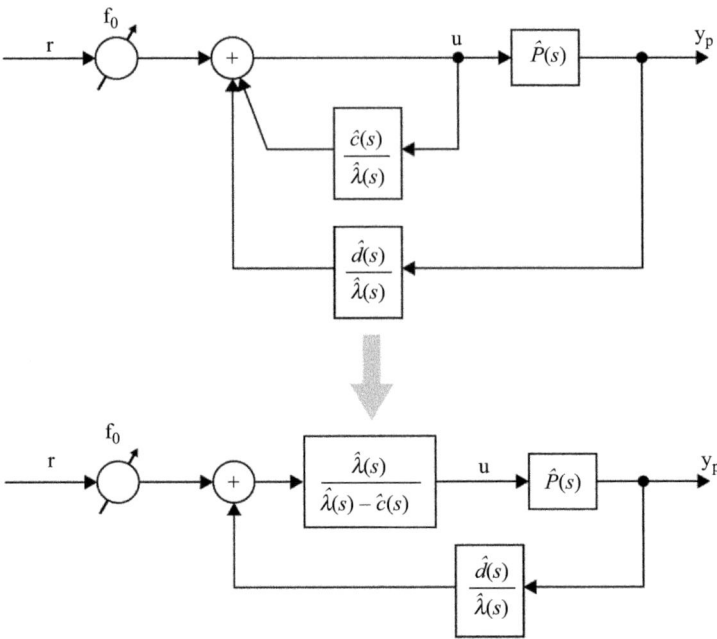

Figure 6.8   Controller structure to stabilize CPL [8]

subscript p is for the plant and subscript m is for the reference model:

$$\frac{\widehat{y}_p(s)}{\widehat{u}(s)} = \widehat{P}(s) = k_p \frac{\widehat{n}_p(s)}{\widehat{d}_p(s)} \tag{6.11}$$

$$\frac{\widehat{y}_m(s)}{\widehat{r}(s)} = \widehat{M}(s) = k_m \frac{\widehat{n}_m(s)}{\widehat{d}_m(s)} \tag{6.12}$$

To achieve our control objective, we consider the controller structure found in Figure 6.8 where f_0 is a scalar, $\widehat{c}(s)$, $\widehat{d}(s)$, and $\widehat{\lambda}(s)$ are polynomials of degrees $n-2$, $n-1$, and $n-1$, respectively.

The closed loop transfer function between the output, y_p, and input, r, can be shown to be (6.13). Note that (6.13b) can be obtained if (6.11) is substituted into (6.13a). It is possible to *make* (6.13) equal to $\widehat{M}(s)$ if and only if (6.14) is satisfied. If satisfied, there exists unique f_0^*, $\widehat{c}^*(s)$, and $\widehat{d}^*(s)$ such that the transfer function in (6.13) equals $\widehat{M}(s)$ described by (6.12), assuming that $\widehat{n}_p$ and $\widehat{d}_p$ are coprime [8].

$$\frac{y_p(s)}{r(s)} = \frac{f_0 \dfrac{\widehat{\lambda}(s)}{\widehat{\lambda}(s) - \widehat{c}(s)} \widehat{P}(s)}{1 - \dfrac{\widehat{d}(s)}{\widehat{\lambda}(s) - \widehat{c}(s)} \widehat{P}(s)} \tag{6.13a}$$

$$\frac{y_p(s)}{r(s)} = \frac{f_o k_p \widehat{\lambda}(s)\widehat{n}_p(s)}{\left(\widehat{\lambda}(s) - \widehat{c}(s)\right)\widehat{d}_p(s) - k_p \widehat{n}_p(s)\widehat{d}(s)} \tag{6.13b}$$

$$\left(\widehat{\lambda}(s) - \widehat{c}(s)^*\right)\widehat{d}_p(s) - k_p \widehat{n}_p(s)d(s)^* = f_o \frac{k_p}{k_m}\widehat{\lambda}_o(s)\widehat{n}_p(s)\widehat{d}_m(s) \tag{6.14}$$

The *unique* solutions of these polynomials can be determined with (6.15). The expression for $\widehat{\lambda}_0(s)$ is an arbitrary Hurwitz polynomial of degree *n-m-1* and $q(s)$ is the quotient obtained when dividing $\widehat{\lambda}_0 \widehat{d}_m$ by $\widehat{d}_p$ of degree *n-m-1*:

$$\widehat{d}^*(s) = \frac{1}{k_p}\left(\widehat{q}(s)\widehat{d}_p(s) - \widehat{\lambda}_o(s)\widehat{d}_m(s)\right) \tag{6.15a}$$

$$\widehat{c}^*(s) = \widehat{\lambda}(s) - \widehat{q}(s)\widehat{n}_p(s) \tag{6.15b}$$

$$f_o^* = \frac{k_m}{k_p} \tag{6.15c}$$

6.1.3.1   Attempt to stabilize second-order plant with model reference control

Consider the second order plant associated with the output impedance of the system of Figure 6.6 listed as (6.16). The plant takes on the form of (6.17):

$$\widehat{P}(s) = Z_{2,out}(s) = K_{DC}\frac{Ls - Y}{LCs^2 - YCs + 1} \tag{6.16}$$

$$\widehat{P}(s) = k_p\frac{\widehat{n}_p(s)}{\widehat{d}_p(s)} = k_p\frac{s - q_o}{s^2 - p_1 s + p_o} \tag{6.17}$$

The zero of (6.17) is not of minimum phase because the zero is located in the right half of the complex plane. The model that is chosen is listed as (6.18). One advantage of the control structure chosen (Figure 6.8) is that the forward block cancels the zeros of $\widehat{P}(s)$ and replaces them with the zeros of $\widehat{M}(s)$:

$$\widehat{M}(s) = k_m\frac{\widehat{n}_m(s)}{\widehat{d}_m(s)} = k_m\frac{s + a_o}{s^2 + b_1 s + b_o} \tag{6.18}$$

Utilizing the procedures outlined in the previous section, polynomials (6.19), (6.20), (6.21), and (6.22) can be written. One will notice that (6.20) introduces negative feedback terms:

$$\widehat{\lambda}(s) = \widehat{\lambda}_o(s)\widehat{n}_m(s) = s + a_o \tag{6.19}$$

$$\begin{aligned}
\widehat{d}^*(s) &= \frac{1}{k_p}(\widehat{q}\widehat{d}_p - \widehat{\lambda}_o\widehat{d}_m) \\
&= \frac{-1}{k_p}[(p_1 + b_1)s + (b_o - p_o)] \\
&= d_1^* s + d_o
\end{aligned} \tag{6.20}$$

$$\widehat{c}^*(s) = \widehat{\lambda}(s) - \widehat{q}\widehat{n}_p = (s + a_o) - (s - q_o) = a_o + q_o \qquad (6.21)$$

The feedforward block of Figure 6.8 is listed as (6.22). Comparing (6.16) and (6.17), $q_o = Y/L$ resulting in an unstable compensator since $q_o > 0$:

$$\frac{\widehat{\lambda}(s)}{\widehat{\lambda}(s) - \widehat{c}(s)} = \frac{s + a_o}{s - q_o} \qquad (6.22)$$

6.1.3.2   Stabilized third order plant with model reference control

Utilizing the short line modeling assumption has resulted in an unstable compensator design. The third order plant model, (6.5), will be used, which introduces additional capacitance into the system. Our plant is now described by (6.23) with all appropriate definitions provided in terms of the system circuit parameters:

$$\widehat{P}(s) = k_p \frac{\widehat{n}_p(s)}{\widehat{d}_p(s)} = k_p \frac{s^2 + q_1 s + q_0}{s^3 + p_2 s^2 + p_1 s - p_0} \qquad (6.23)$$

$$k_p \equiv K_{DC}/C_1 \quad p_0 \equiv 1/(LC_1 C_2 Y)$$

$$q_0 \equiv (Y - R)/(LC_2 Y) \quad p_1 \equiv (C_1 Y - RC_1 + C_2 Y)/(LC_1 C_2 Y)$$

$$q_1 \equiv (RC_2 Y - L)/(LC_2 Y) \quad p_2 \equiv (RC_1 C_2 Y - LC_1)/(LC_1 C_2 Y)$$

To use MRC, the plant must be minimum phase but not necessarily stable. This condition requires that both q_0 and q_1 be greater than 0. $q_0 > 0$ is easily satisfied as the line resistance R is usually small or ignorable. However, the condition that $q_1 > 0$ constrains the system capacitor C_2. This capacitor value should be lower bounded by $L/(RY)$. When these conditions are satisfied, a reference model described by (6.24) is obtained:

$$\widehat{M}(s) = k_m \frac{\widehat{n}_m(s)}{\widehat{d}_m(s)} = k_m \frac{s^2 + a_1 s + a_0}{s^3 + b_2 s^2 + b_1 s + b_0} \qquad (6.24)$$

Using the model reference controller structure as in Figure 6.8 and following the procedures outlined in the latter section, $\widehat{\lambda}(s)$, $\widehat{d}^*(s)$ and $\widehat{c}^*(s)$ can be determined with (6.25), (6.26), and (6.27), respectively:

$$\widehat{\lambda}(s) = \widehat{\lambda}_o(s)\widehat{n}_m(s) = s^2 + a_1 s + a_o \qquad (6.25)$$

$$\begin{aligned}
\widehat{d}^*(s) &= \frac{1}{k_p}(\widehat{q}\widehat{d}_p - \widehat{\lambda}_o \widehat{d}_m) \\
&= \frac{1}{k_p}\left[(p_2 - b_2)s^2 + (p_1 - b_1)s - (b_o + p_o)\right] \\
&= d_2^* s^2 + d_1^* s + d_o
\end{aligned} \qquad (6.26)$$

$$\widehat{c}^*(s) = \widehat{\lambda}(s) - \widehat{q}\widehat{n}_p = s(a_1 - q_1) + (a_o - q_o) = c_1{}^* s + c_o \qquad (6.27)$$

Finally, the compensator for this third-order plant scenario is listed as (6.28) and is clearly stable if the parameters a_o, a_1, b_o, b_1, and b_2 are selected appropriately:

$$\frac{\widehat{\lambda}(s)}{\widehat{\lambda}(s) - \widehat{c}(s)} = \frac{s^2 + a_1 s + a_o}{s^2 + q_1 s + q_o} \tag{6.28}$$

6.1.4   Parameter selection

To pick the reference model parameters, the designer must consider the following criteria. First, the DC gain of the reference model must equal one to ensure that the output voltage will follow the reference voltage with no steady-state error. Second, the reference model must be stable. Third, the selected parameters should allow the system response to behave with sufficiently small rise time, settling time, and percent overshoot.

To meet the first criterion, the reference model's DC gain should satisfy

$$\widehat{M}(0) = k_m \frac{a_o}{b_o} = 1$$
$$\text{Stability}: a_i, b_i > 0$$

If $k_m = 1$, then $a_o = b_o$. For the system to be critically damped, (6.29) must be satisfied. A system that is critically damped will have the desired transient properties as defined in the latter paragraph. To meet the second criterion, guidelines are provided in [9]. Here, it is advised to choose the pole locations as shown in Figure 6.9. The third-order denominator of the reference model is chosen to be composed of a stable real pole and a complex conjugate pair as described by (6.30). It is suggested to not place all the poles in one location in the complex plane. Otherwise, the controlled system will tend to have a slower response and utilize

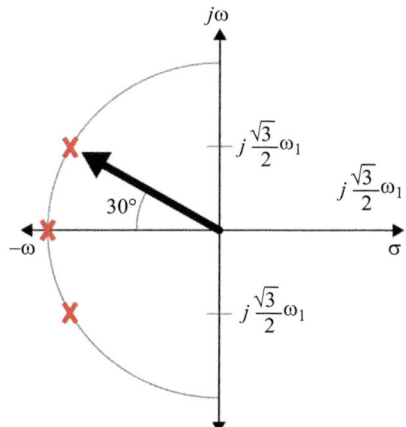

Figure 6.9   Pole locations ensuring a stable and well-behaved reference model

large control signals. In this example, the poles are placed 30 degrees apart along a circle of radius ω_1. Using (6.30), b_o, b_1, and b_2 can be solved for as shown in (6.31):

$$a_1 = 2\sqrt{a_o} \tag{6.29}$$

$$(s + \omega_1)\left(s + \frac{\sqrt{3}}{2}\omega_1 + j\frac{\sqrt{3}}{2}\omega_1\right)\left(s + \frac{\sqrt{3}}{2}\omega_1 - j\frac{\sqrt{3}}{2}\omega_1\right) \tag{6.30}$$

$$(s + \omega_1)(s^2 + \sqrt{3}\omega_1 s + \omega_1^2)$$

$$s^3 + (\sqrt{3}\omega_1 + \omega_1)s^2 + (\omega_1^2 + \sqrt{3}\omega_1^2)s + \omega_1^3 = s^3 + b_2 s^2 + b_1 s + b_o \tag{6.31}$$

After determining the denominator of (6.24), we choose the numerator parameters as shown in (6.32). These parameters ensure that $\widehat{\lambda}(s) = \widehat{\lambda}_0(s)\widehat{n}_m(s)$ is critically damped:

$$a_0 = \omega_1^3, \ a_1 = 2\sqrt{a_0} = 2\omega_1\sqrt{\omega_1} \tag{6.32}$$

6.1.5 Basic verification

The parameters associated with the system under study are provided in Table 6.2. Here, the bidirectional converter is rated for 200 kW corresponding to a transformer leakage inductance of 5.21 mH. Demanded power, P_{rms}, is 100 kW. The critically damped, closed loop response with designed compensators is provided in Figure 6.10. For comparison purposes, if the PD compensator were selected to regulate the plant, the expected closed loop response is also provided in Figure 6.10. The noticeable observation is that the PD controller, no matter how well designed, cannot stabilize the output voltage. Using MRC design ensures that a very smooth, well performing system response is achieved. Hence the power of why these routines were introduced in this part of the chapter.

Table 6.2 System parameters

Parameter	Variable	Value
Transformer turns	n	5
Leakage inductance	L_T	5.21 mH
Duty cycle	d_h	0.15
DC bus voltage	V_{dc}	1 kV
Switching frequency	f_s	3 kHz
Line resistance	R	0.0176 Ω/km
Line inductance	L	0.248 mH/km
Line capacitance	$2C_{1,2}$	0.923 µF/km
Cable distance	–	0.1 km
CPL	Y	10 Ω_{DC}
Line frequency	–	377 rad/s
Complex pole frequency	ω_1	3,500 rad/s

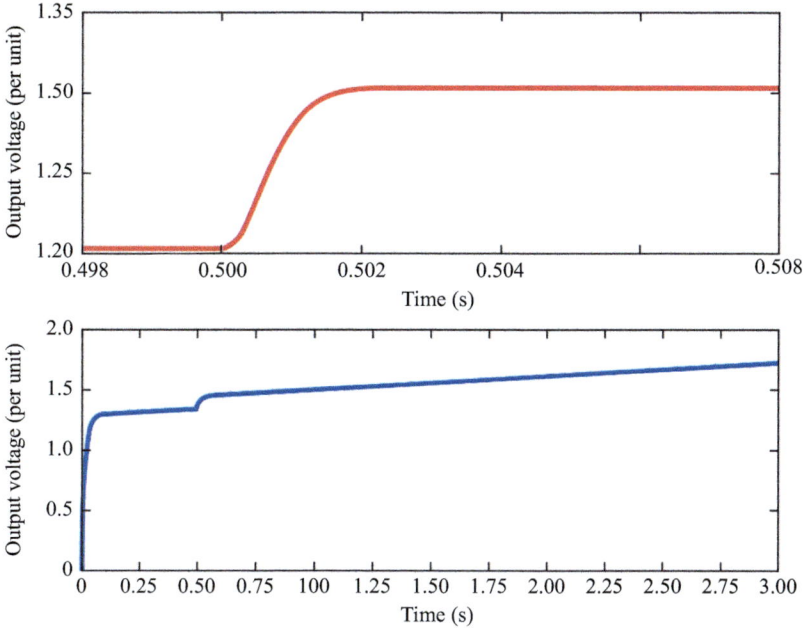

Figure 6.10   Closed loop controller performance based upon MRC design (top) and PD response (bottom)

6.2   Fault tolerant, single input, multiple output, bidirectional converter modeling and design

As the capacity for renewable energy sources within a system continues to develop, it becomes practical to design converters at the MVDC level [10–12]. Unfortunately, issues arise for DC systems at the MV level, especially as the control of multiple converters interact among one another. By incorporating multiple ports into a single converter, the complexity of the multi-bus system can be reduced [13]. While current-fed dual-active bridge (DAB) designs are not as common at low voltage (LV) levels, current-fed designs can surpass voltage-fed designs by offering larger voltage level transformation, lower input current ripple, a multi-port interface, and DC fault ride-through capability [14,15].

Having studied the single input, single output bidirectional DAB in the previous section of this chapter, this section explores the power output of a DAB converter with two unique variations. The first variation is the addition of mutual inductors at the MVDC ports to establish current-fed behavior. The second is using a three-winding transformer to create a three-port converter. Both modifications are incorporated into a single converter, and equations are developed to describe power flow with respect to passive component values and switched voltage waveforms.

6.2.1   Circuit analysis

The circuit schematic for the current-fed three-port DAB converter is shown in Figure 6.11. The converter interconnects one LVDC bus and two MVDC buses. Port 1 is the LVDC port connected to the transformer with an H-bridge. Ports 2 and 3 are MVDC ports with mutual inductor pairs incorporated into each port. The arm modules are made up of full bridge and half bridge submodules as shown in the figure. Each MVDC port also has an associated bus capacitor.

The converter operates by controlling the phase shift between the switching waveforms at each port. Port 1 switching waveform v_1 is a square wave. Ports 2 and 3 switching waveforms can be defined as $v_{armx} = v_{armx-12} - v_{armx-11}$, where x is the port number. A simplification is made in the following analysis that each arm's voltage waveform is equivalent to a square waveform. The switching waveforms for each port are given in Figure 6.12. Note that the duty cycle of the individual arms on a MV port can be in the range of $0 \leq D \leq 1$ but the resulting port switching waveform v_{armx} will remain between 0 and 0.5.

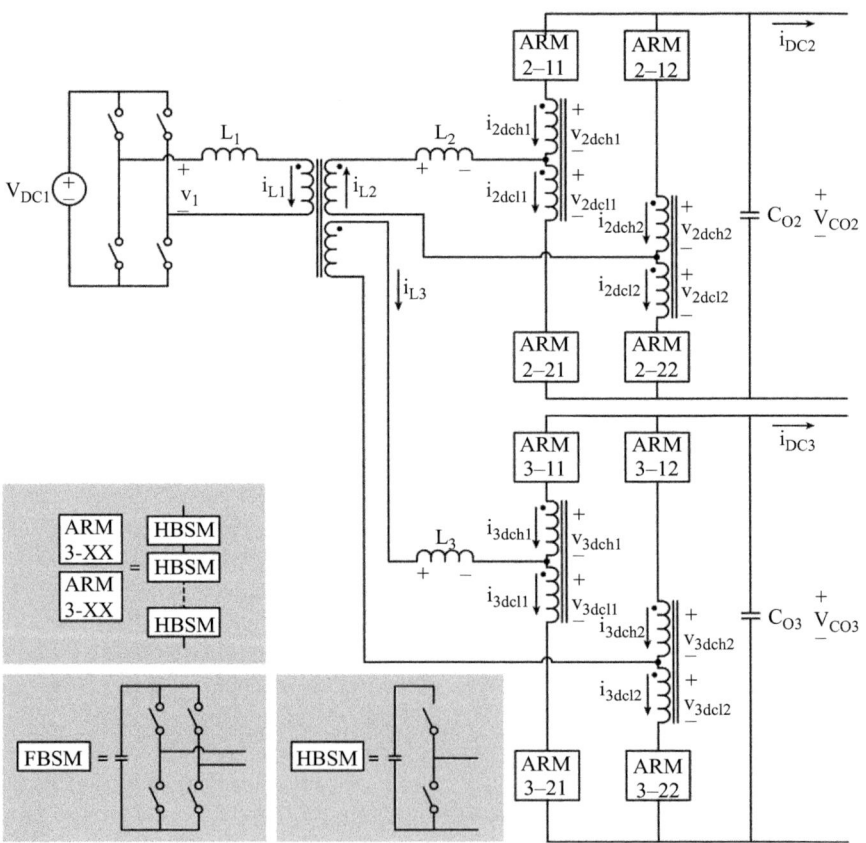

Figure 6.11   Current-fed, multi-port DAB converter circuit schematic

6.2.2   *State variable computation*

The MV ports require a more in-depth analysis than the LV port when defining the state variables. The power flow transferred from one port to another can be computed by the product of the AC port voltage and the transformer leakage inductor current. Figure 6.13 shows the example of key parameters at port 2. In this example, the instantaneous power transferred into port 2 is described by (6.33). Under balanced operating conditions, the average power flowing into port 2 is equal to the power flowing into the DC bus at port 2, since the arm capacitors and mutual

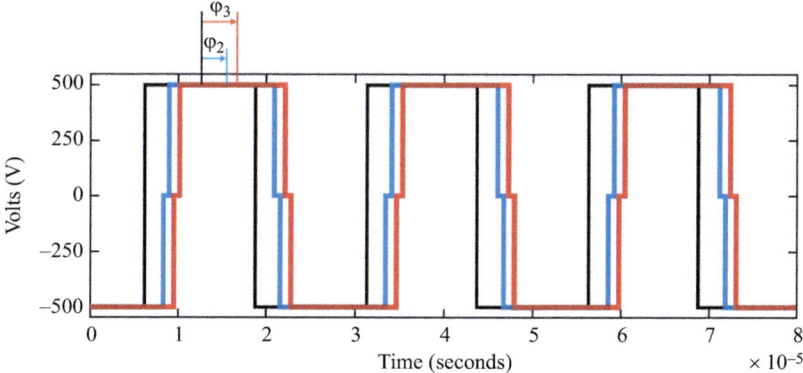

Figure 6.12   Switching waveforms for ports 1, 2, and 3. Also includes phase shift between ports 1 and 2 (ϕ_2) and ports 1 and 3 (ϕ_3)

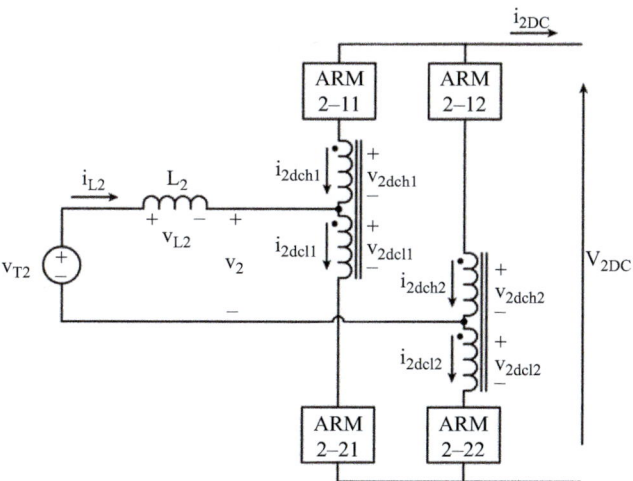

Figure 6.13   Circuit parameters at port 2

inductors in equilibrium have an average power of zero:

$$P_2(t) = v_2(t)i_{L2}(t) \tag{6.33}$$

6.2.2.1 Computation of port voltage v_2 with respect to v_{T2}

The first step in the computation is to define the voltage v_2 in terms of the controllable voltage waveform v_{arm2} and the reflected transformer voltage v_{T2}. To start, Faraday's law for mutual inductors gives the following equations:

$$v_{2dch1}(t) = L_{dc}\frac{di_{2dch1}(t)}{dt} + L_M\frac{di_{2dcl1}(t)}{dt} \tag{6.34}$$

$$v_{2dcl1}(t) = L_{dc}\frac{di_{2dcl1}(t)}{dt} + L_M\frac{di_{2dch1}(t)}{dt} \tag{6.35}$$

$$v_{2dch2}(t) = L_{dc}\frac{di_{2dch2}(t)}{dt} + L_M\frac{di_{2dcl2}(t)}{dt} \tag{6.36}$$

$$v_{2dcl2(t)} = L_{dc}\frac{di_{2dcl2}(t)}{dt} + L_M\frac{di_{2dch2}(t)}{dt} \tag{6.37}$$

Then, through Kirchhoff's Current Law in Figure 6.13, the transformer winding current can be defined in (6.38). From there, its derivative relationship can be established in (6.39):

$$i_{L2}(t) = -i_{2dch1}(t) + i_{2dcl1}(t) = i_{2dch2}(t) - i_{2dcl2}(t) \tag{6.38}$$

$$\frac{di_{L2}(t)}{dt} = -\frac{di_{2dch1}(t)}{dt} + \frac{di_{2dcl1}(t)}{dt} = \frac{di_{2dch2}(t)}{dt} - \frac{di_{2dcl2}(t)}{dt} \tag{6.39}$$

Through Faraday's law for inductors for the transformer leakage inductance, the relationships in (6.40) and (6.41) are found:

$$\frac{v_{L2}(t)}{L_2} = \frac{1}{L_{dc} - L_M}(v_{2dcl1}(t) - v_{2dch1}(t)) \tag{6.40}$$

$$\frac{v_{L2}(t)}{L_2} = \frac{1}{L_{dc} - L_M}(v_{2dch2}(t) - v_{2dcl2}(t)) \tag{6.41}$$

Thus, (6.42) can be obtained from (6.40) and (6.41) and using Kirchhoff's Voltage Law in Figure 6.13, (6.43) can be found from (6.42):

$$v_{2dcl1}(t) - v_{2dch1}(t) = v_{2dch2}(t) - v_{2dcl2}(t) \tag{6.42}$$

$$v_{2dcl2}(t) = v_{2dch1}(t) \text{ and } v_{2dcl1}(t) = v_{2dch2}(t) \tag{6.43}$$

Then, Kirchhoff's Voltage Law can be applied again to solve for the voltage across each mutual inductor as shown in (6.44) and (6.45):

$$v_{2dch1}(t) = (v_{arm2}(t) - v_{T2}(t))\left(\frac{L_{dc} - L_M}{L_2 + L_{dc} - L_M}\right) + v_{2dch2}(t) \tag{6.44}$$

$$v_{2dch2}(t) = -(v_{arm2}(t) - v_{T2}(t))\left(\frac{L_{dc} - L_M}{L_2 + L_{dc} - L_M}\right) + v_{2dch1}(t) \tag{6.45}$$

Finally, the voltages v_{L2} and v_2 and can be defined in (6.46) and (6.47), respectively:

$$v_{L2}(t) = \left(\frac{L_2}{L_2 + L_{dc} - L_M}\right)(v_{T2}(t) - v_{arm2}(t)) \tag{6.46}$$

$$v_2(t) = \left(\frac{L_{dc} - L_M}{L_2 + L_{dc} - L_M}\right)v_{T2}(t) + \left(\frac{L_2}{L_2 + L_{dc} - L_M}\right)v_{arm2}(t) \tag{6.47}$$

It is clear in (6.47) that if the transformer leakage inductance is much larger than the mutual inductance, the voltage v_2 will be equivalent to the voltage v_{arm2}. As the mutual inductor value increases, the reflected voltage across the transformer from the other ports begin to have an increasing effect on the voltage v_2. The effects of inductor size on converter performance will be discussed after computing power flow equations. The same steps can be applied to port 3, yielding (6.48) and (6.49):

$$v_{L3}(t) = \left(\frac{L_3}{L_3 + L_{dc} - L_M}\right)(v_{T3}(t) - v_{arm3}(t)) \tag{6.48}$$

$$v_3(t) = \left(\frac{L_{dc} - L_M}{L_3 + L_{dc} - L_M}\right)v_{T3}(t) + \left(\frac{L_3}{L_3 + L_{dc} - L_M}\right)v_{arm3}(t) \tag{6.49}$$

6.2.2.2 Computation of port voltages v_2 and v_3 with respect to v_1, v_{arm2}, and v_{arm3}

The following section relates the port voltage and leakage inductor voltage of the single port solved above to the remaining ports within the converter. The simplified diagram of the 3-port system is shown in Figure 6.14. Note that v_1 is an equivalent square wave voltage source representing the LV DC source being switched by a full bridge.

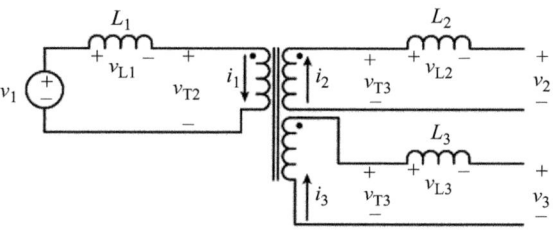

Figure 6.14   Circuit parameters of the 3-port converter

Faraday's Law of mutual induction yields (6.50):

$$\frac{v_{T1}(t)}{n_1} = \frac{v_{T2}(t)}{n_2} = \frac{v_{T3}(t)}{n_3} \tag{6.50}$$

Total MMF in a multi-winding transformer is the magnetic analog to Kirchhoff's current law as shown in (6.51). The derivative can be applied, resulting in (6.52):

$$n_1 i_1(t) = n_2 i_2(t) + n_3 i_3(t) \tag{6.51}$$

$$n_1 \frac{di_1(t)}{dt} = n_2 \frac{di_2(t)}{dt} + n_3 \frac{di_3(t)}{dt} \tag{6.52}$$

Now, applying Faraday's law for inductors to (6.52) results in (6.53):

$$n_1 \frac{v_1(t) - v_{T1}(t)}{L_1} = n_2 \frac{v_{T2}(t) - v_2(t)}{L_2} + n_3 \frac{v_{T3}(t) - v_3(t)}{L_3} \tag{6.53}$$

Next, solving (6.47) and (6.49) for v_{T2} and v_{T3}, respectively, and substituting into (6.53), it is possible to express the voltages v_2 and v_3 in terms of controllable voltage waveforms as shown in (6.54) and (6.55), respectively:

$$v_2(t) = K_{\beta 2} \frac{n_1}{L_1} v_1(t) + \frac{K_{\beta 2} n_2 + L_2}{L_{a2}} v_{arm2}(t) + K_{\beta 2} \frac{n_3}{L_{a3}} v_{arm3}(t) \tag{6.54}$$

$$v_3(t) = K_{\beta 3} \frac{n_1}{L_1} v_1(t) + K_{\beta 3} \frac{n_2}{L_{a2}} v_{arm2}(t) + \frac{K_{\beta 3} n_3 + L_3}{L_{a3}} v_{arm3}(t) \tag{6.55}$$

Likewise, the voltages across each transformer leakage inductance can be computed and are given in (6.56)–(6.58). From Faraday's law for inductors, this implies the derivative of the port currents $\frac{di_1}{dt}$, $\frac{di_2}{dt}$, and $\frac{di_3}{dt}$ are also known. The constant values expressed in (6.54)–(6.58) are listed below:

$$v_{L1}(t) = \frac{L_1 - Kn_1^2}{L_1} v_1(t) - \frac{Kn_1 n_2}{L_{a2}} v_{arm2}(t) - \frac{Kn_1 n_3}{L_{a3}} v_{arm3}(t) \tag{6.56}$$

$$v_{L2}(t) = \frac{L_2}{L_{a2}} \left(\frac{Kn_1 n_2}{L_1} v_1(t) + \frac{Kn_2^2 - L_{a2}}{L_{a2}} v_{arm2}(t) + \frac{Kn_2 n_3}{L_{a3}} v_{arm3}(t) \right) \tag{6.57}$$

$$v_{L3}(t) = \frac{L_3}{L_{a3}} \left(\frac{Kn_1 n_3}{L_1} v_1(t) + \frac{Kn_2 n_3}{L_{a2}} v_{arm2}(t) + \frac{Kn_3^2 - L_{a3}}{L_{a3}} v_{arm3}(t) \right) \tag{6.58}$$

$$L_{a2} = L_2 + L_{dc} - L_M$$

$$L_{a3} = L_3 + L_{dc} - L_M$$

$$K_{\beta 2} = Kn_2 \left(\frac{L_{dc} - L_M}{L_{a2}} \right)$$

$$K_{\beta 3} = Kn_3 \left(\frac{L_{dc} - L_M}{L_{a3}} \right)$$

$$K = \frac{L_1 L_{a2} L_{a3}}{n_1^2 L_{a2} L_{a3} + n_2^2 L_1 L_{a3} + n_3^2 L_1 L_{a2}}$$

6.2.3 *Visualizing multiport, bidirectional DC/DC converter power modes*

With all switching duty cycles equal to $D = 0.5$, the converter will have six distinct switched states within one full switching cycle. To calculate average power flow, the state values will need to be computed before and after the duration of each switched state using the formulas in (6.59). Since the capacitor time constants are designed to be much longer than the duration of the switching cycle, the voltage values can be approximated as remaining constant until the end of each switched state. With unchanging voltage values, the slopes of inductor currents remain constant throughout each switched state. Beginning with zero initial conditions, the final state values can be computed for all six switched states through the entire switching cycle. Last, under steady-state conditions, the average current waveforms within a DAB converter are equal to zero. The steady-state values can be found by taking the initial current values at the fourth step, dividing by two, and using the negative of those values as the initial conditions. With updated initial conditions, the state values for current can be re-solved for each switched state, and from there the average power flow can be computed using the instantaneous power flow equations in (6.60) and (6.61):

$$i_1(t) = \int \frac{v_{L1}(t)}{L_1} dt; \ i_2 = \int \frac{v_{L2}(t)}{L_2} dt; \ i_3 = \int \frac{v_{L3}(t)}{L_3} dt \tag{6.59}$$

$$P_1(t) = v_1(t)i_1(t); \ P_2(t) = v_2(t)i_2(t); \ P_3(t) = v_3(t)i_3(t) \tag{6.60}$$

$$P_1(t) = P_2(t) + P_3(t) \tag{6.61}$$

The average power flow equations were computed using MATLAB's symbolic toolbox, and the equations for the mode when $0 < \phi_2 < \phi_3$ are given in (6.62)–(6.64). Since the time step T_s is small, higher order terms (second order and above) can be neglected. In addition, the higher order phase shift terms (third order and above) are also neglected. Performing this reduction yields the following approximations for power flow:

$$0 < \phi_2 < \phi_3$$

$$P_{1avg} \approx L_1^{-1} L_a^{-1} T_s n_1 V_1 \left[L_2 L_{3a} n_2 V_{Carm2-0} \left(-2\phi_2^2 + \phi_2 \right) \right.$$
$$\left. + L_{2a} L_3 n_3 V_{Carm3-0} \left(-2\phi_3^2 + \phi_3 \right) \right] \tag{6.62}$$

$$P_{2avg} \approx L_a^{-1} T_s n_2 V_{Carm2-0} \left[-2(L_{3a} n_1 V_1 - L_3 n_3 V_{Carm3-0}) \phi_2^2 \right.$$
$$+ (L_{3a} n_1 V_1 + L_3 n_3 V_{Carm3-0}) \phi_2 + L_3 n_3 V_{Carm3-0} \left(2\phi_3^2 - \phi_3 \right)$$
$$\left. - 4 L_3 n_3 V_{Carm3-0} \phi_2 \phi_3 \right] \tag{6.63}$$

$$P_{3avg} \approx L_a^{-1} T_s n_3 V_{Carm3-0} \left[L_2 n_2 V_{Carm2-0} \left(-2\phi_2^2 - \phi_2 \right) \right.$$
$$\left. + (L_{2a} n_1 V_1 + L_2 n_2 v_{Carm2-0}) \left(-2\phi_3^2 + \phi_3 \right) + 4 L_2 n_2 V_{Carm2-0} \phi_2 \phi_3 \right] \tag{6.64}$$

The average power flow formulas were plotted for the full range of ϕ_2 and ϕ_3. The power output for port 2 is shown in Figure 6.15 and the power output for port 3 is shown in Figure 6.16. Each mode is distinguished by a unique color in the 3D plot.

A PLECS simulation was run to confirm the accuracy of the voltage and average power flow equations. The components were selected from a previous experiment of a current-fed DAB converter [15]. Table 6.3 shows the circuit parameters used in the simulation.

Recall that each arm was treated as a single submodule. Therefore, each arm voltage either has its capacitor switched in series (positive voltage) or the capacitor bypassed (zero volts), which results in a square wave. Note that for balanced

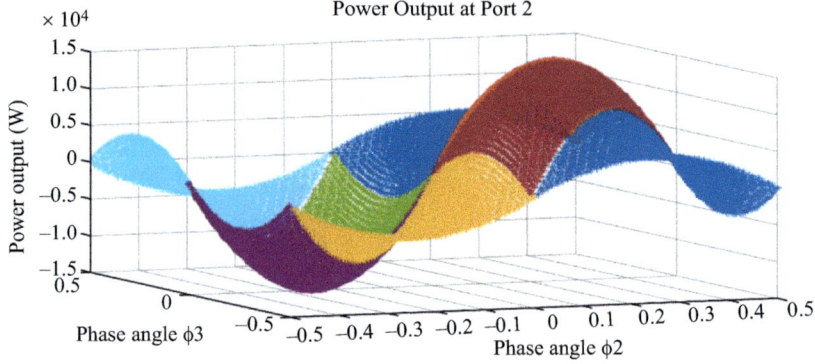

Figure 6.15   Power output of port 2. The surface shows the power output under all possible phase angle combinations assuming a constant duty cycle of D = 0.5 for all switching waveforms

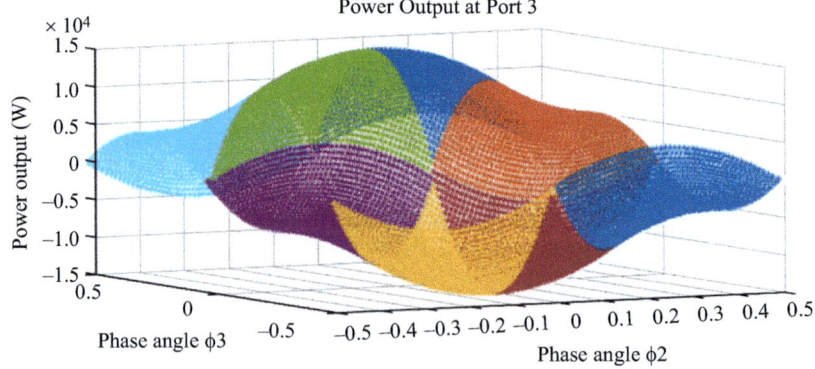

Figure 6.16   Power output of port 3. The surface shows the power output under all possible phase angle combinations assuming a constant duty cycle of D = 0.5 for all switching waveforms

Table 6.3   Circuit parameters

Parameter	Variable	Value	Parameter	Variable	Value
Primary turns ratio	n_1	1	Primary duty cycle	D_1	0.5
Secondary turns ratio	n_2	1	Secondary duty cycle	D_2	0.525
Tertiary turns ratio	n_3	1	Tertiary duty cycle	D_3	0.525
Primary winding inductance	L_1	20 µH	Secondary switching delay (p.u.)	ϕ_2	0.10
Secondary winding inductance	L_2	20 µH	Tertiary switching delay (p.u.)	ϕ_3	0.15
Tertiary winding inductance	L_3	20 µH	Primary DC voltage	V_{1DC}	500 V
Coupled inductor inductance	L_{dc}	100 µH	Secondary DC voltage	V_{2DC}	525 V
Mutual inductance	L_M	80 µH	Tertiary DC voltage	V_{3DC}	525 V
Switching frequency	f_s	40 kHz	Secondary arm voltage	V_{arm2}	500 V
Switching period	T_s	25 µs	Tertiary arm voltage	V_{arm3}	500 V

operation, arm$_{2\text{-}1}$ and arm$_{2\text{-}4}$ switch as a pair; likewise, arm$_{2\text{-}2}$ and arm$_{2\text{-}3}$ switch as a pair. Also, as the duty cycle of the arms extend beyond 0.5, the effective duty cycle of the voltage difference waveform v_{arm2} is equal to $1-D$. To illustrate this example, a secondary and tertiary duty cycle for each arm was selected as $D = 0.525$. The switching waveforms v_{arm2} and v_{arm3} have an effective duty cycle of $D = 0.475$, as was shown originally in Figure 6.12.

An 80% mutual coupling coefficient M was selected to determine the mutual inductance L_M. For simplicity, the turns ratio of the transformer is kept at 1:1:1, despite the desired application of LV – MV - MV 3-port converter. Note that to maintain stable operation at an arbitrary current fed port x, the arm voltage must follow the relationship in (6.65)

$$V_{xDC} = 2D_x V_{armx} \tag{6.65}$$

6.2.4   Mutual inductor effects on converter performance

After verifying the performance of the converter, the effects of the mutual inductor on converter performance were explored by varying the mutual inductor value but holding the transformer leakage inductance constant. In particular, the changing mutual inductance had two main effects. First, the port voltage waveform, typically a square waveform, became distorted as the mutual inductance value approached the magnitude of the transformer leakage inductance (see Figure 6.17). Second, the average power output decreased as the mutual inductance value approached the magnitude of the transformer leakage inductance as shown in Figure 6.18. The ratio of mutual inductor value to transformer leakage inductance was used as the independent variable in the graphs. It can be concluded that the mutual inductors cannot exceed the transformer leakage inductance without significant loss of power flow capability.

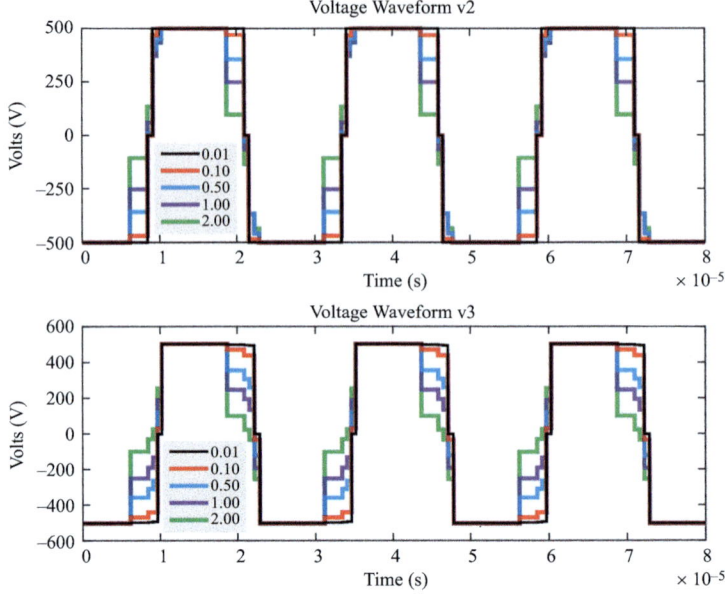

Figure 6.17   Current-fed port voltage waveforms (v_x at port x) with inductance ratio $(L_{dc} - L_M) / L_x$ between 0.01 and 2

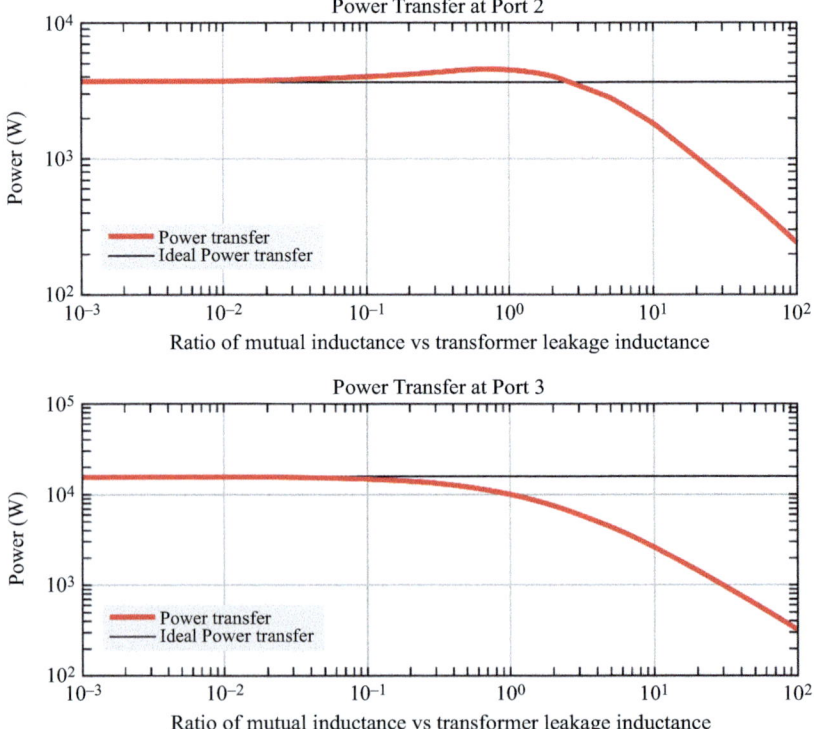

Figure 6.18   Average power transfer at each current-fed port (P_{xavg} at port x) vs the inductance ratio $(L_{dc} - L_M) / L_x$

Further Reading

Beyond some of the key references, the authors paraphrased or extracted information from some of their past work to build this chapter. The first half of the chapter was written as a tutorial for designing a model reference controller design. Research books are great for this type of chapter because information is collected from various sources and put in one central location. For further material and sources that form the foundation of the first half of chapter, readers are encouraged to consult the following dissertation, [16], and reference article [17]. For visualizing the various modes of operation of a multiport, dual active bridge design, readers are encouraged to consult the following thesis, [18], and reference articles, [19–20]. Others can be found by the chapter authors.

References

[1] A. Emadi, A. Khaligh, C. H. Rivetta, and G. A. Williamson, "Constant power loads and negative impedance instability in automotive systems: definition, modeling, stability, and control of power electronic converters and motor drives," *IEEE Trans. Veh. Technol.*, vol. 55, no. 4, pp. 1112–1125, Jul. 2006, doi: 10.1109/TVT.2006.877483.

[2] A. Kwasinski and C. N. Onwuchekwa, "Dynamic behavior and stabilization of DC microgrids with instantaneous constant-power loads," *IEEE Trans. Power Electron.*, vol. 26, no. 3, pp. 822–834, Mar. 2011, doi: 10.1109/TPEL.2010.2091285.

[3] C. N. Onwuchekwa and A. Kwasinski, "Analysis of boundary control for buck converters with instantaneous constant-power loads," *IEEE Trans. Power Electron.*, vol. 25, no. 8, pp. 2018–2032, Aug. 2010, doi: 10.1109/TPEL.2010.2045658.

[4] H. K. Krishnamurthy and R. Ayyanar, "Building block converter module for universal (AC-DC, DC-AC, DC-DC) fully modular power conversion architecture," in *2007 IEEE Power Electronics Specialists Conference*, 2007, pp. 483–489, doi: 10.1109/PESC.2007.4342035.

[5] International Electrotechnical Commission, "Power cables with extruded insulation and their accessories for rated voltages from 1 kV (Um = 1,2 kV) up to 30 kV (Um = 36 kV) - Part 2: Cables for rated voltages from 6 kV (Um = 7,2 kV) up to 30 kV (Um = 36 kV)," 2014.

[6] A. Emadi, B. Fahimi, and M. Ehsani, "On the concept of negative impedance instability in the more electric aircraft power systems with constant power loads," *SAE Trans.*, pp. 689–699, Aug. 1999. doi: 10.4271/1999-01-2545.

[7] J. Glover, M. Sarma, and T. Overbye, *Power System Analysis and Design*, 4th ed. New York: Thomas Learning, 2008.

[8] S. Sastry and M. Bodson, *Adaptive Control: Stability, Convergence and Robustness*. New York: Dover Publications, 2011.

[9] C.-T. Chen, *Linear System Theory and Design*, 4th ed. Oxford, UK: Oxford University Press, 2013.

[10]   S. P. Engel, M. Stieneker, N. Soltau, S. Rabiee, H. Stagge, and R. W. De Doncker, "Comparison of the modular multilevel DC converter and the dual-active bridge converter for power conversion in HVDC and MVDC grids," *IEEE Trans. Power Electron.*, vol. 30, no. 1, pp. 124–137, 2015, doi: 10.1109/TPEL.2014.2310656.

[11]   F. Mura and R. W. De Doncker, "Design aspects of a medium-voltage direct current (MVDC) grid for a university campus," *8th Int. Conf. Power Electron. - ECCE Asia "Green World with Power Electron. ICPE 2011-ECCE Asia*, no. Mvdc, pp. 2359–2366, 2011, doi: 10.1109/ICPE.2011.5944508.

[12]   A. Q. Huang and R. Burgos, "Review of solid-state transformer technologies and their application in power distribution systems," *IEEE J. Emerg. Sel. Top. Power Electron.*, vol. 1, no. 3, pp. 186–198, 2013, doi: 10.1109/JESTPE.2013.2277917.

[13]   M. Michon, J. L. Duarte, M. Hendrix, and M. G. Simões, "A three-port bi-directional converter for hybrid fuel cell systems," *PESC Rec. - IEEE Annual Power Electronics Specialists Conference*, vol. 6, pp. 4736–4742, 2004, doi: 10.1109/PESC.2004.1354836.

[14]   Y. Shi, R. Li, Y. Xue, and H. Li, "Optimized operation of current-fed dual active bridge DC-DC converter for PV applications," *IEEE Trans. Ind. Electron.*, vol. 62, no. 11, pp. 6986–6995, 2015, doi: 10.1109/TIE.2015.2432093.

[15]   Y. Shi, R. Mo, H. Li, and Z. Pan, "A novel ISOP current-Fed modular dual-active-bridge (CF-MDAB) DC-DC converter with DC fault ride-through capability for MVDC application," *2017 IEEE Energy Conversion Congress & Expo (ECCE)*, vol. 2017-January, pp. 4525–4530, 2017, doi: 10.1109/ECCE.2017.8096776.

[16]   B. Grainger, Design and Power Management of an Offshore Medium Voltage DC Microgrid Realized Through High Voltage Power Electronics Technologies and Control. Doctoral Dissertation, University of Pittsburgh, 2014.

[17]   B. M. Grainger, G. F. Reed and Z. Mao, "Model reference controller design for stabilizing constant power loads in an offshore medium voltage DC micro-grid," *2015 IEEE 16th Workshop on Control and Modeling for Power Electronics (COMPEL)*, 2015, pp. 1–8, doi: 10.1109/COMPEL.2015.7236495.

[18]   Z. Smith, Zachary, Analytical Treatment of the Power Transfer Relationships for a Coupled, Current-Fed, Multi-Port Dual Active Bridge Converter. Master's Thesis, University of Pittsburgh, 2018.

[19]   Z. T. Smith and B. M. Grainger, "Analytical treatment of the power transfer relationships for a coupled, current-fed, multi-port dual active bridge con-verter," 2019 IEEE Electric Ship Technologies Symposium (ESTS), 2019, pp. 562–568, doi: 10.1109/ESTS.2019.8847816.

[20]   Z. T. Smith and B. M. Grainger, "Medium Voltage Ring-Bus Grid Design Employing Current-Fed, Three-Port Dual Active Bridge Converters with Average Power Flow Control," 2021 IEEE Electric Ship Technologies Symposium (ESTS), 2021, pp. 1–7, doi: 10.1109/ESTS49166.2021.9512328.

Chapter 7

Medium frequency and medium voltage transformer technology for DC–DC converter applications

Richard B. Beddingfield[1] and Paul R. Ohodnicki Jr.[2]

To describe performance and operation of medium frequency and medium voltage inductors and transformers in DC–DC converter applications, several fundamental concepts must first be defined and understood. Numerous textbooks, for example, [1,2], have been published on this topic and so this review chapter will emphasize the relevant topics and physics from an intuitive perspective while referencing detailed derivations and theoretical descriptions described elsewhere.

7.1 Magnetic domain analog to electric circuit

Because the operational principle of an ideal transformer requires complete coupling of magnetic flux between two or more independent electrical windings, a common starting point for the analysis lies in description of magnetic flux throughout the transformer and surrounding space. As with all electromagnetic components, the foundational behavior is based upon Maxwell's equations considering presence of polarizable media. Specifically, Ampere's law and Gauss's law provide the basis for magnetomotive force (MMF) and flux law, respectively, in Rowland's magnetic equivalent circuit (MEC) analog to corresponding Kirchhoff's voltage and current laws of electrical circuit analysis (Table 7.1) [3]. Treating magnetic domain physics with familiar electric circuit models can be introduced and applied for analysis.

Although these laws could be applied in a continuum sense, practical transformer design is typically employed through development of a network of lumped elements which describe the relationship between the MMF drop across and the flux flowing through the surfaces which bound a volume element. The proportionality between the two are described in terms of the properties of the corresponding media, much like conductivity in electrical circuits. Therefore, a magnetic

[1]Department of Electrical and Computer Engineering, North Carolina State University, Raleigh, NC, USA
[2]Department of Mechanical Engineering and Materials Science, University of Pittsburgh, Pittsburgh, PA, USA

Table 7.1   Comparison of magnetic and electric governing equations

Magnetic domain (Rowland)		Electrical domain (Kirchhoff)	
Name	*Symbol*	*Name*	*Symbol*
Magnetomotive force	$\mathcal{F} = \oint \vec{H} \cdot \vec{dl}$	Electromotive force	$\varepsilon = \int E \cdot dl$
Magnetic field	$H = I_{enclosed} = NI$	Electric field	$\varepsilon = -\frac{d\phi}{dt}$
Magnetic flux	$\Phi_{surface} = \iint\limits_S \vec{B} \cdot \vec{dS}$	Electric current	$I = \iint\limits_S \vec{J} \cdot \vec{dS}$
Reluctance	$\mathcal{R} \equiv \frac{\mathcal{F}}{\Phi_{Surface}} = \frac{Length}{\mu Area}$	Resistance	$R \equiv \frac{\varepsilon}{I} = \frac{Length}{\sigma Area}$

reluctance is analogous to an electric resistance and the inverse, magnetic permeance is analogous to a conductance. To avoid confusion with power, permeance is denoted as $\widehat{P}$.

By describing the corresponding transformer in terms of a number of discrete volumes separated by magnetic nodes (assuming zero MMF drop at the nodes) and represented as lumped reluctance or permeance elements, an analytically tractable description of the relationship between the current applied to the winding and the flux distribution throughout the transformer can be realized. Examples of two different derived MECs are presented in Figure 7.1 for a single-winding inductor and a two-winding transformer, the first neglecting fringing flux at gaps and flux 'leakage' from the magnetic core and the second accounting for these additional details. The importance of fringing and leakage flux in both inductors and transformers derives from relatively weak permeability contrast between high performing magnetic cores and free space (maximum ~10^6) in stark contrast to the electrical conductivity contrast between electrical conductors and free space (~10^{24}) thereby allowing for neglect of 'leakage currents' in analysis of electrical equivalent circuits. An additional subtle difference between magnetic and electrical equivalent circuits is inherent nonlinearity of B–H characteristics for a magnetic material resulting in a corresponding magnetic flux dependent reluctance or permeance in contrast with typical electrical conductors for which E–I characteristics are highly linear. These non-linearities result in MECs which are more computationally challenging to solve. Accurate modelling of all effects may require computationally intensive 3D finite element analysis with variable material properties. However, provided operating conditions sufficiently far from saturation, a reasonable degree of accurate modelling can be achieved with simple MECs as illustrated in Figure 7.1(b) and (f). Further accuracy can be achieved by modelling additional magnetic flux paths such as fringing and leakage as in Figure 7.1(c) and (g).

As discussed above, leakage inductance is a phenomenon where some of the flux, leakage flux, is generated by a coil that flows through the air space around a component instead of within the core material. In the MEC of Figure 7.1(c) and (g),

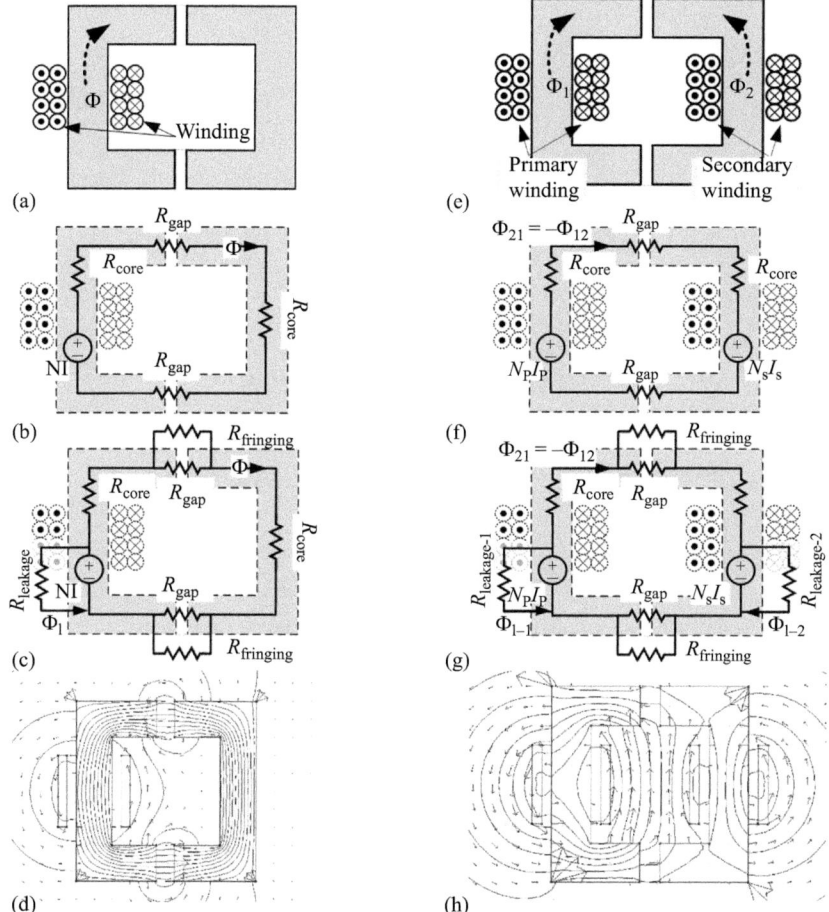

Figure 7.1   *(a) Example single-winding transformer schematic with a gapped magnetic core. (b) Simple MEC neglecting leakage or gap fringing flux. (c) More realistic MEC also accounting for leakage and gap fringing flux. (d) 2D FEA based model of inductor showing flux lines in regions approximated by MEC. (e)–(h) Corresponding two winding transformer schematic and MECs with 2D FEA highlighting leakage flux paths*

this is the reluctance path that is parallel to a MMF source. The MEC model allows one to easily think of the potential leakage flux paths as alternate magnetic flux flows when two MMFs supply opposing flux flows in the core. This line of thinking also provides intuition for the derivation of the ideal transformer. That is, a transformer that requires no magnetizing current, and no magnetic flux in the core, due to perfect coupling between windings.

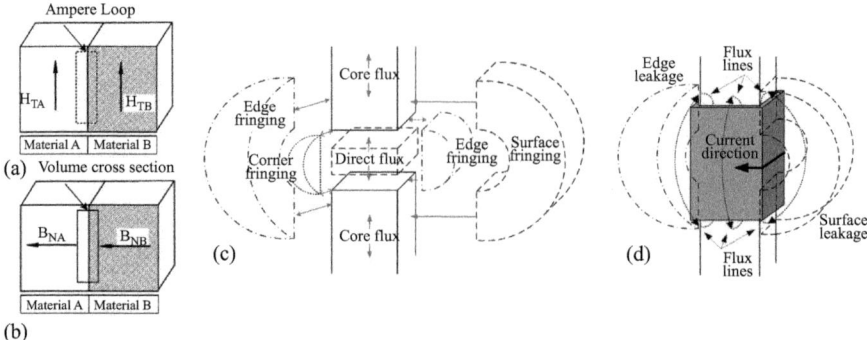

Figure 7.2   Maxwell's boundary conditions for H and B based upon (a) Ampere's law, (b) Gauss's law as well as example flux paths for which flux emanates normal from the magnetic core surface including, (c) fringing flux at a gap, and (d) leakage flux around a winding

7.1.1 Magnetic flux behavior at material interfaces

Analytical calculations of flux that flows within leakage and fringing paths of a transformer can be performed rigorously through applied electromagnetic techniques, however, reasonable approximations can often be derived through a simpler and more intuitive approach guided by the boundary conditions on flux which penetrates across media of differing permeabilities, Figure 7.2(a) and (b). Through the application of Ampere's law and Gauss's law across an interface between a high permeability material and free space, it can be determined that the magnetic flux emanating from a magnetic core is expected to cross the boundary approximately normal to the interface which places constraints on the flux distribution within the fringing and leakage flux paths, Figure 7.2(c) and (d), respectively. Although this approximation is based on simplifying assumptions not strictly valid in all cases (e.g., dissimilar media with magnetic relative permeability that deviate by less than approximately 1–2 orders of magnitude [4]), the conclusions provide a useful guide to interpreting the flow of flux within a transformer or other magnetic component for a given winding and magnetic core configuration. Intuitive approaches to analysis of flux paths within a transformer for a given core and winding configuration can also allow for improved understanding of the dependence of geometrical aspects on the total amount of leakage flux for a given design [5]. For example, Ampere's law with Maxell's boundary conditions at the interface between magnetic core and adjacent conductor, Figure 7.2(d), highlights that magnetic flux produced by the coil is orthogonal to current direction and tangential to core surface.

7.2   Magnetic coupling in multi-winding components

In the case of magnetic components with two or more windings, the fundamental working principle relies on coupled flux between windings which result in an induced electromotive force of the secondary winding upon a time-varying excitation of the primary winding. The coupling of magnetic flux between windings and associated electromotive force is described in terms of a concept known as flux linkage, λ, (7.1), which describes the number of turns, N, of a winding and the total flux flowing through the winding, Φ, and which is also directly related to the concept of inductance, (7.2). The time variation in flux linkage through a winding yields an induced electromotive force (voltage, V) across the winding according to Faraday's law of induction, (7.3). Subscripts i and j are the winding number.

Flux linkage		Inductance		Induced electromotive force	
$\lambda_i \equiv N_i \varnothing_i$	(7.1)	$L_{ij} \equiv \dfrac{\lambda_i}{i_j}$	(7.2)	$V_i = \dfrac{d\lambda_i}{dt} = N_i \dfrac{d\varnothing_i}{dt} = L_{ij} \dfrac{di_j}{dt}$	(7.3)

In cases where (7.2) is referring to flux linkage and excitation current of the same winding, it is a self-inductance, and it is referred to as a mutual inductance in cases where flux linkage and excitation current correspond to two different windings. For an ideal transformer, all flux generated by one coil translates to flux linkage in the other, and hence only a mutual inductance is associated with the primary and secondary windings. In such cases, the voltage ratio between the secondary and primary for an open-circuit secondary with a primary winding excitation can be shown to be simply the turns ratio ($n = N_2/N_1$).

Finite permeability values that can be obtained in practice even in the highest performing magnetic materials result in finite leakage flux and deviation from the ideal transformer approximation. Incomplete coupling between flux flowing through multiple windings and associated deviations from ideal transformer characteristics can be described in part through the concept of an apparent turn ratio. This concept accounts for leakage of magnetic flux from the primary coupling (or magnetizing) path resulting in additional self-leakage inductance with the transformer mutual inductance, and it can be expressed in terms of the ratio of flux flowing through leakage paths, as compared to flux coupling two windings through the magnetizing path. Schematic illustrations of the leakage and magnetizing flux for a two-winding transformer under different excitation conditions are illustrated below in Figure 7.3(a)–(c), where Φ_{L1} and Φ_{21} correspond to the leakage flux and mutual flux, respectively when the primary is excited while Φ_{L2} and Φ_{12} correspond to secondary winding excitation. Figure 7.3(d) illustrates the resulting electrical equivalent circuit.

In case of an ideal transformer, all generated flux on primary and secondary windings are mutually coupled ($\phi_{11} = \phi_{12} = 0$ under all excitation conditions) and hence the coupling coefficient, k, is defined to be equal to 1. In the more general case, the coupling coefficient ranges between a value of 0 and a maximum value of 1 and is defined in (7.4). The coupling coefficient may also be described in terms of the coil self-inductance, L_i, the magnetizing, L_{magi}, and leakage inductances, L_{Li}, (7.5). This assumes that the primary side referred secondary leakage inductance,

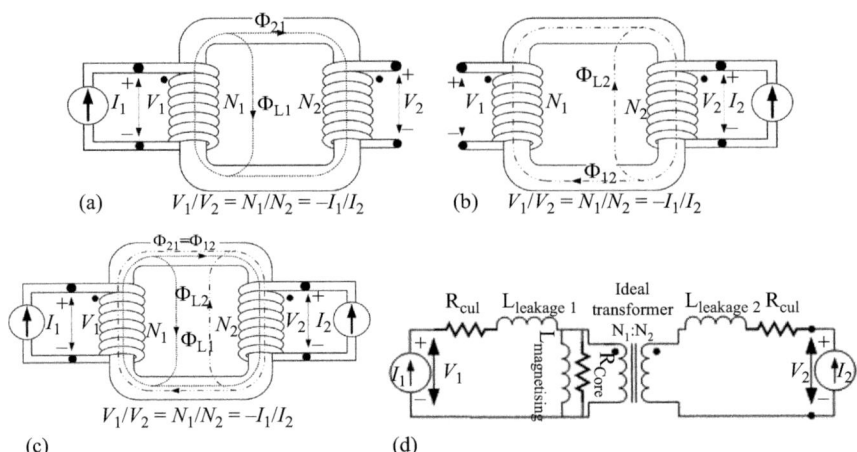

Figure 7.3   *Schematic flux distributions of leakage and magnetizing flux for (a) primary winding excitation and secondary winding open, (b) secondary winding excitation and primary winding open, and (c) both primary and secondary winding excitation, and (d) resulting, electric equivalent model, 'T Model.'*

L'_{L2}, is the same value as the primary leakage inductance, L_{L1}, and that the measured primary side inductance when the secondary is an open circuit is the sum of the primary leakage and magnetizing inductances.

$$k \equiv \sqrt{k_1 k_2}, \; k_1 \equiv \frac{\varnothing_{21}}{\varnothing_{l1} + \varnothing_{21}}, \; k_2 \equiv \frac{\varnothing_{12}}{\varnothing_{l2} + \varnothing_{12}}; \; \text{Coupling coefficient in terms of flux}$$

(7.4)

$$L_{mag} \equiv kL_1; L_{L1} \equiv (1-k)L_1 = L'_{L2}$$

$$\therefore k = \frac{L_{mag1}}{L_m + L_{L1}}; \; \text{Coupling coefficient in terms of primary inductances}$$

(7.5)

   Although the voltage ratio (and current ratio) between transformer windings can simply be described in terms of the nominal turns ratio for an ideal transformer (e.g., $k=1$), the load impedance, Z_{load}, and frequency, f, dependent voltage drop across the effective inductances associated with leakage flux must be accounted for with imperfectly coupled windings. If the conduction and magnetization losses are neglected, this leads to an effective, lossless, voltage gain, G_v, at the output of the 'T' model transformer of (7.6). The apparent turns ratio, which represents the open circuit voltage measured on the secondary for an applied voltage excitation on the primary, is simply the ideal turns ratio reduced by the coupling factor (7.7). The apparent turns ratio incorporates the voltage drop due to uncoupled flux which is represented electrically by the leakage inductance. One could include the voltage gain reduction due to transformer losses; however, they will have minimal impact in a well-designed transformer:

$$G_{v-lossless} = \frac{k}{\frac{2\pi f \sqrt{L_{L1}L_{L2}}}{Z_{load}} + \frac{N_1}{N_2}}; \text{ Lossless voltage gain of imperfectly coupled transformer}$$

(7.6)

$$n_{App} \equiv \lim_{Z_{load} \to \infty} G_{v-lossles} = k\frac{N_2}{N_1} = kn_{ideal}; \text{ Apparent turns ratio} \quad (7.7)$$

As can be seen based upon the definition of the coupling coefficient in (7.7), the greater the leakage of flux on the primary and secondary windings (hence $k <$ 1), the larger the deviation in the apparent turn ratio from nominal. It is pointed out that in some designs, a significant leakage inductance may indeed be desirable to provide an effective inductance to the circuit without the need for additional discrete inductors. As just one example, a minimum level of inductance is required for the application of medium frequency transformers in dual-active bridge converters, to ensure that the power flow can be accurately controlled and to allow for high efficiencies through soft-switching of the semiconductor devices [6]. In the case of resonant converters as in wireless power applications, transformers may exhibit exceedingly high values of leakage inductance and relatively low coupling coefficients.

7.3   Realization of non-ideal magnetic components

Another non-ideality of transformers arises due to the local storage of electrical charge and the associated electric fields which result from the potential differences that exist throughout the transformer. Just as the leakage flux describes stray magnetic flux that does not follow the magnetizing path, the stray electric flux arising from local electrical charges can be described in terms of a parasitic capacitance. The calculation of equivalent capacitance values for a given transformer design is a non-trivial task and it is highly dependent upon the constituent insulation materials and the specific winding configuration as well as the core and winding geometry. An example schematic of the equivalent turn-to-turn capacitance for a coil of wire is illustrated in Figure 7.4(a) along with the equivalent winding resistance and inductance. Figure 7.4(b) shows the lumped electrical equivalent model with Figure 7.4(c) demonstrating how an inductor or an inductive coil of wire ultimately will exhibit capacitive impedance at sufficiently high frequency due to the parasitic capacitance with resonance occurring at $f_r \approx 1/2\pi\sqrt{L_{Inductor}C_{Parasitic}}$, assuming negligible winding resistance (R_s). In a wound core component such as a transformer, several contributions to equivalent capacitance must be considered including (i) turn-to-turn, (ii) turn-to-core, (iii) layer-to-layer, and (iv) interwinding capacitance contributions, Figure 7.5. The calculations of parasitic capacitance can be carried out using several techniques which are analogous to those used in the calculation of leakage inductance as a

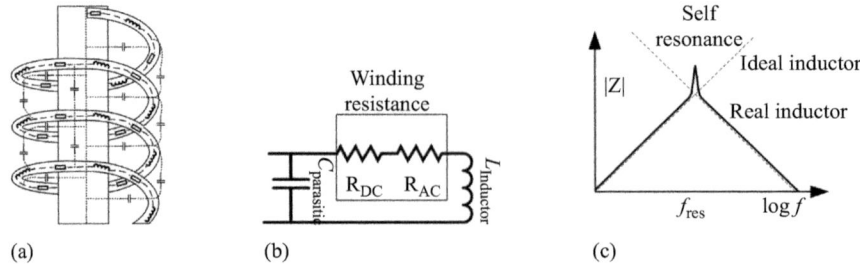

(a)　　　　　(b)　　　　　(c)

*Figure 7.4   (a) Graphical representation of the distributed resistance, inductance,
and capacitance of a coil of conductor. (b) Equivalent circuit for the
wound coil of (a). (c) Resonant frequency (f_r) and effective impedance
of a coil winding or inductor with a negligible winding resistance
showing inductive characteristics at low frequencies and capacitive
characteristics at high frequencies*

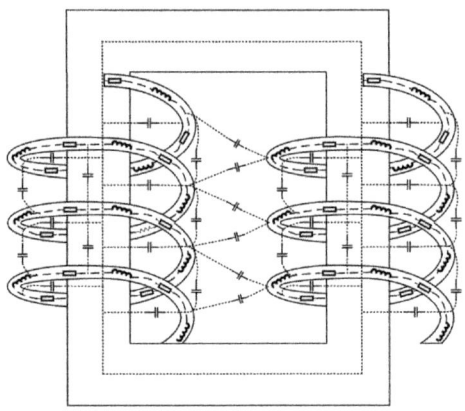

*Figure 7.5   Extension of distributed elements for transformers including
interwinding capacitance elements*

result of duality of the permeance and capacitance expressions if the permeability is
replaced by the permittivity and the path length is replaced by the distance between
two charged surfaces ($C = \epsilon A/d$). In addition to non-linear, undesired resonant
characteristics, parasitic capacitances associated with a transformer design can be
sources for transient, common or differential mode, currents arising from stray
electric fields and local stored charge. These currents are proportional to capacitance, and the voltage and frequency Fourier composition, $i_f \approx C f V_f$. This can
result in safety hazards, electromagnetic interference, and/or failures of switching
devices [7]. These capacitances also provide high frequency electrical pathways
between high voltage power circuitry and low voltage controllers. This is especially
true with fast switching wide bandgap devices which can produce voltage transient
edges of 2–80× of traditional Si devices [8].

7.3.1 Medium frequency resistance

Existing commercial line frequency (50–60 Hz) transformers for medium voltage applications typically show excellent efficiencies of >98%–99%, but higher frequency switching yields additional loss considerations which must be accounted for in the case of medium frequency applications. Overall transformer losses are comprised predominantly of winding and core losses, with typical heuristic designs targeting approximately equal partitioning between the two contributions for conventional transformers. In the case of medium frequency applications, a more rigorous loss analysis is required which includes considerations for (i) core material, (ii) winding material and gauge, (iii) excitation waveform (including frequency, amplitude, and shape), and (iv) transformer core and winding geometry. For line frequency applications the winding losses can be estimated by DC resistance values and depend upon the total cross-sectional area and length of the winding as well as the root-mean squared current according to $P_{Conduction-DC} = I_{RMS}^2 R_{DC}$. For higher switching frequencies, additional AC winding loss contributions resulting from induced eddy currents must be accounted for including non-uniform current distributions within the conductor as well as proximity losses which derive from stray magnetic fields impinging upon the winding. The approximate characteristic winding dimension or alternatively frequency, f, at which AC winding losses become relevant are dictated by the skin depth, δ, of the winding according to (7.8), where ρ_W is the DC resistivity, μ is the magnetic permeability, and ε is the permittivity of the conductor. Figure 7.6 shows current concentrating to the outside of a conductor:

$$\delta = \sqrt{\frac{2\rho}{\omega\mu}}\sqrt{\left(\sqrt{1 + (\rho\omega\epsilon)^2} + \rho\omega\epsilon\right)} \approx \sqrt{\frac{\rho_W}{\pi\mu_0 f}}; \text{Effective skin depth}$$

$$(7.8)$$

$$R_{AC} = R_{DC}\left(Re\left(h\sqrt{\frac{j\omega\mu_0 N_l t}{\rho b}}\coth\left(h\sqrt{\frac{j\omega\mu_0 N_l t}{\rho b}}\right)\right)\right.$$
$$\left. + \frac{(m^2 - 1)}{3}Re\left(2h\sqrt{\frac{j\omega\mu_0 N_l t}{\rho b}}\tanh\left(\frac{h}{2}\sqrt{\frac{j\omega\mu_0 N_l t}{\rho b}}\right)\right)\right) \quad (7.9)$$

At frequencies high enough to yield an effective skin depth on the order of the conductor characteristic dimension, a non-uniform current density that is enhanced at the conductor surface results from eddy currents to yield a reduced effective cross-sectional area. In addition, multi-layer or multi-turn windings will have an

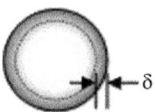

Figure 7.6   Diagram of skin effect in round conductors

additional contribution to AC losses due to proximity effects resulting from the magnetic fields produced by, m, adjacent conductor layers (7.9) [9]. This assumes layers of height, h, that are stacked N_l turns high of thickness, t, in a winding window width of b, Figure 7.7(a). Details of conductor diameter and even the

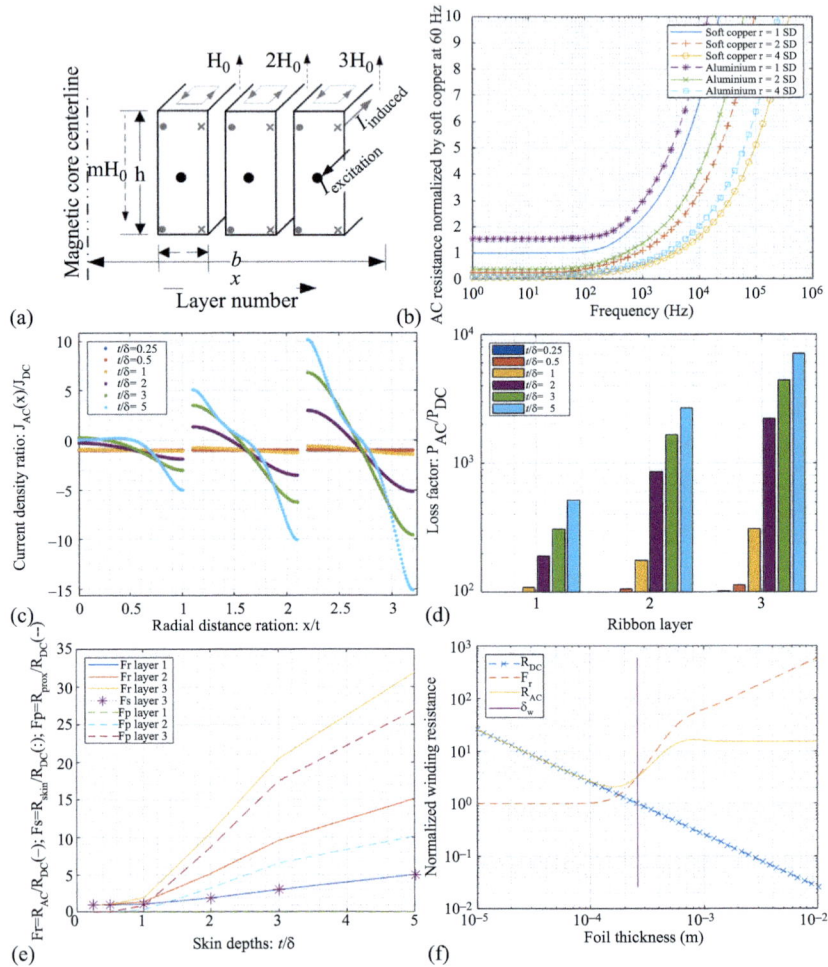

(a)

(b)

(c)

(d)

(e)

(f)

Figure 7.7   *(a) Diagram of Dowel's equation for proximity effect, (b) variation of winding resistance due to skin effect for different materials and thicknesses, (c) distribution of coil density in foil conductors, (d) Ratio of AC to DC conduction loss for foil of various skin depth thicknesses, (e) contribution to total (F_R) conduction loss from skin (F_S), proximity (F_p), for thicknesses normalized by skin depth, and (f) optimizing winding thickness at f=100 kHz for the sixth layer of a multilayer winding coil for minimum resistance (R_W) with DC winding resistance (R_{DC}) and F_R as a function of foil winding thickness*

geometry of the winding configuration therefore become extremely important as AC winding effects begin to become an important contributor to overall losses. For typical copper conductor diameters utilized in medium frequency and medium power transformers, AC winding losses begin having a significant impact for frequencies above the range of approximately 1–10 kHz and must be considered and accounted for in component design. In cases where AC winding effects dominate due to high required switching frequencies, more sophisticated conductor designs of foil or Litz wire may be needed. Litz wire windings are round or square conductors comprised of thin gauge copper strands interleaved in such a way as to minimize proximity losses. Higher costs, complexity, and lower fill factor of Litz wire windings require significant improvements in terms of overall winding losses to justify their utilization. These effects are illustrated in Figure 7.7(b–e) with (f) showing an optimization example.

7.3.2 Magnetic core loss

Magnetic core losses are highly dependent upon the core material selection, with standard electrical steels and bulk crystalline alloys used for line frequency applications showing excessive losses at frequencies of greater than approximately 500 Hz. Total core losses ($P_{TotalCore}$) for a purely sinusoidal excitation waveform can be separated and categorized into three primary categories including hysteretic losses (P_H), classical eddy current losses (P_{Cl}), and anomalous losses (P_A) according to (7.10) where $C1$, $C2$, $C3$, and η are constants specific to a particular core material:

$$P_{TotalCore} = P_H + P_{Cl} + P_A = C_1 f B^\eta + C_2 f^2 B^2 + C_3 f^{1.5} B^{1.5}; Core\ Losses$$

(7.10)

Hysteretic losses are attributed to irreversible, microscopic magnetization processes which can be found even under quasi-static excitation conditions resulting from magnetic domain wall pinning. Classical eddy current losses derive from the assumption of induced eddy currents generated by a homogeneous magnetization process throughout the core material, while anomalous losses provide correction for the eddy currents which result from heterogeneous magnetization processes such as in the vicinity of moving domain walls. Eddy current losses are well-established to be mitigated through both increasing resistivity and decreasing lamination thickness, and they begin to dominate with increasing excitation frequency. For this reason, the primary transformer core materials relevant for medium frequency and medium power applications include: (i) MnZn-ferrites, (ii) amorphous alloys, and (iii) nanocrystalline alloys. In the case of amorphous and nanocrystalline alloys, higher resistivities and thinner laminations ($\sim$110–150 $\mu\Omega$ cm and $\sim$15–25 μm) as compared to bulk crystalline alloys ($\sim$50–80 $\mu\Omega$ cm and $\sim$150–1,000 μm) are effective at mitigating eddy currents at frequencies up to approximately 50 kHz. MnZn-ferrites have dramatically higher resistivities ($\sim$10^6 $\mu\Omega$ cm) such that they can be used without laminations and the losses tend to be dominated by hysteretic loss contributions up to frequencies of approximately 100 kHz. However, the relatively small saturation magnetization of MnZn-ferrites

($B_S \sim 0.4$ T) as compared to the amorphous and nanocrystalline alloys ($B_S \sim 1$–1.4 T) can result in excessively large transformers at the lower end of the medium frequency range ($f \sim 10$ kHz). In the case of NiZn-ferrites which saturate around 0.2 T, the saturation induction is too low for practical applications in the medium frequency range.

7.3.3   Leakage flux and the likely permeance path approximation

Quantitative calculations of leakage inductance are challenging due to dependence on core and winding geometry as well as the operating conditions. Assumptions and simplifications can be made to assist in estimation of leakage inductance for practical components. First, the magnetic flux path exiting the core is perpendicular to core material for typical designs, based upon requirements imposed by the boundary conditions of Maxwell's between two materials with one having a much larger permeability. Second, the magnetic flux path can be assumed to have no normal component with the simplified conductor boundary, rigorously valid in a tightly wound coil that can be homogenized as a simple rectangle or ring. Another assumption is that the shape of the leakage flux space within the window or windings is rectangular and predominately hemispherical when curved around the component exterior.

Based upon such assumptions, one can segment the space within and around a magnetic component into simple and elementary shapes. These various flux paths are all in parallel and for numerical ease, one can determine the magnetic permeance for each space segment where parallel permeances sum. Table 7.2 shows some common elementary shapes that are applicable to leakage inductance calculations. Different shapes connect different parts of the component around an MMF source by either surfaces or edges. The path permeance is derived by dividing the permeance volume by the flux path length to determine the permeance area to path length ratio like the process in Ref. [10]. It should be noted that half annuli with small values of g, $g<3t$, the permeance is $\hat{P}_N = 2\mu dt/\pi(g + t)$. The average arc length in the half cylinder and the 45° spherical sector may be estimated by assuming the flux makes an elliptical path between the boundaries of the permeance geometry which results in an average path length of $l_{elliptical-avg} = \frac{\pi l}{8}\left(2 + \sqrt{2}\,\sinh^{-1}(1)\right)$.

Figure 7.8 provides an example of the segmentation of space into various permeance paths. The solid flux line in the window is a vertical slot leakage permeance. The dashed lines connect core surfaces to core surfaces. Dot dash lines connect various edges of the core or coil. Once all permeance paths are identified and summed, the leakage inductance is easily calculated for a coil by multiplying the number of turns in the coil squared by the total permeance (7.12). One analytical approximation to make in determining permeance paths is that one-third of the coil copper area is included in the permeance area. In adjacent winding configurations, this means that equal areas of the conductive coil have magnetic flux 'entering', 'leaving', and 'not crossing' the coil. In concentric winding configurations, the magnetic flux area is increased by one-third of the width of the coil thickness, accounting for differences in practical and homogenized coils.

Table 7.2 Elementary shapes for magnetic flux in space and their respective permeance

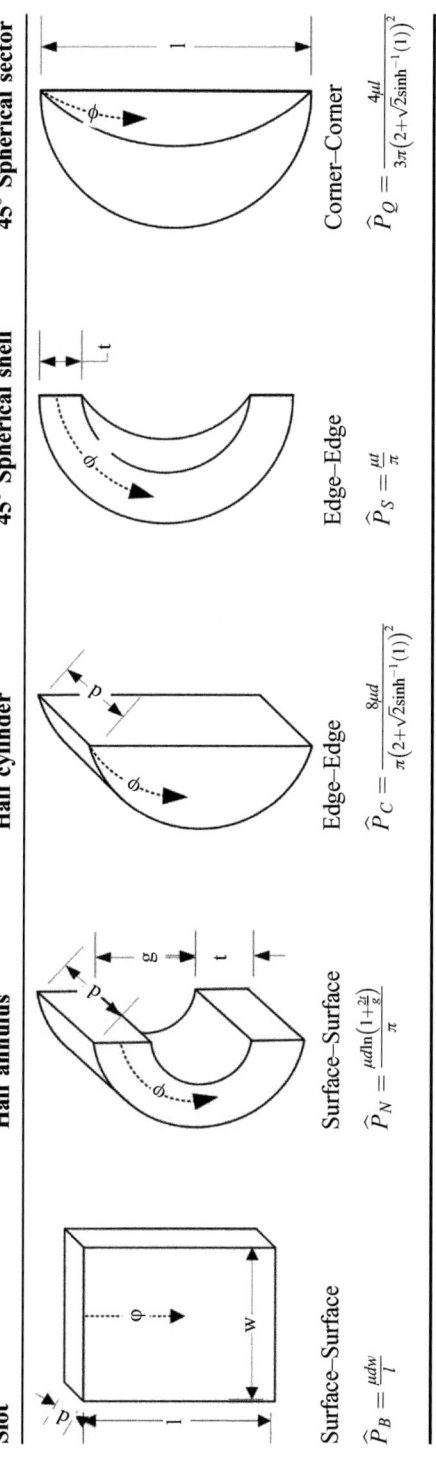

Slot	Half annulus	Half cylinder	45° Spherical shell	45° Spherical sector
Surface–Surface $\widehat{P}_B = \frac{\mu dw}{l}$	Surface–Surface $\widehat{P}_N = \frac{\mu d \ln\left(1+\frac{2t}{g}\right)}{\pi}$	Edge–Edge $\widehat{P}_C = \frac{8\mu d}{\pi\left(2+\sqrt{2}\sinh^{-1}(1)\right)^2}$	Edge–Edge $\widehat{P}_S = \frac{\mu t}{\pi}$	Corner–Corner $\widehat{P}_Q = \frac{4\mu l}{3\pi\left(2+\sqrt{2}\sinh^{-1}(1)\right)^2}$

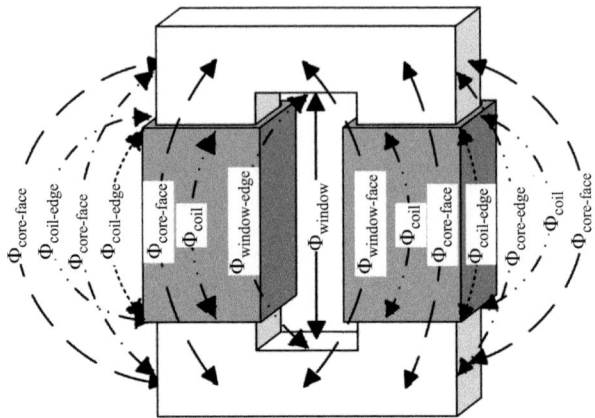

Figure 7.8   Leakage permeance paths around adjacent wound transformer

Another important permeance geometry and leakage path is between two coils wrapped around each other. Depending on the core and winding geometry, the area of permeance can be either a toroid or rectangular. Again, the area of permeance should extend one-third of the winding dimension into the conductor. As an example, the leakage permeance between circular winding is shown below in (7.11) where the radius of the inside edge of the outer conductor of width, w_o, is r_o. Similarly, the outside radius of the inner winding of width, w_i, is r_i. The height of this permeance region is h:

$$\widehat{P}_T = \frac{\mu\pi\left(\left(r_o + \frac{w_o}{3}\right)^2 - \left(r_i - \frac{w_i}{3}\right)^2\right)}{h}; \ Toroid\ Permeance \tag{7.11}$$

Transformer leakage flux is easily estimated once the various leakage flux paths are determined. Permeance building blocks can be assembled in series between various edges and surfaces around the coils of the magnetic component. All these regions are parallel magnetic paths for leakage flux. Thus, the total leakage permeance is simply the sum of all the assembled path regions. The leakage inductance for a given coil is simply the total permeance for the coil multiplied by the turns squared (7.12):

$$L_{leak-i} = N_i^2 \ \widehat{P}_{Total} = N_i^2 \sum \left(\widehat{P}_{B-i} + \widehat{P}_{N-i} + \widehat{P}_{C-i} + \widehat{P}_{S-i} + \widehat{P}_{Q-i} + \widehat{P}_{T-i}\right); \ \text{Leakage Inductance}$$

$$\tag{7.12}$$

7.3.4   Parasitic capacitance and the iso-potential surface approximation

As discussed above, a similar method to that used for determining a magnetic component leakage inductance may also be used to determine the interwinding

Table 7.3   Parasitic capacitance models for medium frequency transformer geometries

Parallel surfaces	Concentric cylinders
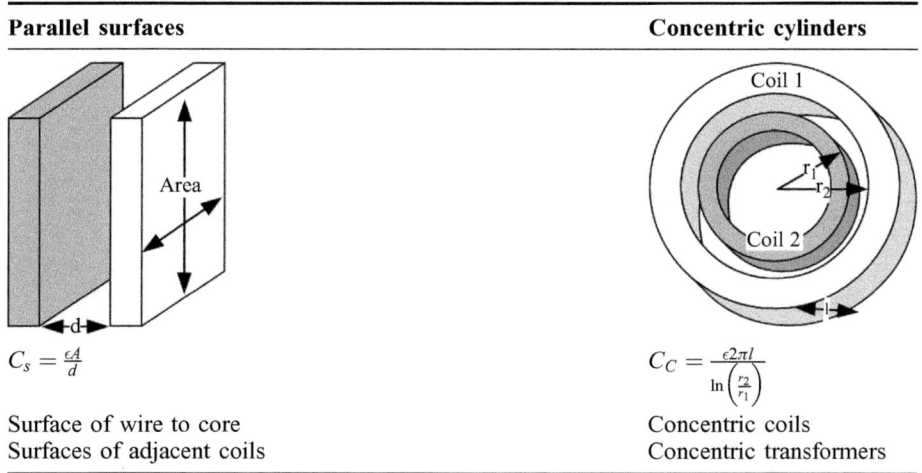	
$C_s = \frac{\epsilon A}{d}$	$C_C = \frac{\epsilon 2\pi l}{\ln\left(\frac{r_2}{r_1}\right)}$
Surface of wire to core Surfaces of adjacent coils	Concentric coils Concentric transformers

parasitic capacitance. When coils of single layers are used the simplified capacitive region shapes of interest are shown in Table 7.3. More detailed approximations can be found in the literature [11].

7.4   Material and magnetic component characterization

The full characterization of medium frequency magnetic components requires several tests. The most relevant characterization method, in terms of understanding operational performance, is characterizing the component in a full-scale converter. However, this approach presents significant challenges due to measurement sensitivity and the co-development of the other converter systems for example, controls, thermals, and protection. As a result, three tests, focusing on different power levels, can provide enough information for an informed design approach given practical constraints of the design process: (i) B–H hysteresis loop and loss performance characterization at relevant excitation levels, (ii) low power impedance analysis to provide a high frequency model, and (iii) series power flow analysis.

7.4.1   Large signal: hysteresis loop analysis

A large signal test, where characterization excitation is at or near operational levels and waveform shapes, can provide insight into soft magnetic material performance generally, or alternatively a magnetic component with an explicit coil structure specifically. The large signal test will determine the operationally dependent effective relative permeability and component losses. These metrics must be characterized at excitation levels applicable to system operation to ensure an appropriate determination of metrics for the final system operation. While characterizing a soft magnetic material generally will provide applicable core material parameters, a final characterization of a magnetic component design may be needed to assess additional geometry and structure dependent performance metrics and losses.

A common approximation for the magnetic core losses under sinusoidal excitation conditions is through use of (7.13) for which three separate fitting constants, k, α, and β, can be empirically determined from measured core losses as a function of frequency and peak induction:

$$P_{Steinmetz} = kf^{\alpha}B^{\beta}; \; Steinmetz \; Expression \; of \; Core \; Losses \qquad (7.13)$$

A specific core material will be described with a unique set of Steinmetz coefficients. Manufacturers generally provide these values for a given sinusoidal excitation in a frequency and flux density range. However, parameters for excitations outside the defined space of the coefficients, for example , different excitation levels, or non-sinusoidal wave shapes, may need another unique set of coefficients. Determining the unique coefficients is achievable through a high power open secondary test. There also exist a large number of alternative models used to describe losses under a range of excitation conditions, with the most widely utilized correcting standard sinusoidal excitation derived coefficients to other waveforms such as Modified Steinmetz Equation (MSE), Generalized Steinmetz Equation (GSE), and Improved Improved Generalized Steinmetz Equation (i²GSE) amongst others [12,13].

The material and component large signal characterization at operationally applicable power levels generates parameters that are extracted from the hysteresis loop for a specific excitation. These loops are measurements of the excitation dependent material behavior contribution of the loss components (7.10), and these parameters are then used to generate performance maps. The hysteresis loop is generated through two coils around the magnetic material or magnetic core of the component. In a transformer, this can be the primary and secondary winding. In an inductor, a second pickup coil must be added for characterization. Multiple winding components can use any pair of coils but may need a complex set of maps that provide characterization of the multiple coupling paths. A two-winding test allows for an explicit measurement of the magnetic core losses when the excitation current is measured and the induced voltage on second winding is measured. The magnetizing field intensity, H, and magnetic flux density, B, are found with (7.14) and (7.15) by relying on the excitation current and turns, I_p and N_p, secondary voltage and turns, V_s and N_s, and the core magnetic path length and cross-sectional area, l_c and A_c. It is important to note that this approach allows measurements of values that are independent of coil parasitic properties. A variety of techniques can be used to excite the medium frequency transformer at relevant excitation levels and with application specific wave shapes. Depending upon the detailed experimental conditions, power amplifiers and specialized power electronics circuits can be utilized. With H and B determined, the excitation specific hysteresis loop of the magnetic core may be developed using the $B(H)$ relationship. This loop allows engineers to determine the effective relative permeability, μ_{re}, and core volume, V_c, specific core loss for this operating point through (7.16) and (7.17), respectively.

Magnetic field strength	Flux density	Effective relative permeability	Hysteresis loop loss
$$H(t) = \frac{N_p i(t)}{l_e}$$ $$(7.14)$$	$$B(t) = \frac{1}{N_s A_e} \int_0^T v_s(t)dt$$ $$(7.15)$$	$$\mu_{re} = \left. \frac{B(t)}{\mu_0 H(t)} \right|_{H_{max}}$$ $$(7.16)$$	$$\frac{P}{V_c} = f \oint H dB$$ $$(7.17)$$

Once operating values are determined from high power tests, values may be mapped for curve fitting. With basic linear algebra for three separate map points (7.18), one can find unique coefficients to the Steinmetz loss model (7.13). Alternatively, curve fitting software can provide coefficients which may lead to a better general fit. This technique is referred to as development of loss maps and Figure 7.9 shows example loss and empirical permeability mapping. Excitation waveform dependency is shown as differences between triangular and trapezoidal performance:

$$\begin{bmatrix} ln\,(P_1) \\ ln\,(P_2) \\ ln\,(P_3) \end{bmatrix} = \begin{bmatrix} 1 & ln(f_1) & ln(B_1) \\ 1 & ln(f_2) & ln(B_2) \\ 1 & ln(f_3) & ln(B_3) \end{bmatrix} \begin{bmatrix} ln(k) \\ \alpha \\ \beta \end{bmatrix} ; \; Linear\; Extraction\; of\; Steinmetz\; Coefficients$$

$$(7.18)$$

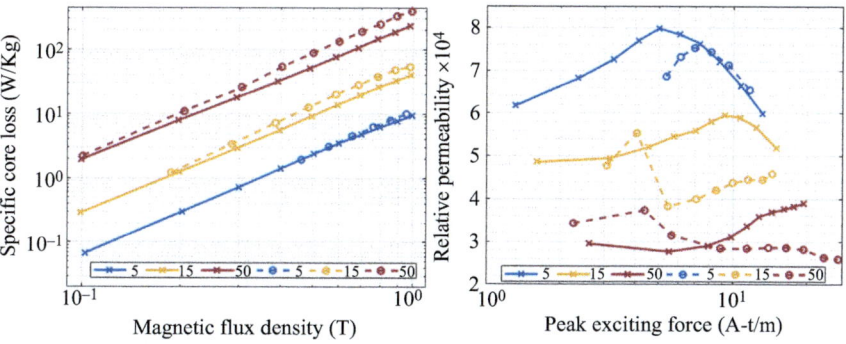

Figure 7.9   *Core loss map (left) and effective permeability map (right) for a MANC core excited by triangular (-x) and trapezoidal (–o) current of various fundamental frequencies 5-50 kHz*

7.4.2   *Small signal: parasitic analysis*

A low power analysis can be used to validate the physics and physical models applicable to practical magnetic components for a small signal model. An LCR meter or impedance analyzer can measure the frequency response of a component for electric equivalent parameter extraction and modeling. A model of a two-port transformer applicable for medium frequencies that captures the dominant parasitic elements listed in Table 7.4 is illustrated in Figure 7.10. This model expands the basic 'T' model to include the transformer high frequency response. More parasitic elements may be needed to capture effects that are relevant for higher frequency, >500 kHz, applications [14,15]. Furthermore, resistances may need to be added in series with capacitors, in series with inductors, and in parallel with inductors to model dissipation losses, conduction losses, and magnetic losses, respectively. This will also contribute to numerical stability in circuit modeling software. Example impedance plots of a transformer tested in an open secondary and shorted secondary configuration are shown in Figure 7.11. The values of the model elements may be extracted from the slopes and resonant frequencies of the impedance plots shown in (7.19)–(7.24). While the medium frequency transformer should be designed to have resonance frequency well above converter switching and fundamental frequencies, high voltage excitation with high dv/dt, square voltages, may produce harmonics of sufficient amplitude as to cause significant current at or beyond the resonant frequencies. This is commonly observed in high frequency

Table 7.4   *Major parasitic elements of MF transformers*

Physical model	Element
Magnetizing inductance	L_m
Primary leakage inductance	L_1
Secondary leakage inductance	L_2
Primary self capacitance	C_1
Secondary self capacitance	C_2
Interwinding capacitance	C_s

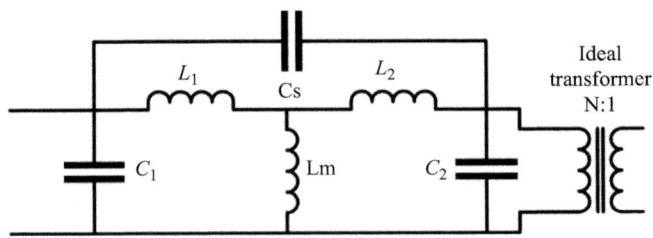

Figure 7.10   Transformer parasitic model

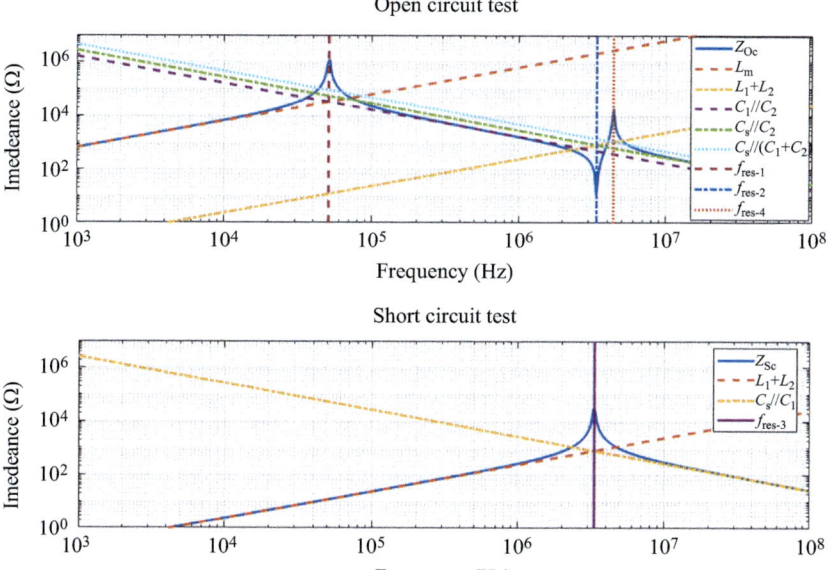

Figure 7.11   *Impedance plots of open and short circuit tests showing resonant frequencies*

common mode currents that pass through the series parasitic capacitance of the transformer.

Small signal magnetizing inductance	Small signal leakage inductances

$$L_m = \frac{dZ_{OC}}{d\omega}\bigg|_{f \ll f_{res-1}} \quad (7.19)$$

$$L_1 + L_2' = \frac{dZ_{SC}}{d\omega}\bigg|_{f \ll f_{res-3}} \quad (7.20)$$

Parallel resonance

Secondary series resonance

$$f_{res-1} = \frac{1}{2\pi\sqrt{L_m(C_1 + C_2)}} \quad (7.21)$$

$$f_{res-2} = \frac{1}{2\pi\sqrt{(L_1 + L_2)(C_s + C_2)}} \quad (7.22)$$

Primary series resonance

Magnetization resonance

$$f_{res-3} = \frac{1}{2\pi\sqrt{(L_1 + L_2)(C_1 + C_s)}} \quad (7.23)$$

$$f_{res-4} = \frac{1}{2\pi\sqrt{(L_1 + L_2)\left(C_s + \frac{C_1 C_2}{C_1 + C_2}\right)}} \quad (7.24)$$

7.4.3   Power flow: series loss analysis

An additional large signal test for transformers is a short circuit test that emulates the operating conditions when power is flowing through a magnetic component. This test is performed by exciting one coil while the second coil is shorted. The power delivered by the excitation circuit, as measured at the component under test terminals, P_{sc}, is the component loss which includes conduction and stray effect losses (7.25). It should be noted that this component loss also includes some core magnetization loss from the short circuit excitation, B_{SC}, which should be corrected using results from prior tests. This analysis provides details of both the conduction losses and the stray flux losses. Stray flux losses are highly sensitive to the core and superstructure material properties and winding configuration, and hence for this test to produce representative loss mapping it should be performed on a completed medium frequency component. This short circuit test can characterize the impact of skin and proximity effect physics, described in (7.8) and (7.9), on the conduction loss of both windings. It can also provide insight into the losses associated with magnetic flux in the leakage path that interacts with conductive materials in the core or superstructure. After a short circuit map is complete, Figure 7.12, one may extract a component specific conduction resistance including skin and proximity effects, R_{AC}, and a component specific model of stray flux losses, a resistance parallel to the leakage inductances, R_{Li}:

$$
\begin{aligned}
P_{SC} &= f \int_0^{f^{-1}} v(t)i(t)dt \\
&= R_{AC1}I_1^2 + R_{AC2}I_2^2 + f^2 \left(\frac{L_{L1}^2}{R_{L1}} I_1^2 + \frac{L_{L2}^2}{R_{L2}} I_2^2 \right) \\
&\quad + kf^\alpha B_{SC}^\beta; \textit{Short Circuit Loss}
\end{aligned}
\tag{7.25}
$$

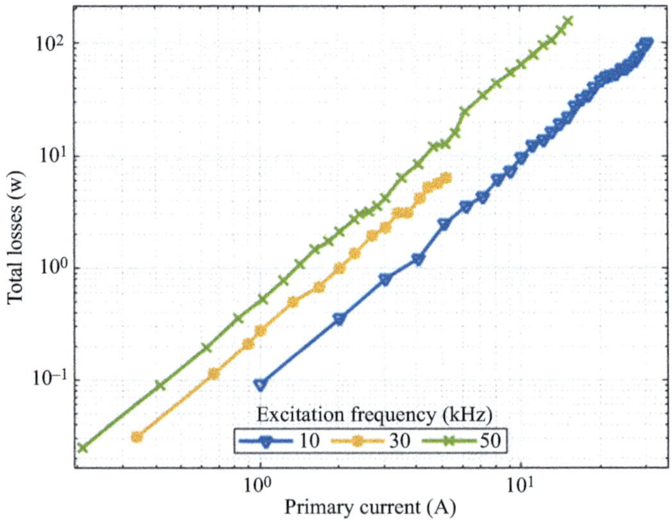

Figure 7.12   Short circuit loss measurements

As a component of the stray flux losses, the short circuit test also captures the losses associated with leakage fluxes. Although leakage flux is typically considered to be losses because free space is a lossless medium, the leakage flux can be a dominant loss component in high leakage inductance transformers and in low effective permeability inductors due to interactions with neighboring conductors and leakage flux induced eddy currents at the surfaces of the magnetic core. For example, nearby conductive materials with large surface areas for example, wide ribbon of tape wound cores, conductive power planes, metal housings, etc. Stray flux encountering this conductive surface and the normal component of flux induces eddy currents around the perimeter. When there are differences in the relative permeability of two materials and the magnetic material has a finite electrical conductivity, that is, air and laminated magnetic ribbon cores, magnetic flux at the interface boundary is near a 90° angle resulting in significant stray flux induced eddy currents, discussed below and shown in Figure 7.13.

Fundamentally, the transformer is a configuration comprised of a core and two or more conductive coils. While air could be the core of choice, as in Inductive Power Technologies for wireless power applications, most cores are soft magnetic materials to control the flow of and amplify the density of magnetic flux. Custom shapes and geometries are available, and most cores will be constructed from elementary 'I' or 'U' shapes if they are from cut pieces. If the core is uncut, it is generally a rectangular shape (i.e., a racetrack core) with corners that can be square or rounded. Isotropic materials are ferrite ceramics or powder cores. Tape wound materials comprised of amorphous and nanocrystalline alloys are fundamentally anisotropic normal to the plane of the strip or ribbon resulting in additional considerations which must be accounted for to achieve optimized medium frequency designs.

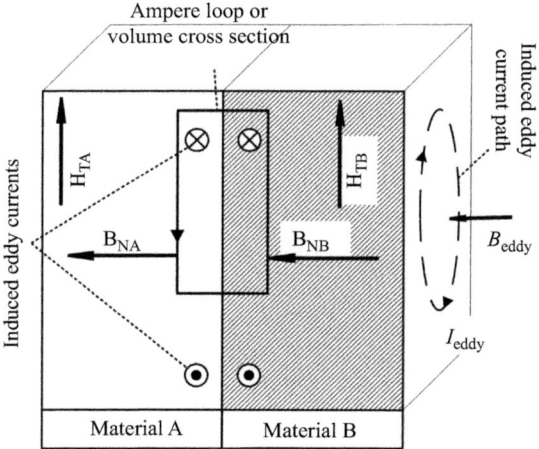

Figure 7.13   Magnetic flux behavior at material interface with normal flux induced eddy currents

7.5   Special considerations for laminated magnetic materials

The basic behavior of the transformer is generally consistent if either isotropic or anisotropic cores are used. However, other finer details such as fringing and leakage flux and thermal profiles with hotspots can have significantly differently behavior. For macro component analysis, it is not practical to explicitly model the thin laminations with thickness dimensions orders of magnitude smaller than the component. Therefore, it is useful to homogenize the laminations into a bulk material model for example [16,17]. However, care must be taken when comparing isotropic and homogenized anisotropic models. Specifically, isotropic cores do not have an orientation and therefore may be assembled without consideration to the material directionality. However, orientation matters for anisotropic and laminated cores and the consideration of this orientation is critical to medium frequency performance. The orientation of laminated cores with respect to the interface of materials, for example, air and core, and assemblies of blocks of is critical. Figure 7.13 shows the definition of boundary conditions at the material interface and, according to Maxwell's equation boundary conditions, that the normal flux density and the tangential magnetic field intensity are equal on either side. In laminated materials where electrical conductivity can be high; this can result in eddy currents being induced in the broad surface of the ribbon [18]. When considering assemblies of laminated cores, there are three principal orientations that are recommended. It is important to avoid configurations where magnetizing flux would be directed into the broad surface of the ribbon as the magnetizing flux can induce excessive eddy currents in the relatively high electrical conductivity of tape-wound cores. The three acceptable configurations are shown in Table 7.5. An example fourth configuration is also shown in Table 7.5 which illustrates an orientation where the magnetizing flux is directed into broad ribbon surfaces which should always be avoided due to exceedingly large magnetizing flux induced eddy current losses which result.

7.6   Winding orientations

For a given core geometry, the method, and details of the coil winding around the core does not change the magnetizing inductance in principle assuming permeability is much higher than free space, the typical case with transformers. However, the winding orientation has significant impact on leakage inductance as the coil orientation and boundaries define leakage flux paths as discussed in detail below. Here we focus on two winding transformers, but the geometries and arrangements are readily extended to multi-winding transformers. While there are several winding options, three common methods can be classified by relative position of the coil rotation axis as illustrated in Table 7.6. The most common orientation used in medium frequency transformers is a concentric winding design in which the axis of coil rotation is colinear and one winding significantly overlaps the other. This

Table 7.5   Laminated core assembly orientations for cut cores from wound or block core pieces

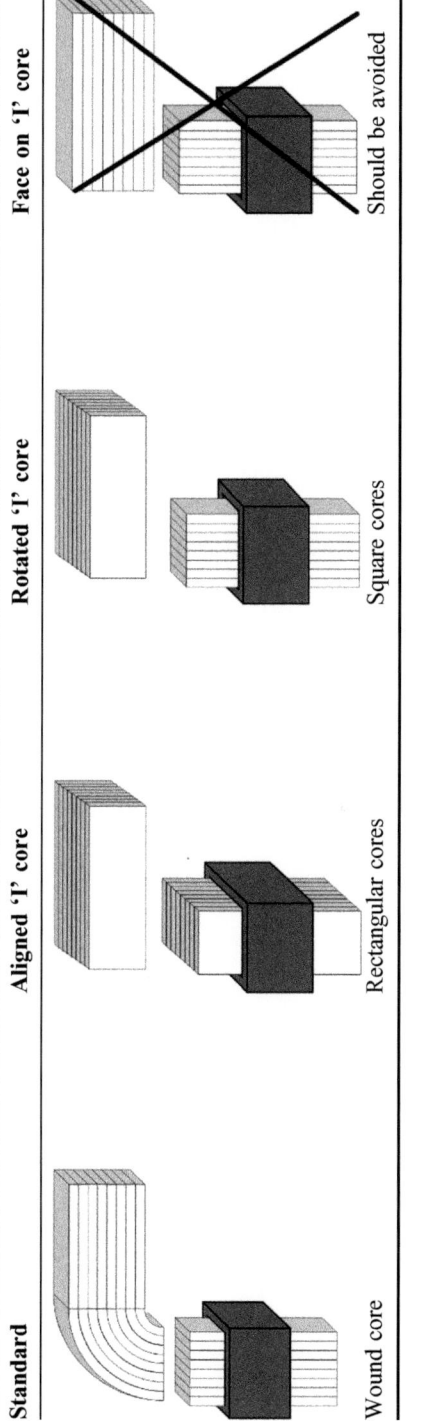

Table 7.6 Winding orientations around basic rectangular core

Label	Concentric wound	Abutting wound	Adjacent wound
Configuration			
Leakage flux space	Between windings	Around windings Along one core limb	Around core limb and yoke Around windings
Inter-capacitance paths	Coil: core Coil: coil	Coil: core: coil Coil edge: coil edge	Coil: core: coil Coil side: coil side

configuration directs much of the leakage flux in between the two coils and results in a relatively low leakage inductance but a high parasitic capacitance. A closely related arrangement is observed with the abutting configuration where the axis of coil rotation is again colinear, but in this case the two coils are physically separated from each other along the core. In this case, the leakage flux is pushed around and concentrated into the space between the two coils resulting in an intermediate level of leakage inductance and parasitic capacitance. A third arrangement is an adjacent winding configuration for which axes of coil rotation are separated onto different regions of the core, resulting in the greatest leakage inductance and lowest effective parasitic capacitance. There are other configurations involving combinations of arrangements or alignment on a per turn basis, for example, two concentric pairs in an adjacent design or an interwoven design of two coils alternating each turn which minimize leakage flux at the expense of greater coil to coil parasitic capacitance.

An additional combination of core and winding arrangement applicable to medium frequency magnetic components is an axial based design. This design is beneficial when very tight tolerances are needed between predicted and measured parasitic elements of the component and in cases where leakage flux normal to core surfaces should be avoided as in tape wound cores. An example axial design is shown in Figure 7.14 where all magnetic permeances paths are toroidal. Furthermore, regions interior to a winding do not contribute to magnetic permeance paths associated with such a winding, resulting in an asymmetric electrical model.

In a full medium frequency transformer design, the selection of core geometry and winding arrangement will set physics boundary conditions and ultimately drive the solution space. The configuration of the core and winding is critically important in determining parasitic aspects such as leakage flux paths and capacitive paths. A detailed review of practical determination of leakage flux and capacitive paths is discussed below.

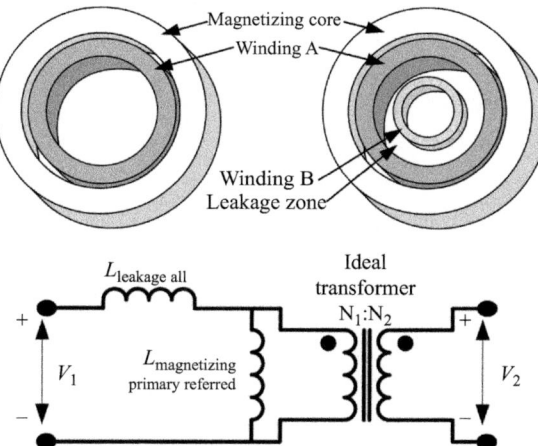

Figure 7.14   Axial inductor (top left), transformer (top right) and transformer equivalent inductances (bottom)

In power converters, a series inductance to the transformer is often desirable. In phase shift converters, this inductance enables the controllability of power flow. It can also be a part of the resonant tank in resonant converters. Further, this inductance impedes common mode current as a high impedance to very high frequency noise. Of course, an external inductor can be added in series with a transformer. However, one can design the transformer with a self-series inductance which can result in an improved power density and higher efficiency in optimized designs.

7.7 Closing remarks

This chapter has provided an overview of the principal concepts and physics that are applicable to magnetic components that are designed for medium to high frequency and medium to high voltage power converters. An emphasis was placed on developing readers intuition for further, in depth study, elsewhere. This overview highlights the fundamental challenge of magnetics for these applications: there are a breadth of physics and applied engineering concepts that practicing engineers need to be aware of and they must know and understand in sufficient depth and detail to properly apply the concepts. Tools such as MECs, large and small signal characterization, and physical analytical models provide the framework for designs that account for the complex interaction between materials, geometry, and excitation dependent component behavior. Readers are encouraged to use this chapter as the starting point for further, in depth, study of the myriad concepts relevant to advanced magnetic components for medium frequency and medium voltage power converter applications.

References

[1] Sudhoff S. *Power Magnetic Devices: A Multi-Objective Design Approach*. 1st ed. Hoboken, NJ: IEEE Press/Wiley, 2013.
[2] Kazimierczuk M. *High-Frequency Magnetic Components*. Chichester. Sussex, UK: Wiley, 2014.
[3] Rowland H. On the magnetic permeability and the maximum of magnetism of iron, steel and nickel. *American Journal of Science*. 1873;s3–6(36):416–425.
[4] Byerly K, Ohodnicki PR, Moon SR, *et al.* Metal amorphous nanocomposite (MANC) alloy cores with spatially tuned permeability for advanced power magnetics applications. *Journal of Materials*. 2018;70: 879–891.
[5] Juds M. *Practical Magnetic and Electromechanical Design*. New Berlin, WI: Self Published, 2020.
[6] De Doncker R, Divan D, and Kheraluwala M. A three-phase soft-switched high-power-density DC/DC converter for high-power applications. *IEEE Transactions on Industry Applications*. 1991;27(1):63–73.
[7] Han D, Li S, Wu Y, Choi W, and Sarlioglu B. Comparative analysis on conducted CM EMI emission of motor drives: WBG versus Si devices. *IEEE Transactions on Industrial Electronics*. 2017;64(10):8353–8363.

[8]   Johannesson D, Nawaz M, and Ilves K. Assessment of 10 kV, 100 A silicon carbide MOSFET power modules. *IEEE Transactions on Power Electronics.* 2018;33(6):5215–5225.

[9]   Dowell P. Effects of eddy currents in transformer windings. *Proceedings of the Institution of Electrical Engineers.* 1966;113(8):1387.

[10] Roters H. *Electromagnetic Devices.* 1st ed. New York: Wiley; 1941.

[11] Biela J and Kolar J. Using transformer parasitics for resonant converters: a review of the calculation of the stray capacitance of transformers. *IEEE Transactions on Industry Applications.* 2008; 44(1):223–233.

[12] Reinert J, Brockmeyer A, and De Doncker R. Calculation of losses in ferro- and ferrimagnetic materials based on the modified Steinmetz equation. *IEEE Transactions on Industry Applications.* 2001;37(4):1055–1061.

[13] Muhlethaler J, Biela J, Kolar J, and Ecklebe A. Improved core-loss calculation for magnetic components employed in power electronic systems. *IEEE Transactions on Power Electronics.* 2012;27(2):964–973.

[14] Fouassier P. *Modelization Electrque des Composants Magnetiques Haute Frequence: Prise en Compte en de la Temperature et Caracterization des Ferrites.* Grenoble, France: Energie électrique. Institut National Polytechnique; 2007.

[15] Schellmanns A, Berrouche K, and Keradec J. Multiwinding transformers: a successive refinement method to characterize a general equivalent circuit. *IEEE Transactions on Instrumentation and Measurement.* 1998;47(5):1316–1321.

[16] Kiwitt J, Huber A, and Rei K. Modellierung geblechter Eisenkerne durch homogene anisotrope Kerne fur dynamische Magnetfeldberechnungen. *Electrical Engineering.* 1999;81(6):369–374.

[17] Wang J, Lin H, Huang Y, and Sun X. A new formulation of anisotropic equivalent conductivity in laminations. *IEEE Transactions on Magnetics.* 2011;47(5):1378–1381.

[18] Beddingfield R, Bhattacharya S, and Ohodnicki P. Shielding of leakage flux induced losses in high power, medium frequency transformers. *IEEE Energy Conversion Congress and Exposition (ECCE).* 2019: 4154–4161.

Chapter 8

MVDC stability: modeling, analysis, and enhancement approaches

Fei (Fred) Wang[1,2], Yaosuo Xue[2] and Le Kong[1]

This chapter presents a general introduction to the fundamental theory and various aspects of stability issues in MVDC systems. The objective is to provide an overview of system stability analysis and physical reasoning to facilitate the design of the circuits, control, and protection of MVDC systems. Starting with the discussion of different types of small-signal stability analysis approaches in Section 8.1, it is shown that the terminal impedance characteristics of each subsystem or equipment in a system seen from the common bus is one of the most important indicators for system stability. The subsystem or equipment impedance can be obtained by mathematical modeling or measurement. However, unlike DC systems of low voltage and power (e.g., a notebook computer power system) in which impedances can generally be measured with commercially available instrumentation, it is difficult to use the commercial instrument in direct impedance measurement for MVDC systems. Therefore, several small-signal impedance measurement techniques proposed in literature for MVDC systems are discussed in Section 8.2. Section 8.3 introduces three commonly used large-signal stability analysis approaches for considering large disturbances that cannot be covered by small-signal stability analysis. Based on the stability analysis results, small-signal and large-signal stability enhancement approaches are summarized in Section 8.4. General stability improvement approaches for DC systems regarding small-signal instabilities and constant power loads (CPLs) induced transient instabilities can be applied for MVDC applications. However, protection strategies to ensure large-signal stability for AC systems or low-voltage DC systems are not suitable for MVDC systems due to large and fast-rising fault current. Therefore, the protection remains as a key issue in MVDC applications. Section 8.5 summarizes this chapter, including modeling, analysis, and enhancement approaches of MVDC stability.

[1]Center for Ultra-wide Area Resilient Electric Energy Transmission Network (CURENT), Min H. Kao Department of Electrical Engineering and Computer Science, University of Tennessee, Knoxville, TN, USA
[2]Electrification and Energy Infrastructures Division, Oak Ridge National Laboratory, Knoxville, TN, USA

8.1   System modeling and small-signal stability analysis

Stability analysis is a vital step in the design stage for any electrical system, including MVDC systems [1]. The stability can be classified as small-signal stability and large-signal stability. Small-signal stability is defined as the ability of a system to maintain steady-state operating points when it is subject to small disturbances such as incremental changes in system loads, while the large-signal stability is defined as the ability in events of large disturbances, such as fault conditions, pulse loads, or load transients, that the system response will be able to return to an acceptable equilibrium range. In both the small-signal and large-signal stability studies, the common practice is to capture system dynamic characteristics via proper analytical models, such as state-space models and impedance models [2].

8.1.1   State-space model-based stability analysis

8.1.1.1   State-space representation and linearization

The dynamic behavior of a system, including a MVDC system, can be described by a set of n nonlinear ordinary differential equations in the state space form as in (8.1),

$$\begin{cases} \dot{x} = f(x,u) \\ y = g(x,u) \end{cases} \tag{8.1}$$

where,

$x = \begin{bmatrix} x_1 & x_2 & \cdots & x_n \end{bmatrix}^T$ is the state variable vector,
$y = \begin{bmatrix} y_1 & y_2 & \cdots & y_m \end{bmatrix}^T$ is the output variable vector, and
$u = \begin{bmatrix} u_1 & u_2 & \cdots & u_r \end{bmatrix}^T$ is the input variable vector.

At equilibrium (or singular) point, (8.2) is satisfied.

$$\dot{x} = f(x_o) = 0 \tag{8.2}$$

If the function f in (8.1) is linear, then the system has only one equilibrium point. If the function f is nonlinear, then there may be more than one equilibrium point. And its linearization form can be obtained under the assumption of a small disturbance, which results in variables changes (Δx, Δy, and Δu) at the equilibrium point (x_o, y_o, and u_o). Neglecting the terms containing second and higher-order powers of Δx and Δu, the system (8.1) can be written as (8.3),

$$\begin{cases} \Delta \dot{x} = A\Delta x + B\Delta u \\ \Delta y = C\Delta x + D\Delta u \end{cases} \tag{8.3}$$

where,

$A = [\partial f/\partial x]_{n \times n}$ is the state or plant matrix,
$B = [\partial f/\partial u]_{n \times r}$ is the control or input matrix,
$C = [\partial g/\partial x]_{m \times n}$ is the output matrix, and
$D = [\partial g/\partial u]_{m \times r}$ is the feed-forward matrix.

Note that all the derivatives in these matrices are evaluated at the initial equilibrium operating point $(x_o, y_o,$ and $u_o)$. By performing a Laplace transform on (8.3), the linearized (8.4) in the frequency domain results, where the initial conditions $\Delta x(0)$ are assumed to be 0:

$$\begin{cases} s\Delta x(s) - \Delta x(0) = A\Delta x(s) + B\Delta u(s) \\ \qquad \Delta y(s) = C\Delta x(s) + D\Delta u(s) \end{cases} \tag{8.4}$$

8.1.1.2   Eigen properties and small-signal stability

Eigenvalues and eigenvectors

The poles of the linearized system represented by $\Delta x(s)$ and $\Delta y(s)$ are the roots of the characteristic equation of matrix A as determined in (8.5):

$$\det(sI - A) = 0 \tag{8.5}$$

These roots are also called eigenvalues of matrix A. The eigenvalues $\lambda = [\lambda_1 \quad \lambda_2 \quad \cdots \quad \lambda_n]^T$ are given by the values for which there exist non-trivial solutions to the (8.6):

$$\det(A - \lambda I) = 0 \tag{8.6}$$

Each eigenvalue is defined as $\lambda_i = \alpha_i + j\omega_i$ $(i = 1, 2, \cdots, n)$ and associated with that, there are two eigenvectors: one is the right eigenvector Φ_i and the other is the left eigenvector Ψ_i.

The right eigenvector $\Phi_i = [\Phi_{1i} \quad \Phi_{2i} \quad \cdots \quad \Phi_{ni}]^T$ (i.e., a column vector) satisfies (8.7):

$$A\Phi_i = \lambda_i\Phi_i, i = 1, 2, \cdots, n \tag{8.7}$$

Similarly, the left eigenvector $\Psi_i = [\Psi_{i1} \quad \Psi_{i2} \quad \cdots \quad \Psi_{in}]$ (i.e., a row vector) satisfies (8.8):

$$\Psi_i A = \lambda_i\Psi_i, \ i = 1, 2 \cdots, n \tag{8.8}$$

Correspondingly, two matrices are introduced as (8.9) and (8.10):

$$\Phi = [\Phi_1 \quad \Phi_2 \quad \cdots \quad \Phi_n]_{n\times n} \tag{8.9}$$

$$\Psi = [\Psi_1{}^T \quad \Psi_2{}^T \quad \cdots \quad \Psi_n{}^T]^T{}_{n\times n} \tag{8.10}$$

The above matrices are orthogonal matrices and satisfy $\Psi\Phi = I$ (identity matrix).

Small-signal stability analysis

By investigating the characteristics of λ_i, of which the real part indicates system damping and the imaginary part gives oscillation frequency, system stability can then be determined as follows:

For eigenvalues with only real parts, the system is in a non-oscillatory mode. A negative eigenvalue means a decaying mode and a positive eigenvalue indicates aperiodic unstable phenomena.

For complex eigenvalues, each conjugate pair corresponds to an oscillatory mode. If all the eigenvalues have negative real parts, then the system will be asymptotically stable with a damped oscillation. And if at least one eigenvalue has a positive real part, then the system will be unstable with amplitude-increasing oscillations. The oscillation frequency f in Hz and the damping ratio ζ are given by (8.11) and (8.12), where α is the real part of the eigenvalue and ω is the imaginary part:

$$f = \frac{\omega}{2\pi} \tag{8.11}$$

$$\zeta = \frac{-\alpha}{\sqrt{\alpha^2 + \omega^2}} \tag{8.12}$$

Participation factor
In a system with poor damping, it is of significance to identify the related state variables so that a proper solution can be designed accordingly. A concept called participation factor P_{ki} as shown in (8.13) is then introduced as a measure of the correlation between system oscillation modes and state variables of a linear system:

$$P_{ki} = \Phi_{ki}\Psi_{ik} \tag{8.13}$$

The participation matrix P of size $n \times n$, which consists of the participation factors, is given as (8.14):

$$P = \begin{bmatrix} \Phi_{11}\Psi_{11} & \Phi_{12}\Psi_{21} & \cdots & \Phi_{1n}\Psi_{n1} \\ \Phi_{21}\Psi_{12} & \Phi_{22}\Psi_{22} & \cdots & \Phi_{2n}\Psi_{n2} \\ \vdots & \vdots & \cdots & \vdots \\ \Phi_{n1}\Psi_{1n} & \Phi_{n2}\Psi_{2n} & \cdots & \Phi_{nn}\Psi_{nn} \end{bmatrix} \tag{8.14}$$

The participation factors generally represent the relative participation of the states to the corresponding mode. The larger the participation factor is, the more participation of the respective state is to the critical mode. Therefore, the participation factors can be adopted to identify root causes of system instabilities, and stabilization methods can then be designed accordingly.

8.1.2   Impedance-based stability analysis

8.1.2.1   Small-signal impedance model

Apart from the conventional linearized state-space model-based stability analysis approach, the impedance model-based approach is also prevailing since it only requires the terminal characteristics of each subsystem which can be obtained either through analytical modeling or practical measurement of the port behaviors. Note that a subsystem in a MVDC system can be a piece of equipment such as a DC/DC or AC/DC power electronics converter, or a group of equipment connected together that can be characterized from a terminal.

The most popular impedance-based approach is based on the concept called impedance ratio. For instance, a system can be considered as a piece of equipment connected to the rest of the DC system, which can be modeled as two individually

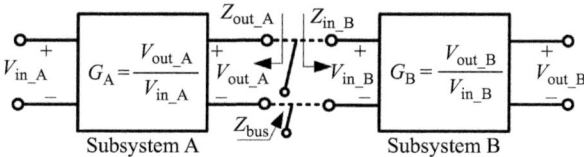

Figure 8.1   Interconnection of two stable independent subsystems in an MVDC system [3]

stable subsystems A and B as shown in Figure 8.1. Its total small-signal input-to-output transfer function G_{AB} can be derived as shown in (8.15), with an introduction of minor loop gain T_{MLG}. Since the transfer functions G_A and G_B are stable, the entire system stability will then only be determined by the characteristics of the impedance ratio of Z_{out_A} and Z_{in_B}:

$$G_{AB} = \frac{V_{out\_B}}{V_{in\_A}} = G_A G_B \frac{Z_{in\_B}}{Z_{out\_A} + Z_{in\_B}} = G_A G_B \frac{1}{1 + T_{MLG}} \tag{8.15}$$

where

$$T_{MLG} = \frac{Z_{out\_A}}{Z_{in\_B}}.$$

In addition to the impedance-ratio-based minor loop gain model, the system stability can also be analyzed through the impedance-sum model $Z_{sum} = Z_{out\_A} + Z_{in\_B}$ [4] or the bus impedance Z_{bus}, which is the equivalent impedance seen from the bus port, i.e., $Z_{bus} = Z_{out\_A} // Z_{in\_B}$ [5].

8.1.2.2   Impedance-based stability criteria
Many stability criteria have been applied with system impedance models to investigate the small-signal stability of MVDC systems, including Middlebrook criterion, forbidden region criteria, passivity-based criteria, and Nyquist stability criterion and its extensions.

Middlebrook criterion and forbidden region
To guarantee system stability, Middlebrook first applied the minor loop gain term into the design of input filters of DC/DC converters in 1976 by confining the ratio of filter output impedance and converter input impedance within the unit circle on the complex plane [6], i.e., $|T_{MLG}| \ll 1$ for all frequencies. Later, the Middlebrook criterion was adopted for stability analysis in distributed DC power systems. For example, in the system as shown in Figure 8.1, if $|Z_{in\_B}| \gg |Z_{out\_A}|$ for all frequencies, then the loading effect of the subsystem B is negligible, and the system will be stable. This approach provides an intuitive but conservative way for defining the impedance specification to assure system stability.

To relax the conservativeness of the Middlebrook criterion, several stability criteria were developed to define a forbidden region in the polar plots of T_{MLG}, such as the gain/phase margin (GMPM) criterion [3], the opposing argument (OA) criterion [7],

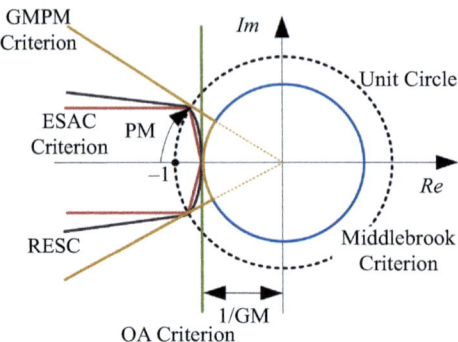

Figure 8.2   Boundaries of the stability criteria on the complex plane [10]

the energy source analysis consortium (ESAC) criterion [8], the root exponential stability criterion (RESC) [9], etc. Compared with the forbidden region defined by the
Middlebrook criterion which is outside the solid circle shown in Figure 8.2, the GMPM
criterion allows $|Z_{in\_B}| \ll |Z_{out\_A}|$ at some frequencies, i.e., the Nyquist contour can be
outside the unit circle, while ensuring the desired gain margin and phase margin to not
encircle $(-1, j0)$ point. The forbidden region specified by GMPM criterion in the s-
plane for T_{GLM} is located to the left of the defined boundary shown in Figure 8.2. The
OA criterion is an extension of the GMPM criterion for systems with multiple paralleled load converters. The forbidden region defined by OA criterion is to the left of the
vertical line that intersects the real axis at $-1/GM$, where the encirclement of $(-1, j0)$
point is avoided. The ESAC criterion can further reduce the forbidden region, which is
to the left of the defined boundary, but keeps the same values of gain margin and phase
margin as in GMPM criterion. As an extension of ESAC, the RESC has a larger
forbidden region i.e., to the left of defined boundary, but it is more robust to the
numerical variations of the load impedances. Yet, all those stability criteria are sufficient but not necessary conditions. Additionally, the forbidden-region-based stability
criteria with only considering avoidance of encirclements of $(-1, j0)$ point is based on
a prerequisite, i.e., all the poles of $T_{MLG}(s)$ are in the left-half plane. If the open-loop
right-half plane (RHP) poles exist, those stability criteria cannot work.

Passivity-based stability criterion
In addition to the minor-loop-gain-based stability criteria, passivity-based stability
criterion (PBSC) has also been proposed for single-bus DC systems [10, 11]. For a
linearized single-input–single-output (SISO) system which is described by a
rational transfer function $Z(s)$, it is passive if it satisfies:

1. $Z(s)$ is stable (no RHP poles), and
2. $Re(Z(j\omega)) \geq 0$ or $-90° \leq arg(Z(j\omega)) \leq 90°$ for any ω.

Applying the passivity-based stability concept to the example system as shown
in Figure 8.1, if the bus impedance Z_{bus} satisfies the passivity conditions, then the
overall system is stable. This also implies that the Nyquist contour of $Z_{bus}(j\omega)$ must

lie in the RHP. The PBSC can also be applied for single-bus DC systems with multiple source converters and load converters, where $Z_{bus}(s)$ will be the parallel combination of all the converters' input/output impedances. However, like Middlebrook Criterion and other forbidden region-based stability criteria, the PBSC can only provide a sufficient but not necessary condition.

Nyquist stability criterion

A sufficient and necessary condition for small-signal stability of MVDC systems can be obtained based on the Nyquist stability criterion. For example, as indicated in (8.15), the stability of the interconnected system in Figure 8.1 is determined by the locations of zeros of $(1+T_{MLG}(s))$ under stable G_A and G_B. Namely, if there are one or more RHP zeros of $(1+T_{MLG}(s))$, meaning there are one or more RHP poles in G_{AB}, then the system is unstable. The number of RHP zeros (N_Z) in $(1+T_{MLG}(s))$ equals to the sum of N_P and N_{CW}, where N_P is defined as the number of RHP poles of $T_{MLG}(s)$; and N_{CW} is defined as the number of clockwise encirclements of $(-1, j0)$ in the Nyquist contour of $T_{MLG}(j\omega)$ in polar plots, which has three possibilities [12]:

1. $N_{CW} = 0$: there is no encirclement of $(-1, j0)$ point.
2. $N_{CW} < 0$: there are one or more counterclockwise encirclements of $(-1, j0)$ point.
3. $N_{CW} > 0$: there are one or more clockwise encirclements of $(-1, j0)$ point.

Therefore, based on the Nyquist stability criterion, for conditions 1 and 2 above, the system is stable if $N_Z = N_P + N_{CW} = 0$; otherwise, the system is unstable. For condition 3, the system is unstable. The common practice for power converter-based DC system stability study with Nyquist criterion is to formulate an impedance ratio according to the types of converters in the system to avoid RHP poles so that the system stability depends upon the Nyquist contour of $T_{MLG}(j\omega)$ only. For instance, the numerator of the impedance ratio should be the impedance of the voltage-type converters, and the denominator should be the impedance of current-type converters [13]. However, if the numerator of the minor loop gain can only be selected as the output impedance of the source-side current-type converter, there will be open-loop RHP poles. Consequently, much numerical calculation of the number of RHP poles may be required. The graphical analysis on the Nyquist contour may be cumbersome as well.

The Nyquist stability criterion can also be extended to impedance-sum models of the system by checking encirclement types of the Nyquist contour with the origin as the critical point. That is, a system with two individually stable subsystems is stable, if and only if the impedance sum of the two subsystems does not encircle the origin clockwise [4].

Inverse Nyquist stability criterion

The inverse Nyquist stability criterion is an alternative way for system analysis that needs much calculation on the RHP poles of the impedance-ratio model. Its equivalence to the Nyquist stability criterion can be proved as follows.

Take the system in Figure 8.1 as an example, where subsystem A is at the source side with V_{in_A} connected to an ideal DC voltage, and subsystem B is at the load side with V_{out_B} connected to a resistive load. The transfer function from input to output of

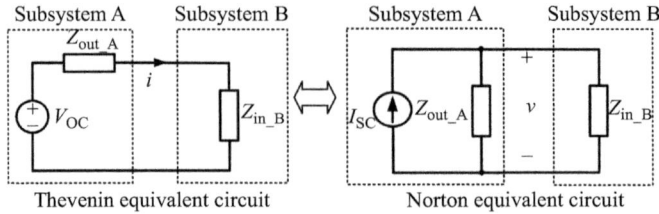

Figure 8.3   Equivalence of small-signal representation of the system with two subsystems using Thévenin and Norton equivalent circuit [14]

subsystem A without the loading effect is modeled as H_A and that of subsystem B with the resistor load is modeled as H_B. Both H_A and H_B are assumed to be stable. Then the entire system can be modeled either as a Thévenin circuit or a Norton circuit as shown in Figure 8.3 since the Thévenin and Norton circuits are exchangeable according to the theory of linear circuit analysis. Both the Thévenin and Norton equivalent circuits can generate a transfer function as given in (8.16),

$$G_{AB} = \frac{V_{out\_B}}{V_{in\_A}} = H_A H_B \frac{1}{1 + T_{MLG\_Inv}} \tag{8.16}$$

where

$$T_{MLG\_Inv} = \frac{Z_{in\_B}}{Z_{out\_A}}.$$

However, as shown in Figure 8.1, the transfer function for the overall system from source-side input to the load-side output can be derived as given in (8.15), where the minor loop gain T_{MLG} is defined as Z_{out_A}/Z_{in_B}. Although the minor loop gain terms defined in (8.15) and (8.16) are different, the same results can be obtained as shown by the following three scenarios.

Scenario I: *Applying Nyquist stability criterion to T_{MLG}*

In this scenario, the Nyquist stability criterion is applied to the minor loop gain $T_{MLG}(s)$ derived based on Figure 8.1, where the numerator is Z_{out_A} and the denominator is Z_{in_B}. The number of RHP poles N_{P_SI} of T_{MLG} is calculated first, which equals to the sum of the number of RHP poles in Z_{out_A} and the number of RHP zeros in Z_{in_B}. The system stability can then be determined by checking whether the Nyquist contour counterclockwise encircles $(-1, j0)$ N_{P_SI} times.

Scenario II: *Applying Nyquist stability criterion to T_{MLG_Inv}*

In scenario II, the Nyquist stability criterion is applied to the minor loop gain T_{MLG_Inv} derived based on Figure 8.3, where the numerator is Z_{in_B} and the denominator is Z_{out_A}. The number of RHP poles N_{P_S2} in $T_{MLG\_INV}(s)$ equals the number of RHP poles of Z_{in_B} plus the number of RHP zeros of Z_{out_A}. The system will be stable if and only if the counterclockwise encirclements of $(-1, j0)$ point of T_{MLG_Inv} is N_{P_S2} times.

Scenario III: *Applying inverse Nyquist stability criterion to T_{MLG_Inv} (s)*

The Nyquist stability criterion can also be applied to the inverse polar plots, which is considered as the inverse Nyquist stability criterion. It may be stated as follows [12]:

> "For a closed-loop system to be stable, the encirclement, if any, of the $(-1, j0)$ point by the $1/T_{MLG\_Inv}(s)$ Nyquist locus must be counter-clockwise, and the number of such encirclements must be equal to the number of poles of $1/T_{MLG\_Inv}(s)$, i.e., the zeros of $T_{MLG\_Inv}(s)$, that lie on the RHP."

Accordingly, apply the Nyquist stability criterion to the inverse polar plots of $T_{MLG\_Inv}(s)$, i.e., $1/T_{MLG\_Inv}(s)$. The number of RHP poles N_{P_S3} of $1/T_{MLG\_Inv}(s)$ is examined first, which is the sum of the number of RHP zeros in Z_{in_B} and the number of RHP poles in Z_{out_A}. If the Nyquist contour of $1/T_{MLG\_Inv}(s)$ counter-clockwise encircles the $(-1, j0)$ point N_{P_S3} times, then the system will be stable. The stability determination in scenario III is equivalent to that in scenario II based on the equivalence of applying the Nyquist criterion to the polar plots and its inverse plane. Additionally, applying the inverse Nyquist stability to $T_{MLG\_Inv}(s)$ in scenario III is equivalent to applying the Nyquist stability to $T_{MLG}(s)$ because $T_{MLG}(s) = 1/T_{MLG\_Inv}(s)$.

Therefore, it can be concluded that by applying the Nyquist stability criterion to either Z_{out_A}/Z_{in_B} or Z_{in_B}/Z_{out_A}, the system stability can all be predicted, although the shapes of the Nyquist plots might be different. The impedance ratio could be selected in a way to ensure that there are no open loop RHP poles in the minor loop gain term to simplify the graphical analysis on Nyquist plots.

Example 1 (Impedance ratio-based stability analysis with Nyquist stability criterion): In this example, a simplified DC system that consists of a source subsystem with an ideal voltage source and current-controlled load subsystem is given as shown in Figure 8.4.

The output impedance of source converter Z_1 and the input impedance of the load converter Z_2 are derived as $Z_1 = (sL_1 + R_1)/(s^2 L_1 C_1 + sR_1 C_1 + 1)$ and

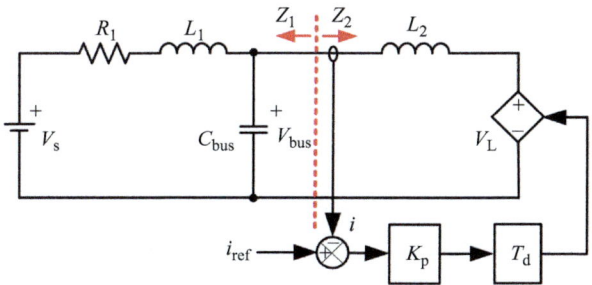

Figure 8.4   A DC system with source subsystem and load subsystem

$Z_2 = sL_2 + G_iG_d$, where G_i is the transfer function of the current controller, and G_d is the transfer function of time delay. The circuit parameters are: $L_1 = 1$ mH, $R_1 = 0.1\ \Omega$, $L_2 = 5$ mH, current controller $K_p = 30$, switching frequency $f_{sw} = 10$ kHz, control delay $T_d = 1.5/f_{sw}$, and $C_{bus} = 20\ \mu F$ for Case I or $5\ \mu F$ for Case II. From the time-domain simulations, it is known that Case I is stable and Case II is unstable.

Applying the Nyquist stability criterion and inverse Nyquist stability criterion to the impedance ratio, we have $T_1 = Z_1/Z_2$ and $T_2 = Z_2/Z_1$. The Nyquist plots of T_1 and T_2 under different cases are given in Figure 8.5.

Then, system stability can be examined with T_1 and T_2 separately, considering the encirclement types of Nyquist plots N_{CW} and the number of open-loop RHP poles N_P as in Table 8.1.

Therefore, it can be seen from the example system, both $T_1 = Z_1/Z_2$ and $T_2 = Z_2/Z_1$ can predict the system stability correctly since they are intrinsically equivalent.

Generalized Bode criterion

In addition to Nyquist plots, the Bode diagram is also a commonly used graphical representation for the frequency response of a system. The main advantage of using the Bode diagram is that the impedance ratio term can be separated into the impedance of each subsystem so that interactions between different subsystems can be more intuitively analyzed. The commonly used Bode criteria in power

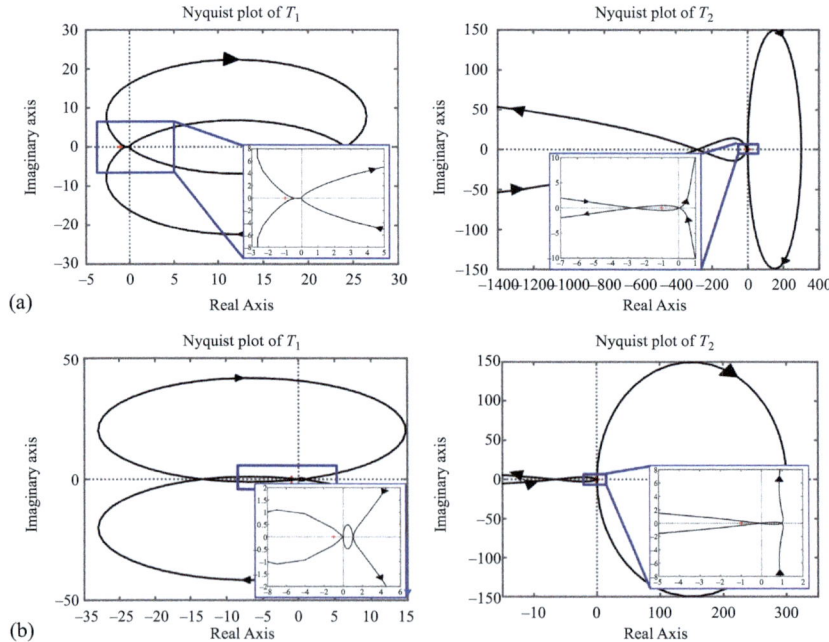

Figure 8.5   Impedance ratio-based stability analysis with Nyquist stability criteria: (a) Case I: $C_{bus} = 20\ \mu F$ and (b) Case II: $C_{bus} = 5\ \mu F$

Table 8.1 System stability analysis with Nyquist criterion applying to T_1 and T_2

Cases	Case I ($C_{bus} = 20\ \mu F$)		Case II ($C_{bus} = 5\ \mu F$)	
	T_1	T_2	T_1	T_2
Minor loop gains				
Number of open-loop RHP poles N_P	0	0	0	0
Number of clockwise encirclements of $(-1, j0)$ N_{CW}	0	$2 + (-2)$	2	2
Number of closed-loop RHP zeros N_Z	0	0	2	2
Prediction of system stability	Stable	Stable	Unstable	Unstable

electronics systems, including the classic Bode criterion, the revised Bode criterion I, and the revised Bode criterion II, are to check if there are enough gain margin and phase margin of the open-loop transfer function under certain conditions [15]. However, these criteria are based on some specific cases of Nyquist criterion, which can only provide a sufficient condition for stability; e.g., the classic Bode criterion is only suitable for minimum phase system without RHP poles or RHP zeros, the revised Bode criterion I can be extended to systems with multiple gain margins and phase margins but with no RHP poles, and the revised Bode criterion II can be applied to systems with RHP poles but with no more than two integrators in the transfer function. Therefore, none of these Bode criteria are like the Nyquist criterion that can apply to any types of systems. Hence, a generalized Bode criterion which is equivalent to the Nyquist stability criterion for arbitrary systems is needed.

The generalized Bode stability criterion can be derived from Nyquist stability criterion with system minor loop gain and formulated on Bode diagrams with impedances of each subsystem by mapping between Nyquist and Bode plots. An example is shown in Figure 8.6, where the minor loop gain is assumed as $T_1 = Z_1/Z_2$ with one magnitude crossing of Z_1 and Z_2 [15–17]. And the goal is still to check the encirclement type of the critical point $(-1, j0)$ of the Nyquist contour of T_1.

General rules of using Bode plots for system stability analysis can also be summarized as in Table 8.2, where relationships of the magnitude, phase, and phase derivatives of the impedances are all needed to determine system stability.

For cases with multiple crossings of the magnitude of the impedances, general rules using Bode plots to identify the encirclements of the critical point of Z_1/Z_2 consist of four steps [17]:

1. Find all the exterior regions (ERs) outside the unit circle of the Nyquist diagram of T_1, where the numerator impedance Z_1 is larger than the denominator impedance Z_2.
2. Check whether there are crossings of the critical boundaries (CBs) and phase of Z_2 within the ERs, where the CBs are defined as the shifted phase of Z_1, i.e., $\angle Z_1 \pm 180°$.
3. Determine the encirclement types if there is phase crossing. At each crossover frequency, if $(\angle Z_1)' < (\angle Z_2)'$, the crossing type will be clockwise crossing; if $(\angle Z_1)' > (\angle Z_2)'$, it will be counterclockwise encirclement.

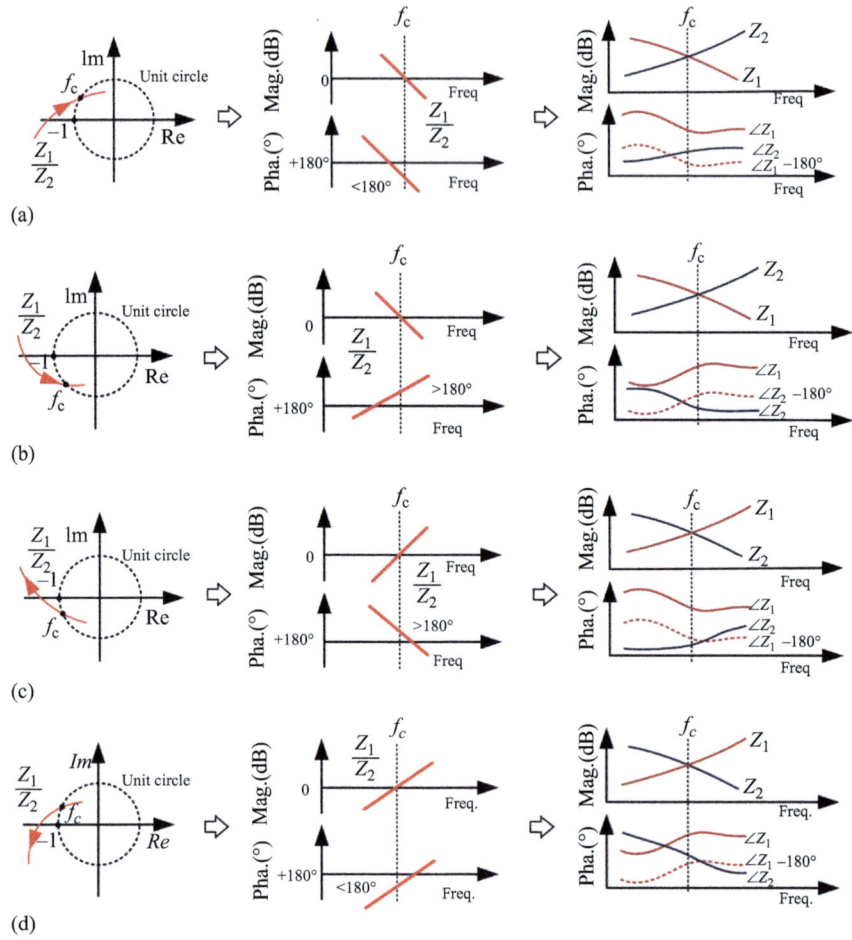

Figure 8.6   Mapping from Nyquist to Bode diagrams: (a) into the unit circle clockwise, (b) into the unit circle counterclockwise, (c) out of the unit circle clockwise, and (d) out of the unit circle counterclockwise [16, 17]

4. Calculate the number of crossings. If the crossover frequency is not 0 Hz, then the crossing number should be doubled, whereas for a crossover frequency equal to 0 Hz, the number should be counted separately.

Example 2 (Stability analysis with generalized Bode criterion): Stability analysis on the system in Example 1 can also be conducted with the generalized Bode criterion.

Table 8.2   General rules of using Bode plots for system stability analysis [16]

Case	Magnitude (lower than f_c)	Phase* at f_c	Encirclement type	Encirclement condition
(a)	$\lvert Z_1 \rvert > \lvert Z_2 \rvert$	$(\angle Z_1)'^{\dagger} < (\angle Z_2)'$	Clockwise	$\angle Z_1 - \angle Z_2 < 180°$
(b)	$\lvert Z_1 \rvert > \lvert Z_2 \rvert$	$(\angle Z_1)' > (\angle Z_2)'$	Counterclockwise	$\angle Z_1 - \angle Z_2 > 180°$
(c)	$\lvert Z_1 \rvert < \lvert Z_2 \rvert$	$(\angle Z_1)' < (\angle Z_2)'$	Clockwise	$\angle Z_1 - \angle Z_2 > 180°$
(d)	$\lvert Z_1 \rvert < \lvert Z_2 \rvert$	$(\angle Z_1)' > (\angle Z_2)'$	Counterclockwise	$\angle Z_1 - \angle Z_2 < 180°$

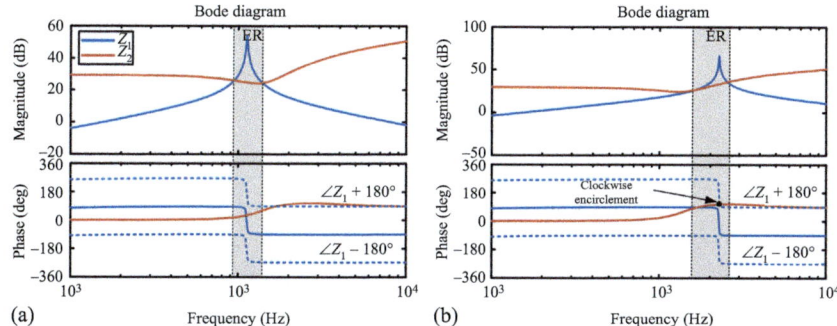

Figure 8.7   Impedance-based stability analysis with generalized Bode criterion: (a) Case I: $C_{bus} = 20\ \mu F$ and (b) Case II: $C_{bus} = 5\ \mu F$

The minor loop gain term T_1 is decomposed as Z_1 and Z_2 and the Bode diagrams of both Z_1 and Z_2 are given in Figure 8.7. For Case I as shown in Figure 8.7(a), there are two crossings of the magnitude of Z_1 and Z_2 which is defined as one ER region. In this region, it can be seen that there is no crossing of CBs and phase of Z_2. Therefore, the system is stable since there is no encirclement $(-1, j0)$ and no open-loop RHP poles. For Case II as shown in Figure 8.7(b), there is also one ER region. In this region, it is seen that there is one crossing of the CB ($\angle Z_1 + 180°$) and phase of Z_2 ($\angle Z_2$). Additionally, the phase derivative of Z_1 is smaller than that of Z_2, i.e., $(\angle Z_1)' < (\angle Z_2)'$. Therefore, it can be inferred that there will be two clockwise encirclements of the critical point of the Nyquist contour of T_1, and the system is predicted to be unstable. These analysis results match with the results using the Nyquist criterion.

Generalized Nyquist stability criterion
The impedance-based stability analysis approach above is mainly illustrated with a SISO system as an example. The method can also be extended to multi-input–multi-output

---

The phase difference $\angle Z_1 - \angle Z_2$ is within $[0°, 360°)$.
$^{\dagger}(\angle Z_1)'$ is the phase derivative of Z_1.

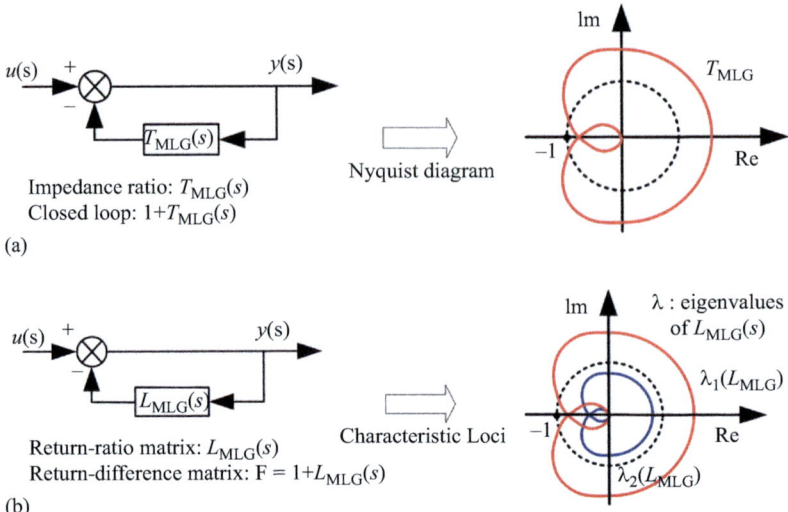

Figure 8.8   Impedance-ratio-based system stability analysis with Nyquist stability criterion: (a) for a SISO system and (b) for a MIMO system

(MIMO) systems, such as multi-port MVDC distribution networks for all-electric ships [18], with the Generalized Nyquist stability criterion.

In a SISO system, the system stability is dependent on the minor loop gain term T_{MLG} as shown in Figure 8.8(a). Similarly, in a MIMO system, a return-ratio matrix $L_{MLG}(s)$ of the system can be derived based on the component connection method (CCM) in the frequency domain with converters separated from the passive connection network [19]. Then the generalized Nyquist stability criterion is applied to analyze the characteristic loci of the eigenvalues of $L_{MLG}(s)$ as shown in Figure 8.8(b).

Additionally, the generalized Bode criterion can also be extended to MIMO systems, where the system stability is determined by analyzing the eigenvalue transfer functions of the return-ratio matrix.

8.2   Small-signal impedance measurement

The state-space model-based small-signal stability analysis requires detailed information of each converter in the system. However, the small-signal stability analysis of a MVDC system is usually performed by system designers, who may not have access to internal architectures of the converters. In this situation, the impedance-based stability analysis approach can be adopted since it only requires the terminal characteristics of each piece of equipment (e.g., converter) or subsystem, which can be acquired through measurement or provided by the vendors. For an equipment vendor, the impedance models can be derived based on the internal structure of the equipment and then validated through measurement. Or for some complex systems, vendors or system integrators can only get the impedance models by measurement. Therefore, measurements of the input and output impedances of connected subsystems play an important role in system stability analysis and design [20].

For DC systems of low voltage and power, commercial products for impedance measurement are available, including the frequency response analyzer (FRA) and input modulators (e.g., the power amplifier (PA) or the line injector). An example circuit of the impedance setup using an FRA and a PA is shown in Figure 8.9, where the FRA normally injects perturbation signals (either sinusoidal or square signals) into the system, and the PA is to provide power up to several hundred watts into the equipment under test (EUT). Both FRA and PA have isolated output so that they can be directly connected to the system at any potential.

However, no commercial product has been released for impedance measurement for MVDC applications [21]. The challenge is mainly on how to generate and inject needed perturbation signals that should be effective and measurable under medium-voltage, high-power conditions. Additionally, isolation issues also need to be considered. For the commercially available FRA and PA, the isolation voltage is usually around a few hundred volts. Therefore, if applying them to MVDC systems, transformers with high bandwidth will be required, otherwise the measurable frequency range will be limited.

To cope with these issues, converters in the system under test can be used without the need for special power hardware. The perturbation signals can be integrated into the control signal of the converter, then the converter output will contain the desired perturbation power [22]. However, this kind of approach requires access to the internal control of the converter. Also, the converter usually has a low-pass filter at the output side, which may filter out the high-frequency perturbation signals. Another approach is to add a converter of the same power and voltage rating as the converter to be measured into the system, which works as the perturbation generation unit [23]. The added converter is controlled to inject perturbations into the system under test. Then, the DC impedances are extracted from the measured voltage and current waveforms. This is an easy-to-use solution, but it requires an additional converter, and the added converter needs to be guaranteed to have no impact on the original operating conditions of the system under test.

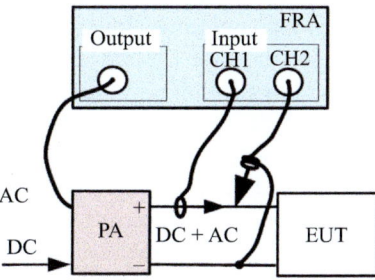

Figure 8.9   Example of impedance measurement setup with an FRA and a PA

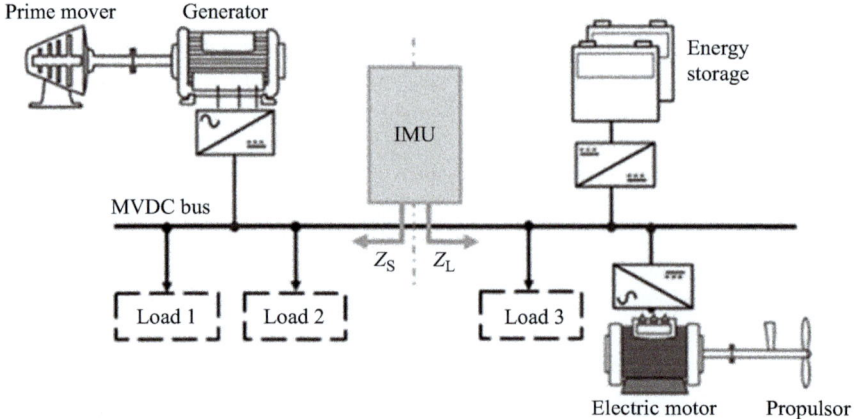

Figure 8.10   Illustration showing IMU insertion into the all-electric ship MVDC distribution system (simplified) [24]

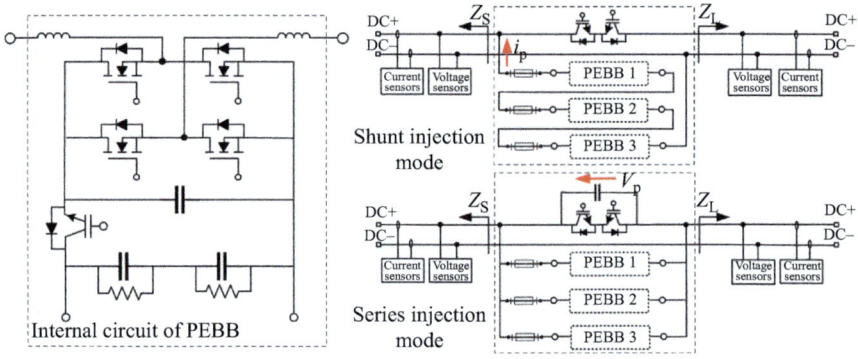

Figure 8.11   Perturbation injection circuits in MVDC system using PEBBs [21]

Another solution is to use an impedance measurement unit (IMU) to inject perturbations into MVDC systems as shown in Figure 8.10 [24]. Its purpose is to measure the upstream and downstream impedances in order to capture the dynamic effect of different control loops of power converters in the system. The IMU consists of several power electronics building block (PEBB) subsystems. In each PEBB, high voltage silicon-carbide (SiC) H-bridge modules are used to reduce the number of PEBBs and achieve a high switching frequency for higher control bandwidth as shown in Figure 8.11 [21]. The output inductor is to limit the current ripple in the current injection mode. The capacitor bank provides the energy buffer to avoid large voltage ripple when the generated perturbations are at very low frequency. The Si IGBT inserted on the DC source side in the internal circuit of PEBB is to protect in case of overcurrent (O/C) faults. Besides, the PEBBs can also be designed to get energy from the system under test without the need for an extra DC power supply. The DC side of the PEBB is floating. And since the injected power is reactive, only losses in the PEBB are active, which is negligible compared to the power rating of the system under test. In this example, three PEBBs form an IMU, which can operate in either shunt injection mode or series injection mode. In the shunt injection mode, the PEBBs are connected in series to form an IMU which is to sustain system voltage and inject perturbation currents into the DC-link of the system under test. And the anti-series IGBTs in the path from the source converter to the load converter are always on. In the series injection mode, the PEBBs are in parallel to sustain the system current and inject voltage perturbations into the system under test. The anti-series IGBTs are initially on, and after the IMU is started, they are turned off. The capacitor in parallel with the anti-series IGBTs is to maintain the perturbation voltage.

8.3 Large-signal model and stability analysis

Many small-signal techniques have been developed to address the instability phenomena in MVDC systems by linearizing the system around the equilibrium point. However, these analysis results are only applicable to the system under small disturbances. While in the operation of an actual MVDC system, large disturbances (e.g., load transients, pulse loads, or fault events) may occur and inflict the system into the transient response stage. In this case, the intrinsic nonlinearity of the system cannot be neglected, and a large-signal model can be applied to capture the underlying nonlinear behaviors. Therefore, stability studies under large disturbances are required.

8.3.1 Large-signal model of MVDC systems

The MVDC systems normally consist of distributed power generations, power loads, connecting lines, and energy storage systems. To investigate large-signal stability issues in the MVDC system, the system is normally simplified as the equivalent circuit shown in Figure 8.12. The source-side converter is modeled as a controlled current source with output capacitor C_o, which is regulated by the droop control with a droop gain m_d ($=1/R_d$) and a current control loop T_{cl}. The RLC filter

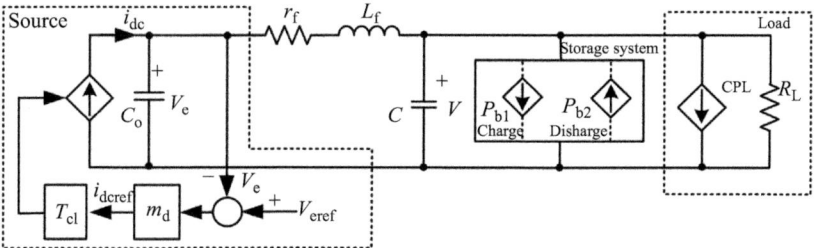

Figure 8.12   The simplified schematic of a MVDC system

(r_f, L_f, and C) represents the actual physical elements, such as the output filter of the front-end converter or the cable impedance. The load side usually consists of resistive loads and power-electronics-interfaced loads which usually behave as CPLs. Additionally, the storage system can work as an inertia element on the DC bus to help stabilize the voltage. When the source output power is more than the load side needed, the storage system works as CPLs in recharging mode; when the source side cannot provide enough power for the load, the storage system acts as a source in discharging mode.

The simplified circuit can represent a single-bus DC system with one source and one load. It can also be extended to systems with multiple sources in parallel and multiple loads in parallel by adding the equivalent circuit blocks in the source side and load side, separately. While for a meshed DC system, the system connectivity can be modeled by the incidence matrix to describe the nodes and distribution lines in the system. The sources and loads can be modeled as the same circuit in the single-bus case. Then, the state-space model for the entire system can be derived [25].

8.3.1.1 Modeling of load dynamics

The dynamic behavior of CPLs under linear assumptions would affect the small-signal stability of a system, it is also the main cause of large-signal instability issues of systems [26, 27]. The CPL can be modeled as a current source with the current drawn by the CPLs from the DC bus equal to $I_{CPL} = P/V$, where P is the load power and V is the load voltage. A typical I/V curve of a CPL is shown in Figure 8.13. It can be seen that an instantaneous impedance is positive ($V/I > 0$), while the incremental impedance is negative ($\Delta V/\Delta I < 0$). The equation representing the straight-line tangent to the I/V curve at the equilibrium point is

$$i \approx -\frac{P}{V^2}v + 2\frac{P}{V} \tag{8.17}$$

Therefore, the CPL can be modeled as a constant current source in parallel with a negative resistor. The negative incremental resistor R_{CPL} at (I_0, V_0) is defined by

$$R_{CPL} = -\frac{V_0^2}{P} \tag{8.18}$$

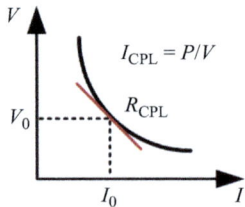

Figure 8.13   I/V curve of a CPL

8.3.1.2 Modeling of fault dynamics

Not only are the large-signal stability issues caused by the tightly regulated load dynamics, but the fault dynamics also have impacts on system large-signal stability. Compared with conventional AC systems, the line impedances and converter impedances in DC system are much lower. In addition, the DC-link capacitor-discharge component, cable-discharge component, and source-side component can all feed current to the fault location. Hence, the inrush fault current will be much larger in DC than in the AC system, for instance, when there is a pole-to-pole fault on the DC bus, all the capacitors in the converters will discharge immediately and feed a large inrush current into the fault location as shown in Figure 8.14.

There are many types of faults in MVDC systems, such as open-circuit fault, short-circuit fault (i.e., phase-to-phase in AC side or pole-to-pole on DC bus), ground fault (i.e., phase-to-ground in AC side or pole-to-ground on DC side), over-voltage fault, etc. [28–30]. In a physical system, fault detection and classification are needed first to distinguish different types of faults [30]. Then, to analyze these fault events and the impact on system stability, the fault currents must be calculated. It can be done by starting from the *RLC* circuit at the instant of the fault, then establishing the pre-fault system matrix and during-fault matrix such that proper protection can be put into action accordingly [28]. Also, the domain of asymptotic stability can be analyzed as the example shown in Figure 8.15, where the hybrid AC/DC power system can be simplified as the equivalent circuit model with a short circuit fault across the load. To obtain such a result with the domain of attraction, large-signal stability analysis techniques need to be applied.

8.3.2 Large-signal stability analysis techniques

Compared to small-signal stability analysis methods, publications on large-signal stability techniques of MVDC systems are limited. The large-signal stability

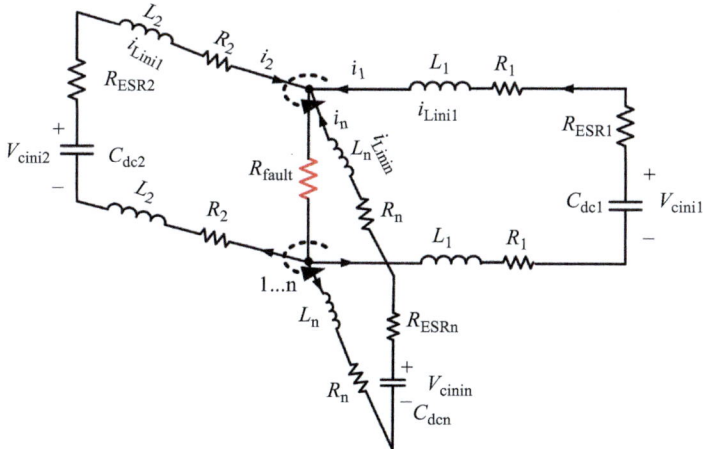

Figure 8.14 RLC circuit of a pole-to-pole fault in DC systems [28]

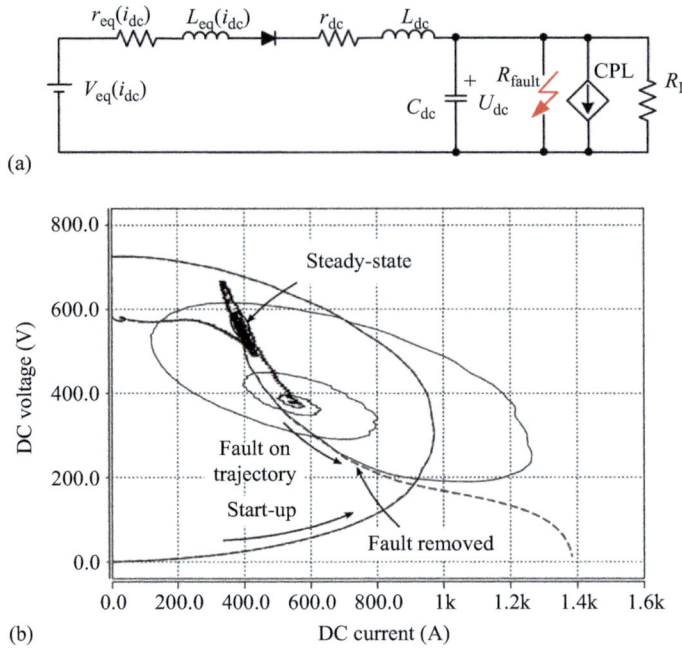

(a)

(b)

Figure 8.15　Large-signal stability analysis of a hybrid AC/DC system with CPLs [31]: (a) equivalent circuit and (b) system trajectory on the phase plane

studies are mostly based on simulations rather than analytical theory. Numerical or circuit models of the system need to be built first, and system performance is then evaluated under certain disturbances. The simulation results can provide insights into the nonlinear or transient behaviors in MVDC systems. However, the simulation process might be computationally intensive to cover all different scenarios and the results are not in the closed-form [32]. To obtain the closed-form solutions and estimate the region of attraction (ROA) i.e., defined as the set of initial states from which the system can converge to the equilibrium point, several mathematical tools for large-signal stability analysis are commonly used, including the Lyapunov direct method, Takagi–Sugeno (TS) fuzzy-model-based method, mixed potential method, etc. [33]. Such knowledge can be used to facilitate the design of the ride-through capability of the MVDC system under load or fault transients.

8.3.2.1 Lyapunov direct method

The Lyapunov direct method, which is also called the second method of Lyapunov, can be used to determine the stability of a system. The general idea of this method is that if there is a measure of energy in the system, which is defined as the Lyapunov function, then system stability can be ascertained by studying the rate of change of energy of the system. Therefore, the key is to find a Lyapunov function and verify that if it meets the required properties.

In an electrical system, such as a MVDC system, the inductor currents i_L and capacitor voltages u_C are normally selected as state variables x, and then the system can be expressed as (8.19) and (8.20):

$$\dot{x} = \Phi(x) \tag{8.19}$$

$$\dot{\Phi}(x) = \frac{\partial \Phi(x)}{\partial x} \dot{x} = J\Phi(x) \tag{8.20}$$

Then, the Lyapunov function can be constructed as (8.21),

$$V(x) = \dot{x}^{\mathrm{T}} P \dot{x} = \Phi(x)^{\mathrm{T}} P \Phi(x) \tag{8.21}$$

where P is positive definite, therefore, $V(x)$ is also positive definite.

If Lyapunov function $V(x)$ satisfies:

1. $V(x)$ is positive definite, and
2. $\dot{V}(x)$ is negative definite

then, the system is asymptotically stable, where $\dot{V}(x)$ is the full derivative of $V(x)$. This means the energy defined by V is always dissipated, except at $x = 0$. Furthermore, if when $\|x\| \to \infty$, $V(x) \to \infty$, then the system is globally asymptotically stable. Note that the Lyapunov direct method can only give sufficient conditions for system large-signal stability.

8.3.2.2 Takagi–Sugeno fuzzy model method

The TS fuzzy model is combined with the Lyapunov theorem to generate a candidate Lyapunov function for system large-signal stability evaluation by completing the following steps [26, 32].

The first step is to build the state-space model in a matrix form of the system while shifting the equilibrium points to the origin for convenience, as $\dot{x} = A(x)x$. If the state matrix A has nonlinearities, then each nonlinear term can be assumed to take the minimum and maximum values. For instance, if there is one nonlinear term $f_1(x_1)$ in state matrix $A(x)$, then the premise variable $\varkappa_1$ can be selected as the function of state variables for simplicity, i.e., $\varkappa_1 = x_1$ in this case. Under $x_1 \in [x_{1min}, x_{1max}]$, $\varkappa_1$ can be further represented by the membership functions M_1 and M_2, where $M_1 + M_2 = 1$, as given in (8.22):

$$\varkappa_1(t) = x_1(t) = M_1(\varkappa_1(t)) \cdot x_{1max} + M_2(\varkappa_1(t)) \cdot x_{1min} \tag{8.22}$$

The membership functions can be expressed as (8.23) and (8.24):

$$M_1(\varkappa_1(t)) = \frac{\varkappa_1(t) - x_{1min}}{x_{1max} - x_{1min}} \tag{8.23}$$

$$M_2(\varkappa_1(t)) = \frac{-\varkappa_1(t) + x_{1max}}{x_{1max} - x_{1min}} \tag{8.24}$$

Then, the final fuzzy model can be written as (8.25):

$$\dot{x}(t) = \sum_{i=1}^{2} H_i(z(t))A_i x(t) \tag{8.25}$$

where, $H_i(z(t))$ is the normalized weighting functions with $H_i(z(t)) = M_i(z_1(t))$, and A_1 and A_2 are the corresponding system matrices under x_{1max} and x_{1min}, separately. If the system state matrix A has r nonlinearities, there would be 2^r matrices A_i, making this TS fuzzy-model-based method ineffective for a higher-order nonlinear system.

The system would be asymptotically stable if the Linear Matrix Inequality (LMI) as given in (8.26) holds:

$$\begin{cases} M = M^{\mathrm{T}} > 0 \\ A_i^{\mathrm{T}} M + MA_i < 0 \end{cases} \tag{8.26}$$

where, M is a symmetric positive definite matrix that needs to be calculated according to the system model A_i. Note that all A_i matrices must be Hurwitz stable (i.e., a square matrix A_i is Hurwitz stable matrix if every eigenvalue of A_i has strictly negative real part) and $A = \sum A_i$ must be a Hurwitz stable matrix to provide a necessary but not sufficient condition for the matrix M to exist.

Once M is calculated, the Lyapunov function can then be expressed as (8.27), and system stability can be determined by the Lyapunov theorem. The TS model-based method also gives conservative stability conditions:

$$V(x) = x^{\mathrm{T}} Mx \tag{8.27}$$

8.3.2.3 Mixed potential function and stability criterion

The mixed potential method was first proposed by Brayton and Moser for the study of nonlinear system stability in 1964. The main idea is to build a power-related scalar function to model the system, which includes the voltage potential function and current potential function. The procedures of constructing the mixed potential function are as follows [27, 33].

First, the differential equation of nonlinear circuit can be derived based on Kirchhoff's law as given in (8.28):

$$\begin{cases} L\dfrac{di_\rho}{dt} = \dfrac{\partial P(i, v)}{\partial i_\rho} \\ C\dfrac{dv_\sigma}{dt} = -\dfrac{\partial P(i, v)}{\partial v_\sigma} \end{cases} \tag{8.28}$$

where, i_ρ are the inductor currents, v_σ are the capacitor voltages, and $P(i,v)$ is the mixed potential function. Note that $P(i,v)$ is also a Lyapunov-type energy function.

Then, the mixed potential function can be obtained in the unified form as (8.29):

$$P(i, v) = -A(i) + B(v) + (i, \gamma v - \alpha) \tag{8.29}$$

where $A(i)$ is the current potential function, $B(v)$ is the voltage potential function, γ is the circuit structure-related function, and α is a constant vector.

If the parameters satisfy (8.30) and (8.31),

$$\mu_1 + \mu_2 \geq \delta > 0, \tag{8.30}$$

$$P^*(i, v) = \left(\frac{\mu_1 - \mu_2}{2}\right) P(i, v) + \frac{1}{2}\left(P_i, L^{-1}P_i\right) + \frac{1}{2}\left(P_v, C^{-1}P_v\right) \rightarrow \infty, \tag{8.31}$$

as $|i| + |v| \rightarrow \infty$, then, the system operating points would move to the steady-state equilibrium point after large disturbances as $t \rightarrow \infty$, where

μ_1 is the minimum eigenvalue of $\quad L^{-\frac{1}{2}}A_{ii}(i)L^{-\frac{1}{2}}$,
μ_2 is the minimum eigenvalue of $\quad C^{-\frac{1}{2}}B_{vv}(v)C^{-\frac{1}{2}}$,

$$A_{ii}(i) = \partial^2 A(i)/\partial i^2, B_{vv}(v) = \partial^2 B(v)/\partial v^2,$$

$$P_i = \partial P(i, v)/\partial i, P_v = \partial P(i, v)/\partial v.$$

The Brayton–Moser's mixed potential theorem is incomplete in some sense [34]. First, the mixed potential function P^* should be defined as $P^*: R^n \rightarrow R$ and be of the class C^1, C^2. Second, the Brayton–Moser's theory can provide sufficient conditions for the convergence to the set which includes all the equilibrium points. However, it cannot indicate the specific equilibrium point that the system will converge to.

These large-signal stability criteria can determine the safe operation regions of the system. The Lyapunov direct method is the most widely used large-signal stability analysis approach. However, the derived stability criterion is usually conservative. The TS fuzzy-model-based approach shares the same conservativeness issue since it is based on the Lyapunov direct criterion. Additionally, the order of the TS model increases with the number of system nonlinearities, which makes the model not suitable for high-order systems considering computational complexity. The mixed potential function-based stability criterion is also a popular approach in recent studies, but its criterion is incomplete in some cases. Practically, system simulation is needed in evaluating the large-signal stability of a MVDC system.

8.4 Methods of improving system stability

8.4.1 Small-signal stability enhancement

The small-signal stability issues in DC systems are usually due to the insufficient damping of system oscillations, which is mainly caused by the negative incremental impedances from CPLs in a small-signal sense. Therefore, the key for DC bus stabilization is to limit the impacts of CPLs. For example, to assure a stable DC system, the specification constraint on the load admittance for system stability can be derived to size the physical elements and to design control parameters on the load-side converter [8]. The system can also be stabilized by adding extra damping for the system impedances (source impedances, load impedances, or bus impedances) through either passive or active means.

The passive damping methods for stabilization of systems with CPLs include adding R, adding RC parallel damping, RL parallel damping, and RL series damping on the source side converter as shown in Figure 8.16. By adding these passive circuits, the output impedance of the source-side converters can be reshaped to meet the small-signal stability criteria. However, when considering other practical design factors, such as weight, size, cost, efficiency, etc., one method may be better than the others. Thus, when applying the passive stabilization method, the application and performance requirements (e.g., system size, weight, cost, and power dissipation) need to be taken into consideration in the system design.

Passive stabilization methods usually introduce extra power dissipation for the system. Hence, the active stabilization method is preferred in some cases, which is generally realized by a feedback control loop with certain control strategies to reshape the system impedances. According to the locations where the stabilization methods are applied, these methods can be further classified as at the source side, at the load side, and at the intermediate circuitry side. The source-side stabilization method modifies the output impedance of the source equipment such as a converter, the load-side method changes the input impedance of loads, and the intermediate circuits reshape the impedance of either the source or the load. Most of the active stabilization methods are implemented on the source side. For example, the state-space small-signal stability analysis can be conducted to determine the control parameters in the source side to improve system damping [1, 37]. Others such as power shaping stabilization control [38], small-signal voltage injection control [39], virtual phase lead impedance stability control [40], etc. are all effective methods for system stabilization by modifying the output impedance of source-side

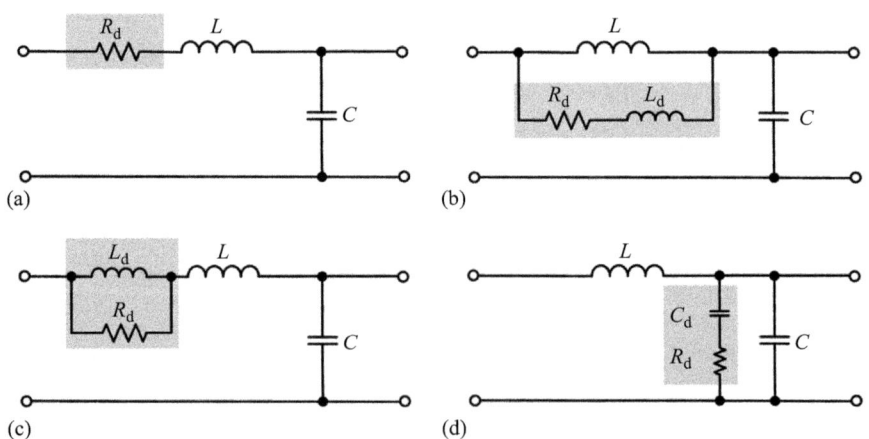

Figure 8.16   Methods of passive damping: (a) adding R damping, (b) adding RL parallel damping, (c) adding RL series damping, and (d) adding RC damping [35, 36]

converters. Similarly, there are some active stabilization methods implemented on the load side to reshape the input impedance, such as the virtual impedance control [41]. However, it is usually challenging to regulate the load impedance for system stability and achieve good load dynamics at the same time since the negative incremental impedance of the load converter is good for load dynamics, while bad for system stability. The stabilization method can also be implemented at the intermediate circuitry side, such as adaptive capacitor converter control [42] or energy-buffering control [43], which can maintain the modularity of converter-based DC systems without any modification on source-side or load-side converters.

All the active stabilization methods can have a minimal change to the system hardware and almost no extra power dissipation is introduced. However, they may also have some disadvantages. First, the stabilization control may contradict other control objectives, such as load dynamics. Second, the control strategies may only be compatible with specific converters. Third, an extra voltage or current sensing circuit may be needed to implement the stabilization control.

8.4.2 Large-signal stability enhancement

Large-signal stability issues in MVDC systems are mainly caused by the presence of load dynamics or faults. Therefore, to improve system large-signal stability, the key is to stabilize CPLs and adopt appropriate protection strategies.

8.4.2.1 Stabilization of constant power loads

Similar to small-signal stabilization methods, existing methods to solve the CPL-induced issues in a large-signal sense can be also categorized as hardware-based methods and control-based methods.

The hardware-based methods add extra damping resistors or capacitor banks to compensate for the negative incremental impedance. For example, storage systems as shown in Figure 8.12 can help improve system large-signal stability if the charging and discharging power of storage systems meet the stability criteria [44]. Nevertheless, the added physical elements will reduce efficiency and increase weight and size of the system, which may not be applicable for some size/weight-sensitive applications, such as the shipboard MVDC power system.

The control-based methods adopt different control strategies on the load side or source side to stabilize the system [45, 46]. The load side stabilizing control strategies usually include the state of the bus voltage to generate reference power for the system, in which the basic principle is state variable-dependent adaptive virtual impedance control (i.e., virtual capacity or virtual resistance). The stabilization from the load side can also be obtained by real-time load shedding since load shedding is also an important backup solution during inadequate generation or voltage swings to ensure stable and continuous operation of the online loads. For the load side control, since each load normally needs a special converter, stabilizing control methods need to be designed accordingly and may have to be recalculated if the system is reconfigured. In addition, in some applications such as systems with pulse loads, a stiff DC bus voltage is important for system operation. Therefore, the stabilization control needs to be implemented in the source side to save efforts on developing the control methods for all different types of loads and help enhance bus voltage regulation capability. The large-signal stability stabilization methods in the source side have been developed based on nonlinear control theorems, such as passivity-based control, sliding-mode control, active damping, state-feedback control, or boundary control.

8.4.2.2 Protection in MVDC systems

Different characteristics of fault currents in DC systems compared with that in AC, such as no zero-crossing point, high amplitude, and fast-rising rate, make protection a challenging task for the stable operation of MVDC systems, including fault detection and fault interruption devices/methods [47, 48]. Compared with AC systems, fault protection methods in the MVDC systems are still in the early stages of development.

For fault detection in DC systems, there are some challenges compared with that in AC systems. First, there is no frequency and phasor information in DC systems, which makes many methods in AC systems not applicable. Second, the limited time-domain information and the response speed requirements in DC systems make it hard to use some digital-signal-processing method with the measurement voltage/current data. The existing fault detection methods in DC systems can be categorized into four types [47]: current-based detection, distance protection, differential protection, and signal processing-based methods. Due to the simplicity, the current-based detection method with O/C relays is popular in AC systems and small DC systems. However, it may not provide adequate selectivity for multi-terminal or meshed DC systems. Establishing communications between O/C relays can help improve the protection coordination and selectivity. Note that a complete selectivity means that protection strategies can minimize the effect of fault events on power systems. The distance protection scheme is suitable for MVDC networks, with the fault distance being estimated using two DC measuring elements installed on the protected lines. The differential protection has little dependency on the fault current level, the existence of distributed generations, fault resistance, or the rate of change of the current, making it one of the best options for DC protection. The signal processing-based methods provide selective protection for DC systems, such as the wavelet-transform-based method or traveling-waves-based method.

The main challenge of fault interrupting schemes in DC systems is to find a fast and cost-effective solution. The existing fault removal methods in DC systems can be classified as the following [47]: converter blocking only, converter blocking with AC-side circuit breakers (CBs), converter blocking with AC-side CBs and DC isolator switches, fault current limiters (FCLs), and the DC circuit breakers (DCCBs). The converter blocking only scheme is to turn off the main converter switches through the converter control after a fault occurrence. Note that depending on the converter topology, there may or may not be fault current paths after the converter is blocked. Therefore, the converter blocking only scheme can just apply to converters with fault circuit current blocking topologies. The converter blocking only protection can de-energize the line when there is a fault within the line in a two-terminal DC system. However, it may cause unnecessary outages of normal loads and sources, especially, in multi-terminal DC systems. The converter blocking with AC-side CBs are suitable for converters without fault current blocking capability, and can provide an economic solution for DC grid protection. The CBs can prevent the fault current contribution from the AC side. However, they are generally very slow and may negatively impact the system, e.g., requiring high fault current capability for converters, or causing large-signal-stability issues. Without DC fault isolation, it is difficult to provide adequate protection coordination and selectivity, especially, for meshed or multi-terminal MVDC systems. A fast DC isolator switch can be integrated with the converter blocking with AC-side CBs to solve the selectivity issues. However, the discharge current of capacitors is still allowed to flow into the fault, which may damage the system equipment without proper design and protection. Fault current limiters such as DC FCLs with superconducting windings, can be added to protect sensitive devices in the system, which, however, is costly. DCCBs are designed based on the specification of DC fault currents, e.g., a solid-state DCCB with adequate power ratings in series with a DC-link capacitor can limit and stop the capacitor discharging currents from flowing. Although DCCBs can interrupt the major portion of the fault current, i.e., the DC-link capacitor discharging elements, they may also affect the bus voltage and system power quality by uncoordinated tripping of the DC-link capacitors. MVDC system protection and its impact on system stability remain a topic for further research.

8.5 Summary

The small-signal and large-signal stability analyses are essential to the design and operation of electrical systems, including MVDC systems. Therefore, in order to cover the basic knowledge for MVDC system stability studies, this chapter presents some modeling methods for system characterization, analytical tools for stability analysis, and approaches for stability improvement.

Small-signal stability analysis for MVDC systems, like for other dynamical systems, is conducted based on the linearization of the system around a certain operating point. The linearized system can be represented by various small-signal models, e.g., the state-space model, the impedance model. The state-space models are best for global stability analysis that can be analyzed through the eigenvalue analysis and participation factors calculation. The impedance models are more suitable for local stability analysis by checking the impedance interaction at the connecting terminal with the stability criteria, such as the Middlebrook criterion, the forbidden-region-based criteria, passivity-based criterion, Nyquist stability criterion, generalized Bode criterion, etc. One can choose a stability criterion for system analysis and design based on the needs, e.g., existence of open-loop RHP poles, simplicity of using the method, sufficiency and necessity of the analysis results, or suitability for controller design. The impedance model-based stability analysis method also shows advantages in cases when the system characteristics can only be obtained at the terminal through modeling or measurement. However, there is no commercially available impedance measurement instrument yet for medium voltage, high power applications due to the difficulty in generating the injected perturbation signals. Therefore, the impedance measurement units to generate the effective and measurable injected perturbation signals are needed, such as PEBB-based IMU.

The small-signal stability studies are only valid under linear assumptions, so large-signal stability studies are needed for stability analysis considering non-linearities in the system. Large-signal stability issues in power systems are mainly caused by load dynamics or fault events. To see the ability of a system to return to an acceptable steady-state operating region under large disturbances, both time-domain simulations and nonlinear mathematical analysis methods can be adopted. Time-domain simulations can provide insights into the transient performance of the system. But they cannot give a closed-form solution for the system stable operation. Therefore, mathematical analysis methods are needed, e.g., Lyapunov direct method, TS fuzzy-model-based method, mixed potential method, etc. These stability analysis methods can be used to estimate the region of attraction of the system so that it is possible to quantify the allowed disturbance magnitude or component sizing to avoid large-signal instability issues. Additionally, simulation tools can complement well with the mathematical analysis methods for more accurate and efficient large-signal stability analysis of the system.

Stability issues in DC systems including MVDC systems are mainly caused by the negative incremental impedance induced by well-regulated CPLs in both small-signal sense and large-signal sense. To enhance the small-signal stability of the system, both passive damping (e.g., *RLC* damping) and active damping (e.g., virtual impedance control) methods can be adopted to reshape impedances at either source subsystem or load subsystem to provide enough damping for system stabilization. Similarly, for the improvement of CPLs-induced large-signal stability, hardware-based methods (e.g., adding damping resistor) and control-based methods (e.g., virtual resistance or virtual capacity) are normally adopted. While for fault-induced large-signal instabilities, proper protection strategies are required, including fault detection and fault interrupting schemes. Transient currents and voltages in different types of fault events need to be sensed or estimated correctly so that effective protection schemes for the MVDC system can be designed. Additionally, semiconductor protection devices with adequate power ratings should be selected to limit fault current and isolate the fault location from the rest of the system.

Acknowledgments

This work is partially supported by the Engineering Research Center Program of the National Science Foundation and the Department of Energy under NSF Award Number EEC1041877 and the CURENT Industry Partnership Program. This work is also partially supported by the US Department of Energy, Office of Electricity, Advanced Grid Modeling Program under contract DE-AC05-00OR22725.

References

[1] G. O. Kalcon, G. P. Adam, O. Anaya-Lara, S. Lo, and K. Uhlen, "Small-signal stability analysis of multi-terminal VSC-based DC transmission systems," *IEEE Transactions on Power Systems*, vol. 27, no. 4, pp. 1818–1830, 2012.

[2] P. Simiyu, A. Xin, G. T. Bitew, M. Shahzad, W. Kunyu, and P. M. Kamunyu, "Small-signal stability analysis for the multi-terminal VSC MVDC distribution network: a review," *The Journal of Engineering*, vol. 2019, no. 16, pp. 1068–1075, 2019.

[3] C. M. Wildrick, F. C. Lee, B. H. Cho, and B. Choi, "A method of defining the load impedance specification for a stable distributed power system," *IEEE Transactions on Power Electronics*, vol. 10, no. 3, pp. 280–285, 1995.

[4] Q. Zhong and X. Zhang, "Impedance-sum stability criterion for power electronic systems with two converters/sources," *IEEE Access*, vol. 7, pp. 21254–21265, 2019.

[5] J. Siegers, S. Arrua, and E. Santi, "Allowable bus impedance region for MVDC distribution systems and stabilizing controller design using positive feed-forward control," in *Energy Conversion Congress and Exposition (ECCE)*, Milwaukee, WI, USA, 2016: IEEE, pp. 1–8.

[6] R. Middlebrook, "Input filter considerations in design and application of switching regulators," in *IEEE Industry Applications Society Annual Meeting*, 1976, pp. 366–382.

[7] X. Feng, J. Liu, and F. C. Lee, "Impedance specifications for stable DC distributed power systems," *IEEE Transactions on Power Electronics*, vol. 17, no. 2, pp. 157–162, 2002.

[8] S. D. Sudhoff, S. F. Glover, P. T. Lamm, D. H. Schmucker, and D. E. Delisle, "Admittance space stability analysis of power electronic systems," *IEEE Transactions on Aerospace and Electronic Systems*, vol. 36, no. 3, pp. 965–973, 2000.

[9] S. D. Sudhoff and J. M. Crider, "Advancements in generalized immittance based stability analysis of DC power electronics based distribution systems," in *IEEE Electric Ship Technologies Symposium*, Alexandria, VA, USA, 2011, pp. 207–212.

[10] A. Riccobono and E. Santi, "Comprehensive review of stability criteria for DC power distribution systems," *IEEE Transactions on Industry Applications*, vol. 50, no. 5, pp. 3525–3535, 2014.

[11] A. Riccobono and E. Santi, "Stability analysis of an all-electric ship MVDC power distribution system using a novel passivity-based stability criterion," in *IEEE Electric Ship Technologies Symposium (ESTS)*, Arlington, VA, USA, 2013, pp. 411–419.

[12] K. Ogata, *Modern Control Engineering*. Prentice Hall PTR, 2001.

[13] J. Sun, "Impedance-based stability criterion for grid-connected inverters," *IEEE Transactions on Power Electronics*, vol. 26, no. 11, pp. 3075–3078, 2011.

[14] B. Wen, D. Boroyevich, R. Burgos, P. Mattavelli, and Z. Shen, "Inverse Nyquist stability criterion for grid-tied inverters," *IEEE Transactions on Power Electronics*, vol. 32, no. 2, pp. 1548–1556, 2017.

[15] D. Lumbreras, E. L. Barrios, A. Urtasun, A. Ursúa, L. Marroyo, and P. Sanchis, "On the stability of advanced power electronic converters: the generalized Bode criterion," *IEEE Transactions on Power Electronics*, vol. 34, no. 9, pp. 9247–9262, 2019.

[16] Y. Liao and X. Wang, "General rules of using bode plots for impedance based stability analysis," presented at the IEEE Workshop on Control and Modeling for Power Electronics (*COMPEL*), Padova, Italy, 2018.

[17] Y. Liao and X. Wang, "Impedance-based stability analysis for inter-connected converter systems with open-loop RHP poles," *IEEE Transactions on Power Electronics*, vol. 35, no. 4, pp. 4388–4397, 2020.

[18] U. Javaid, F. D. Freijedo, W. v.d. Merwe, and D. Dujic, "Stability analysis of multi-port MVDC distribution networks for all-electric ships," *IEEE Journal of Emerging and Selected Topics in Power Electronics*, vol. 8, no. 2, pp. 1164–1177, 2020.

[19] W. Cao, Y. Ma, F. Wang, L. M. Tolbert, and Y. Xue, "Low-frequency stability analysis of inverter-based islanded multiple-bus AC microgrids based on terminal characteristics," *IEEE Transactions on Smart Grid*, vol. 11, no. 5, pp. 3662–3676, 2020.

[20] M. Hiermeier, J. Pforr, M. Mürken, and T. Hackner, "Measurement technique to determine the impedance of automotive energy nets for stability analysis purpose based on a floating capacitor H-bridge converter," in *IEEE Energy Conversion Congress and Exposition (ECCE)*, Milwaukee, WI, USA, 2016, pp. 1–8.

[21] Z. Shen, M. Jaksic, I. Cvetkovic, R. Burgos, and D. Boroyevich, "Small-signal impedance measurement in medium-voltage DC power systems," in *International Conference on Electrical Systems for Aircraft, Railway, Ship Propulsion and Road Vehicles (ESARS)*, Aachen, Germany, 2015, pp. 1–5.

[22] M. Shirazi, J. Morroni, A. Dolgov, R. Zane, and D. Maksimovic, "Integration of frequency response measurement capabilities in digital controllers for DC–DC converters," *IEEE Transactions on Power Electronics*, vol. 23, no. 5, pp. 2524–2535, 2008.

[23] L. Kong, N. Praisuwanna, L. Qiao, and F. Wang, "Development of a two-Level VSC based DC impedance measurement unit," in *IEEE Energy Conversion Congress and Exposition (ECCE)*, Detroit, MI, USA, 2020, pp. 2939–2944.

[24] F. Wang, Z. Zhang, T. Ericsen, R. Raju, R. Burgos, and D. Boroyevich, "Advances in power conversion and drives for shipboard systems," *Proceedings of the IEEE*, vol. 103, no. 12, pp. 2285–2311, 2015.

[25] N. H. v. d. Blij, L. M. Ramirez-Elizondo, P. Bauer, and M. T. J. Spaan, "Design guidelines for stable DC distribution systems," in *IEEE Second International Conference on DC Microgrids (ICDCM)*, Nuremburg, Germany, 2017, pp. 279–284.

[26] H. Kim, S. Kang, G. Seo, P. Jang, and B. Cho, "Large-signal stability analysis of DC power system with shunt active damper," *IEEE Transactions on Industrial Electronics*, vol. 63, no. 10, pp. 6270–6280, 2016.

[27] J. Jiang, F. Liu, S. Pan, X. Zha, W. Liu, C. Chen *et al.*, "A conservatism-free large signal stability analysis method for DC microgrid based on mixed potential theory," *IEEE Transactions on Power Electronics*, vol. 34, no. 11, pp. 11342–11351, 2019.

[28] C. Li, C. Zhao, J. Xu, Y. Ji, F. Zhang, and T. An, "A pole-to-pole short-circuit fault current calculation method for DC grids," *IEEE Transactions on Power Systems*, vol. 32, no. 6, pp. 4943–4953, 2017.

[29] X. Shi, Z. Wang, B. Liu, Y. Li, L. M. Tolbert, and F. Wang, "DC impedance modelling of a MMC-HVDC system for DC voltage ripple prediction under a single-line-to-ground fault," in *IEEE Energy Conversion Congress and Exposition (ECCE)*, Pittsburgh, PA, USA, 2014, pp. 5339–5346.

[30] W. Li, A. Monti, and F. Ponci, "Fault detection and classification in medium voltage DC shipboard power systems with wavelets and artificial neural networks," *IEEE Transactions on Instrumentation and Measurement*, vol. 63, no. 11, pp. 2651–2665, 2014.

[31] A. Griffo and J. Wang, "Large signal stability analysis of 'more electric' aircraft power systems with constant power loads," *IEEE Transactions on Aerospace and Electronic Systems*, vol. 48, no. 1, pp. 477–489, 2012.

[32] M. Kabalan, P. Singh, and D. Niebur, "Large signal Lyapunov-based stability studies in microgrids: a review," *IEEE Transactions on Smart Grid*, vol. 8, no. 5, pp. 2287–2295, 2017.

[33] W. Xie, M. Han, W. Cao, J. M. Guerrero, and J. C. Vasquez, "System-level large-signal stability analysis of droop-controlled DC microgrids," *IEEE Transactions on Power Electronics*, vol. 36, no. 4, pp. 4224–4236, 2021.

[34] F. Chang, X. Cui, M. Wang, W. Su, and A. Q. Huang, "Large-signal stability criteria in DC power grids with distributed-controlled converters and constant power loads," *IEEE Transactions on Smart Grid*, vol. 11, no. 6, pp. 5273–5287, 2020.

[35] A. M. Rahimi and A. Emadi, "Active damping in DC/DC power electronic converters: a novel method to overcome the problems of constant power loads," *IEEE Transactions on Industrial Electronics*, vol. 56, no. 5, pp. 1428–1439, 2009.

[36] M. Cespedes, L. Xing, and J. Sun, "Constant-power load system stabilization by passive damping," *IEEE Transactions on Power Electronics*, vol. 26, no. 7, pp. 1832–1836, 2011.

[37] S. R. Rudraraju, A. K. Srivastava, S. C. Srivastava, and N. N. Schulz, "Small signal stability analysis of a shipboard MVDC power system," in *IEEE Electric Ship Technologies Symposium*, Baltimore, MD, USA, 2009, pp. 135–141.

[38] J. Wang and D. Howe, "A power shaping stabilizing control strategy for DC power systems with constant power loads," *IEEE Transactions on Power Electronics*, vol. 23, no. 6, pp. 2982–2989, 2008.

[39] A. Aldhaheri and A. Etemadi, "DC distributed systems stabilization and performance improvement using small-signal voltage injection," in *IEEE Applied Power Electronics Conference and Exposition (APEC)*, San Antonio, TX, USA, 2018, pp. 3481–3485.

[40] W. Wu, Y. Chen, L. Zhou, X. Zhou, L. Yang, Y. Dong *et al.*, "A virtual phase-lead impedance stability control strategy for the maritime VSC-HVDC system," *IEEE Transactions on Industrial Informatics*, vol. PP, no. 99, pp. 1–1, 2018.

[41] X. Zhang, X. Ruan, and Q. C. Zhong, "Improving the stability of cascaded DC/DC converter systems via shaping the input impedance of the load converter with a parallel or series virtual impedance," *IEEE Transactions on Industrial Electronics*, vol. 62, no. 12, pp. 7499–7512, 2015.

[42] X. Zhang, X. Ruan, H. Kim, and C. K. Tse, "Adaptive active capacitor converter for improving stability of cascaded DC power supply system," *IEEE Transactions on Power Electronics*, vol. 28, no. 4, pp. 1807–1816, 2013.

[43] M. Gutierrez, P. Lindahl, A. Banerjee, and S. B. Leeb, "Controlling the input impedance of constant power loads," in *IEEE Applied Power Electronics Conference and Exposition (APEC)*, San Antonio, TX, USA, 2018, pp. 3452–3458.

[44] X. Liu and Y. Bian, "Large signal stability analysis of the DC microgrid with the storage system," in *IEEE International Conference on Electrical Machines and Systems (ICEMS)*, Sydney, NSW, Australia, 2017, pp. 1–5.

[45] M. Cupelli, F. Ponci, G. Sulligoi, A. Vicenzutti, C. S. Edrington, T. El-Mezyani *et al.*, "Power flow control and network stability in an all-electric ship," *Proceedings of the IEEE*, vol. 103, no. 12, pp. 2355–2380, 2015.

[46] E. Hossain, R. Perez, A. Nasiri, and S. Padmanaban, "A comprehensive review on constant power loads compensation techniques," *IEEE Access*, vol. 6, pp. 33285–33305, 2018.

[47] M. Monadi, M. Amin Zamani, J. Ignacio Candela, A. Luna, and P. Rodriguez, "Protection of AC and DC distribution systems embedding distributed energy resources: a comparative review and analysis," *Renewable and Sustainable Energy Reviews*, vol. 51, pp. 1578–1593, 2015.

[48] J. Duan, Z. Li, Z. Wei, and W. Lu, "A line accelerated protection scheme of flexible MVDC distribution system based on transient current derivative," *Electric Power Systems Research*, vol. 183, p. 106269, 2020.

Chapter 9

Overview of protection technologies in MVDC system

Xiaoqing Song[1], Pietro Cairoli[1] and Marco Riva[2]

9.1 Introduction

9.1.1 Challenges of protection in MVDC system

Medium voltage direct current (MVDC) systems act as a layer of infrastructure between transmission and distribution to facilitate the installation of renewable resources and DC loads like wind farms, solar farms, electrical vehicles (EVs), etc. With reduced stages of power conversion, MVDC systems potentially feature higher efficiency compared to conventional AC systems. However, MVDC systems also propose new challenges, especially coming from the fault protection.

9.1.1.1 No current zero crossing, and difficult to quench the arc

The first challenge is how to interrupt the DC current without natural current zero crossing. To force the DC current to zero, a counter voltage with higher amplitude than the system voltage needs to be created by circuit breakers, from either mechanical circuit breakers (MCBs) arc voltages or energy arrestors like metal oxide varistors (MOVs). For MCBs, some auxiliary circuits like inductor–capacitor resonant circuits are usually required to force the arc current to zero and extinguish the arc. In solid state circuit breakers (SSCBs) and hybrid circuit breakers (HCBs), the current is commutated from power semiconductor devices to MOVs, and then forced to zero by the MOV clamping voltage. However, a high current turn-off capability is needed for the power semiconductor devices to safely interrupt the potentially high fault current.

9.1.1.2 High interruption speed requirement due to high fault current rising di/dt

Second challenge in MVDC system protection comes from the high fault current rising rate (di/dt), especially in short circuit fault, because of the low system impedance and relatively high DC bus voltage. This requires the protection equipment to have fast response time and high current interruption speed. If the short circuit

[1]ABB US Research Center, Raleigh, NC, USA
[2]ABB S.p.A, Dalmine, Italy

current cannot be interrupted quickly, the peak fault current could rise to a dramatically high level, exceeding the current interruption capacity of circuit breakers.

9.1.1.3 High magnetic energy to be absorb during fault current interruption

The high peak current level during a short circuit fault means a large amount of energy stored in the system loop inductances. The great amount of energy also challenges the energy absorbing components like varistors, in SSCBs and HCBs, because all the magnetic energy needs to be absorbed by varistors or other energy storage components like capacitors during the fault current interruption. A carefully selection and design of energy absorbing components is needed to address this challenge.

9.1.1.4 High thermal stresses for solid state devices

Another challenge is specially for SSCBs and HCBs. In SSCBs, the conduction resistances of power semiconductor devices are usually several times or tens of times higher than that of mechanical switches (MSs), which propose a great challenge for SSCBs to address the high conduction losses. During a fault, the current could rise to several times of the nominal current. With all the fault current conducting through the power semiconductor devices, the device junction temperature could rise fast and even exceed the device temperature limit. Especially, the power semiconductor devices are required to withstand the high fault current for a certain period according to some standards and the selected fault tripping curves. Power semiconductor devices with lower conduction resistances are preferred and optimal cooling system needs to be carefully designed to address the high thermal stresses.

9.1.2 Classification of MVDC circuit breakers

Circuit breakers are the primary apparatus used to protect against overload or short circuit faults in MVDC systems. Based on composition and the current interruption mechanism, circuit breakers can be classified into MCBs, SSCBs, and HCBs. MCBs contain only mechanical switches for current conduction and interruption, while SSCBs are realized by power semiconductor switches for current conduction and interruption. HCBs contain both mechanical switches and power semiconductor switches, with the current conducted mainly through mechanical switches and interrupted by power semiconductor switches.

One major advantage of MCBs is low conduction losses, thanks to the low contact resistance, which is usually less than 1 mΩ, while the conduction resistances of power semiconductor switches vary from a few milliohms to hundreds of milliohms. One major disadvantage of MCBs is the long fault current break time, which varies from tens of milliseconds to hundreds of milliseconds. Another disadvantage is the short lifetime and frequent maintenance required. During a fault current interruption event, the arcing etches and causes damages to the contacts of the mechanical switch and maintenance is needed every a few fault clearing events.

Contrast to MCBs, SSCBs are famous for its ultra fast current interruption speed, which is several orders of magnitude faster than that of MCBs. With the fast current

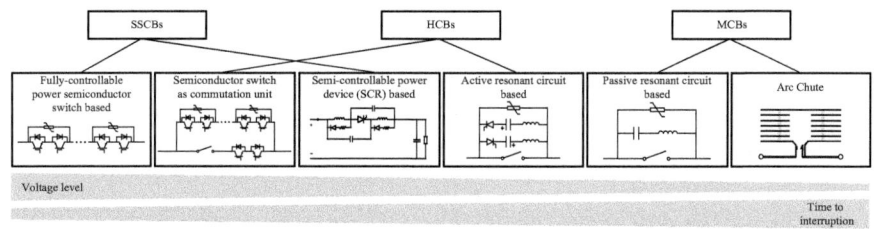

Figure 9.1   Classification of MVDC circuit breakers

interruption speed, the fault current is interrupted before rising to a high level, and the energy in the system parasitic inductances which needs to be absorbed, is also reduced due to lower peak fault current. One major disadvantage of SSCBs is high conduction losses, requiring bulky cooling system to address the thermal stress. The high cost of SSCBs is another major concern, although the cost of some power semiconductor devices like SiC metal oxide semiconductor field effect transistors (MOSFETs) keeps decreasing.

HCBs combine the advantages of both SSCBs and MCBs: low conduction losses and relatively fast current interruption speed. With the current mainly conducted through mechanical switches, conduction losses are reduced, and the fault current is interrupted by solid state switches with fast current interruption speed. The major concern of HCBs is the relatively high cost due to more components and more complicated circuits.

Figure 9.1 shows a classification of the different MVDC circuit breaker technologies. The fully controllable power semiconductor switch-based SSCBs have the fastest current interruption speed and can achieve high voltage level operation by series connection of multiple fundamental cells. The semi-controllable power device like silicon-controlled rectifiers (SCRs)-based SSCBs have slower current interruption speed and less scalability for series connection due to the resonant circuits needed to create zero current crossing. HCBs with semiconductor switches as commutation unit can achieve relatively fast current interruption speed, depending on the separation speed of the fast disconnect mechanical switch. The active resonant circuit-based HCBs have similar interruption speed as the SCR-based SSCBs, as the current interruption mechanism is very similar. MCBs have the slowest current interruption speed and lowest operation current levels compared to the SSCBs and HCBs, although the passive resonant circuit-based MCBs can interrupt the current faster compared to the arc chute MCBs.

9.2   Mechanical circuit breaker

9.2.1   Introduction to mechanical circuit breaker

When interrupting the current, circuit breakers always need to create a counter voltage higher than the system voltage to force the current to zero. The counter voltage can be created by either the mechanical switch arc voltages or surge arresters like MOVs. Figure 9.2 shows an arc chute-based MCB which utilizes the arc voltage to achieve the high counter voltage. To increase the arc voltage, the arc is split into several series connected arcs through the arc chute. To avoid the bulky apparatus, this technology is more feasible for relatively low voltage (below 5 kV) applications, like traction and photovoltaic systems [1,2].

To increase the operational voltage level and the current interruption speed, MCBs used in MVDC protection usually need some auxiliary components/circuits, which create a current zero crossing during current interruption to facilitate the arc extinguishing. Two methods are commonly used to create a current zero crossing for MCBs: passive resonant and active resonant. The following two sections will discuss these two methods respectively.

9.2.2   Passive resonant DC mechanical circuit breaker

Figure 9.3 shows a typical passive resonant mechanical DC circuit breaker topology [3]. The nominal current conducts through the AC MS with low conduction losses and the voltage across the capacitor C is zero. When there is a fault, the MS is opened, and the arc voltage commutates the current from MS to the LC resonant branch. The current starts to oscillate between the inductor L and the capacitor C with exponentially increasing current amplitude. Once the resonant current in the LC branch rises to the fault current level, a current zero-crossing is created for MS and the arc is extinguished. Meanwhile the MS contacts keep separating and the gap distance needs to be large enough to withstand the system voltage when the current zero-crossing occurs. After the MS is fully opened, all the current is

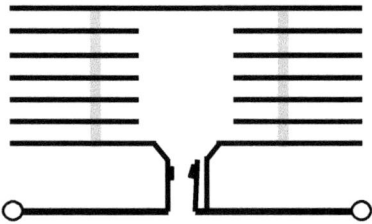

Figure 9.2   Arc chute-based MCBs

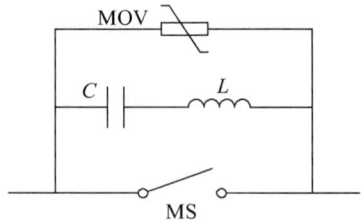

Figure 9.3   Passive resonant DC MCB topology

commutated to the LC branch and continuously resonates. When the voltage amplitude resonates to the MOV clamping voltage, the current starts to commutate to the MOV branch, and eventually decreases to zero with the residual energy in line inductances absorbed by MOVs.

The passive resonant DC MCB is easy to realize by simply adding a LC circuit in parallel with the conventional AC MCB. The fault interruption time is relatively long (several tens of milliseconds) due to the LC circuit resonance time for the current to create a current zero-crossing. To limit the fault current rising rate, a large inductance L is usually required and the LC circuit time constant should be carefully tuned based on the MS contacts separation speed to achieve large enough dielectric strength between the MS contacts at the current zero-crossing transient.

9.2.3 Active resonant DC mechanical circuit breaker

Figure 9.4 shows a typical active resonant mechanical DC circuit breaker topology, which is very similarly to the passive resonant DC MCB in Figure 9.3, except the pre-charged capacitors C1 and C2. Two LC resonant branches are paralleled with the main MS MS1 to realize the bidirectional current interruption. Two auxiliary MSs, MS2 and MS3, in each LC branch are used to excite the resonance. When there is a fault, MS1 is opened and the corresponding auxiliary switch MS2 or MS3 (depending on the current direction) is closed to excite the LC resonance. A zero current-crossing is created for the MS1 when the LC current resonates to the fault current level, and the arc in the MS MS1 is extinguished. Similarly, to the passive resonant DC MCB, the LC branch current starts to commutate to MOV when the voltage reaches the MOV clamping voltage. The MOV absorbs the residual energy in the line inductances and the current eventually decays to zero. The auxiliary MS MS2 or MS3 is opened when the current goes to zero and the fault is isolated.

Compared to the passive resonant DC MCB in Figure 9.3, the pre-charged capacitor provides a fast rising discharge current and the active resonant DC MCB's current interruption speed is much faster due to the shorter LC resonance time needed. Reference [4] presented an active resonant DC MCB, which has successfully interrupted up to 10.5 kA within 5 ms. It is obvious that the circuit of active resonant DC MCB is more complex compared to the passive resonant DC MCB, and two LC resonant branches are required for bidirectional current

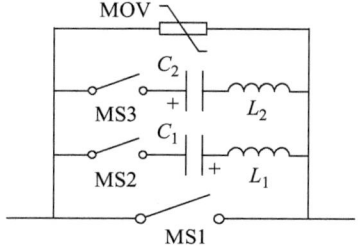

Figure 9.4 Active resonant DC MCB topology

interruption. Additional charging circuits are also needed for capacitors C1 and C2, and the right control logic and timing for all the MSs is necessary.

9.3   Solid state circuit breaker

Solid-state circuit breakers (SSCBs) are power semiconductor-based protection apparatuses, with no moving parts for fault current interruption, renowned for their ultra-fast fault current tripping speed and arc-less current interruption.

9.3.1   *Introduction to solid state circuit breaker*

9.3.1.1   **Fundamentals**

Figure 9.5 shows the key consisting components for a typical SSCB. The power semiconductor devices (or solid state devices) are used to conduct the load current and interrupt the fault current with the change of gate driver signal. Different types of power semiconductor devices can be adopted for SSCBs: based on the semiconductor conduction current modularity mechanism, it can be classified into unipolar power devices (like the MOSFET) and bipolar power devices [like the Insulated gate bipolar transistors (IGBTs), integrated gate-commutated thyristors (IGCTs)]; based on the semiconductor materials, it can be classified into conventional Silicon (Si) devices and the emerging wide band-gap (WBG) power devices like Silicon Carbide (SiC) or Gallium Nitride (GaN).

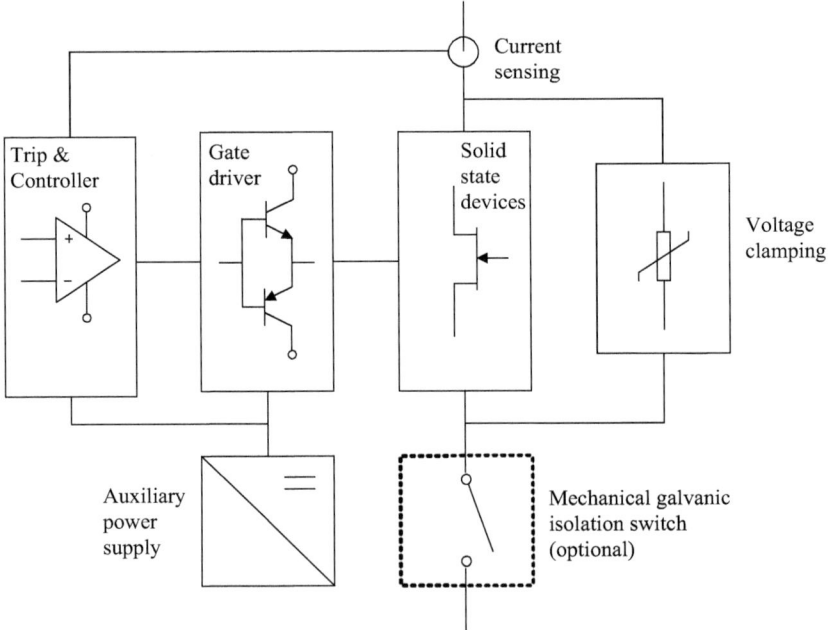

Figure 9.5   The key consisting components for SSCB

The gate driver unit is used to provide enough voltage or current to the gate of the power semiconductor device to turn on or off the conduction current. Different from the power electronics converters, the fast switching speed is not crucial in SSCBs because of limited switching times. Sometimes, a lower switching speed is even preferred to avoid the over-voltage stress across the power semiconductor switches.

The trip and controller unit are needed to monitor the load current and send out the tripping signal to the gate driver circuit when there is a fault. In some cases, the characteristics of the power semiconductor devices (like the conduction voltage drops) could be used as the current measurement method to eliminate the necessity of current sensors and save cost.

All the gate driver circuits, sense and trip electronics, etc. are powered by the auxiliary power supplies, which usually needs to be electrical isolated from the main power branch.

Another difference in SSCBs with the power electronics converters is the necessity of the voltage clamping circuit in parallel with the power semiconductor switches, which has two functions: (a) clamping the voltage across the power semiconductor switches to avoid over-voltage damage during current interruption; (b) absorbing the residual energy in the system parasitic inductances after the power semiconductor switch turn-off.

The mechanical galvanic isolation switch is connected in series with the power semiconductor devices, to provide the galvanic isolation when the breaker is in OFF state for increased safety during service, maintenance, etc. The galvanic isolation switch could also reduce the power semiconductor device's leakage current during OFF state.

9.3.1.2   Advantages

The first and foremost advantage of the SSCB is the ultra-fast fault tripping speed, which is several orders of magnitude faster than that of the conventional MCBs, as shown in Figure 9.6. There are two reasons behind this: (a) SSCBs have no moving parts like the MCB contacts, which means there is zero mass and no inertia when changing circuit breaker state from ON to OFF; (b) the traditional tripping mechanism in electro mechanical breakers inherently lead to a slower response time. The electromechanical breakers historically have two tripping mechanisms: an electromagnetic action for short-circuit protection and a bimetallic action for overload current protection. The electromagnetic action for short-circuit detection can only respond within a few milliseconds to hundreds of milliseconds due to its electromechanical nature. The SSCB's tripping speed is determined by the current sensor delay time and the power semiconductor switch current interruption speed. The current sensors could response as fast as within a few microseconds and the power semiconductor switches could also turn off the current in a microsecond or a few microseconds, depending on what types of power semiconductor technologies are adopted.

The direct benefit of the ultra-fast fault current tripping speed is the much lower peak fault current level as shown in Figure 9.7. SSCBs typically are designed

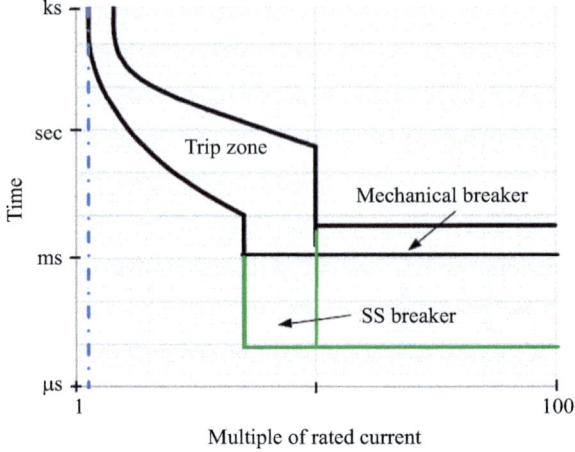

Figure 9.6 Tripping curve comparison between the electromechanical breaker and SSCB

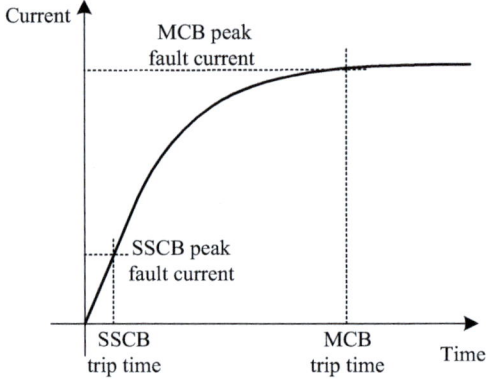

Figure 9.7 Trip time comparison between the SSCB and MCB. The SSCB could trip in microseconds, so that the fault current is interrupted before it rises to a high current level

to respond and interrupt the fault current before the fault current rises to a high level, greatly reducing the fault energy. In many cases, SSCBs do not rely on the current crossing the zero value to interrupt; with the right tripping electronics, they can therefore be used in both DC or AC applications.

The second advantage of the SSCB is the arc-free current interruption. Unlike MCBs, which rely on mechanical contact separation for current interruption, semiconductor devices rely on electric charges movement inside the semiconductor structure to turn on and turn off and forming space charge area to withstand the OFF-state voltage. There is no arcing and no noises generated during the entire process thanks to the elimination of the moving parts.

The operation numbers for SSCBs could be much higher than that of MCBs, which can be translated into a longer lifetime, although there may be some limitations from the voltage clamping components like MOVs. The mechanical breaker contacts gradually wear out with every time of fault current interruption due to the arcing. Compared to MCB, the number of current interruption times of the power semiconductor switches could be considered as unlimited. For example, the same power semiconductor switches are also used in power electronics converters, whose switching frequency varies from tens of Hertz to even Megahertz. With their outstanding electrical life, SSCBs can reduce the maintenance cost and extend the period between maintenance events as compared to the MCBs.

9.3.1.3 Challenges

On the economical side, the cost of SSCBs are deemed to be several times higher than that of MCBs, mainly due to the relatively high cost of the power semiconductor switches. However, it is promising to notice the continuous price dropping of the WBG power semiconductor devices, like SiC MOSFETs, which are very suitable for SSCB applications due to the very low conduction losses and high operational voltages. With the SiC MOSFET market volume continuously increasing because of its wide application in EVs, its price is supposed to reduce even further. Also the power system level benefits need to be considered when SSCBs are used. For example, the SSCB technology is one of the key enabling technologies for the high efficiency DC electric ships and marine power systems, which reduces the power system shut-down time and the system recovery time after a fault. The maintenance service could be reduced as well due to the longer lifetime of the SSCBs.

On the technical side, the power semiconductor switches have relatively high conduction resistances and cause much higher losses in SSCBs compared to mechanical CBs. The mechanical CB's contact resistance is usually much less than 1 mΩ, while the conduction resistances of SSCBs varies from several milliohms to tens of milliohms depending on the current and voltage levels. For example, the ABB Emax E2 circuit breaker generates only 50 W losses at 1,000 A conduction current, while the SSCB could easily generate up to 1,000 W at the similar conduction current level. These losses decrease the overall efficiency of the electrical distribution system, leading to additional operational costs. Of course, the conduction resistance or losses of SSCBs could be reduced or mitigated by adopting lower resistance power semiconductor switches or paralleling more power semiconductor switches, but this could drive the SSCB's cost to an even higher level, or an unacceptable level.

Another technical challenge of SSCBs coming out from the abovementioned high conduction losses is how to dissipate the high losses. While MCB's losses could be easily dissipated through the bus bars or wires thanks to the limited loss amount, the SSCB's conduction losses need to be dissipated by carefully designed cooling system. Also, the required cooling system for the SSCBs result in a relatively large size or form factor of the SSCB. The thermal management challenge does not only apply in the steady-state (nominal current) case but also in transient overload or short circuit fault. During a fault, the potentially high currents flowing through the breaker can lead to significant heating of the semiconductor devices. To ensure that the temperature stays below the safe limit of the semiconductor, it is necessary to interrupt the fault extremely rapidly. The hallmark ultrafast interruption of the SSCB makes this possible, but this comes at a cost: selectivity is more challenging to ensure when an upstream breaker needs to interrupt rapidly to protect itself from overheating.

9.3.1.4   Requirements on power semiconductor devices

Power semiconductor devices in SSCBs share some common requirements as those in power electronics converters, like high enough blocking voltages, conduction loss constraints, etc. Meanwhile, some specific requirements for power semiconductor devices are unique in SSCBs. The following section will elucidate the typical requirements of power semiconductor devices in SSCBs.

Blocking voltage requirement: The power semiconductor device's blocking should be higher than the system voltage and the transient interruption voltage (peak clamping voltage from the voltage clamping components). For bidirectional circuit breakers, bidirectional voltage blocking is required. Symmetrical SCRs and some reverse blocking power semiconductor devices usually have bidirectional voltage blocking capability, however, most power semiconductor switches like MOSFETs, IGBTs need additional diodes or two devices anti-series connection to realize bidirectional voltage blocking.

Conduction resistance requirement: unlike the power electronics converters where power semiconductor devices have both conduction losses and switching losses, the majority losses generated by the power semiconductor devices in SSCBs are conduction losses. Power semiconductor devices with low conduction resistances are preferred for SSCBs and it's common to parallel multiple power devices to reduce conduction losses, especially for power semiconductor devices with good current sharing performances (positive temperature coefficient).

Surge current capability: When there is overload or short circuit fault, the current can rises to several times of the nominal current and surge current capability is required for the power semiconductor switches. Semiconductor device saturation under fault current needs to be avoided, otherwise, conduction voltage drop increases dramatically and significant amount of heat is generated, which may cause device overtemperature failure. Compared to unipolar devices like MOSFETs, junction field effect transistors (JFETs), etc., bipolar semiconductor devices like IGBTs, thyristors, IGCTs, etc. are deemed to have better surge current capability due the strong conductivity modulation mechanism.

Current turn-off capability: Another essential requirement for power semiconductor switches in SSCBs is the current turn-off capability as the potentially high fault current is interrupted by the power semiconductor switches. Fully controllable power semiconductor switches (which can be controlled to turn on and off by gate signals) are preferred in SSCBs due to simplicity in circuit control and utilization. Semi-controllable power switches like SCRs can only be fired to turn on but cannot turn off the current. To realize current interruption, forced current commutation techniques with additional circuits are needed. Unlike the power electronics converter, the high switching speed (e.g., high dv/dt of turn-off voltage) is not favored in SSCBs, as circuit breakers seldomly need to frequently turn on and off. In opposite, a lower switching speed with the help of capacitor snubber circuits are beneficial to reduce the voltage spikes and enhance the power semiconductor device's current turn-off capability.

9.3.2 Classification of solid state circuit breakers

There are different criteria to classify SSCBs. Based on the current conduction and voltage blocking directions, SSCBs can be classified into unidirectional SSCBs and bidirectional SSCBs. Unidirectional SSCBs can only interrupt the current and block the voltage in single direction, while bidirectional SSCBs interrupt the current and block the voltage in both directions. As unidirectional SSCBs can be easily obtained from bidirectional SSCBs by removing the anti-parallel branch or the anti-series connected power switch, this chapter mainly focuses on bidirectional SSCBs, which are more widely used in MVDC systems.

SSCBs can also be classified based on the types of power semiconductor technologies: Semi-controllable power semiconductor devices-based SSCBs and fully controllable power semiconductor device-based SSCBs. The SCR as a typical example for the semi-controllable power semiconductor devices has many beneficial features for SSCBs, making it a very attractive candidate, although additional circuits are needed to assist the SCRs turning off the current. In the next sessions, the SCR-based SSCBs will be further classified and discussed. Fully controllable power semiconductor devices (like MOSFETs, IGBTs, IGCTs, etc.) are easy to control its ON and OFF, making the SSCB design and circuit topology greatly simplified compared to the SCR-based SSCBs.

9.3.3 Unidirectional solid state circuit breakers

As mentioned above, unidirectional SSCBs can only interrupt the current and block the voltage in single direction while the current may conduct in both directions with the help of the anti-paralleled diodes or body diodes in MOSFETs. Two quadrant current-bidirectional power semiconductor switches like IGBTs, MOSFETs can be used to realize unidirectional SSCBs by selecting proper device blocking voltages or using multiple devices in series for MVDC systems, as shown in Figure 9.8, where the SiC MOSFET is taken as an example.

While the circuit topology is simple, some variances are can be found in the gate driver design for the series connected power semiconductor devices. Instead of using separated isolated gate drivers for each series connected power semiconductor switch in Figure 9.8, the single gate driver control for the series connected SiC MOSFET is proposed [5,6], as shown in Figure 9.9. The single gate driver control simplifies the circuit design and is more user-friendly. More importantly, the number of auxiliary power supply needed for the gate driver is reduced from N to 1.

To be mentioned that, SiC MOSFETs in Figures 9.8 and 9.9 can be replaced by other power semiconductor switches, like IGBTs, JFETs, etc. The normally-ON

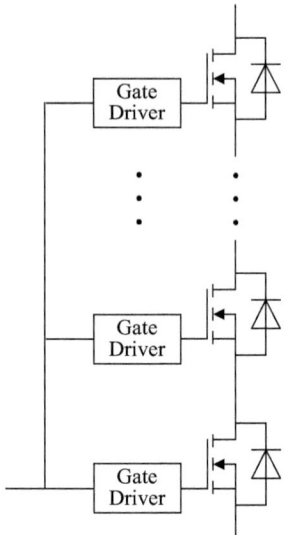

Figure 9.8   Unidirectional SSCB based on SiC MOSFET connected in series with separate gate drivers for each MOSFET

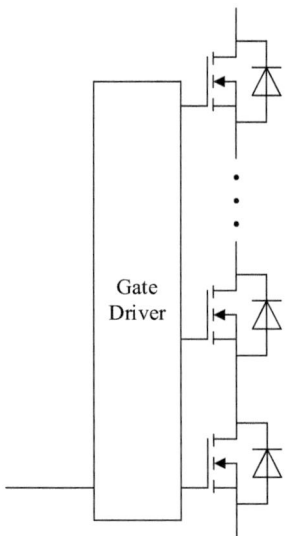

Figure 9.9   Unidirectional SSCB based on SiC MOSFET connected in series with single gate driver circuit for all MOSFETs

SiC JFETs which maintain ON state with zero volt gate voltage, as shown in Figure 9.10, can further facilitate the single input gate driver design as no positive gate voltage is needed to maintain the switches in ON state [7,8]. Another popular

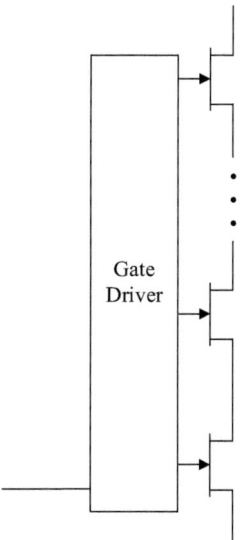

Figure 9.10   Unidirectional SSCB based on SiC JFET connected in series with single gate driver circuit for all JFETs

configuration is called super-cascode [9] as shown in Figure 9.11, where the bottom switch is MOSFET and the upper switches are SiC JFETs which can utilize the lower switch's drain to source voltage to generate a negative gate voltage to turn off the upper switches.

One thing to be noted is the voltage balancing for the series connected power semiconductor devices needs to be carefully considered. Snubber circuits (like capacitor snubber) or voltage clamping components (like MOVs) in parallel with each power semiconductor device are adopted to realize dynamic voltage balancing.

9.3.4   Bidirectional solid state circuit breakers

Most power semiconductor devices like MOSFETs, JFETs, IGBTs, A-IGCTs, emitter turn-off thyristors (ETOs) are two quadrant current-bidirectional power semiconductor switches which can only block the voltage in one direction. To realize bidirectional SSCBs, either anti-series connection of two power semiconductor switches or utilizing diodes to block reverse voltage are needed. In the following topologies, the IGBT, one of the most commonly used power semiconductor switches, will be taken as an example to illustrate different bidirectional SSCB topologies. The IGBTs in the following topologies can also be replaced by MOSFETs (especially SiC MOSFET for medium voltage application), JFETs, ETOs, etc.

Figure 9.12 shows the most widely used topology for the bidirectional SSCB [10], which has two switches connected in anti-series as one module, then multiple modules are connected in series to reach the required voltage level.

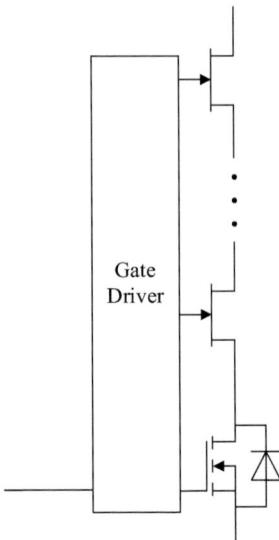

Figure 9.11   Unidirectional SSCB based on SiC JFET in cascode configuration with single gate driver circuit for all devices

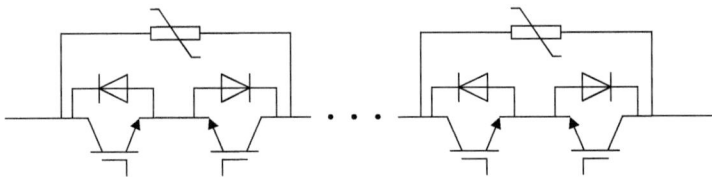

Figure 9.12   Bidirectional SSCB topology based on IGBTs connected in anti-series

In this topology, the number of isolated gate driver circuit needed is half of the number of IGBTs, as every two IGBTs are connected with common emitter configuration, which can share the same gate driver circuit. The dynamic voltage balancing is realized by MOVs in parallel with every two anti-series connected IGBTs.

Figure 9.13 shows another bidirectional SSCB topology [11,12], which used a medium voltage diode to block the reverse voltage. Multiple IGBTs are connected in series in each direction to reach the targeted voltage level. More diodes in series could be adopted to increase its operational voltage level.

The major advantage of this topology is relatively low conduction losses, compared to the topology in Figure 9.12, due to less number of power devices in the current conduction path.

However, more isolated gate drivers are required compared the topology in Figure 9.12, as each IGBT needs an isolated gate driver circuit. The challenges of dynamic voltage balancing between the series connected IGBTs may exist, as

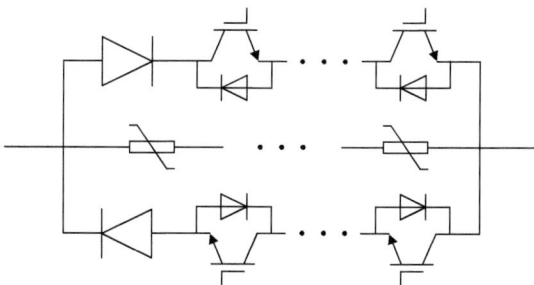

Figure 9.13   Bidirectional SSCB topology based on IGBTs connected in series with diodes

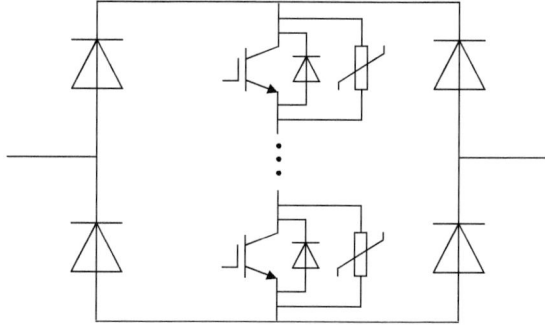

Figure 9.14   Bidirectional SSCB topology based on diode bridge and IGBTs connected in series

MOVs are connected between the input and output, instead of paralleled with each IGBT. Additional capacitor snubber circuit in parallel with each IGBT may be needed to realize the dynamic voltage balancing.

Figure 9.14 shows another way to realize bidirectional SSCBs, by using a diode bridge and connecting IGBTs inside the diode bridge [12]. This topology can reach a trade-off between the topologies in Figures 9.12 and 9.13. It needs the same number of isolated gate drivers as Figure 9.12, but has lower conduction losses. The voltage balancing can also be easily realized by connecting MOVs in parallel with each IGBT.

9.3.5   Thyristor-based solid state circuit breakers

The thyristor technology, developed as early as 1960s, are still the main workhorse in high power applications such as HVDC, FACTS, etc. due to its large power handling capability [13,14]. After a survey of the state-of-art commercial controllable power devices in terms of the upper boundary of the voltage and current ratings achieved in a single-packaged device [13], it can be found that the thyristors

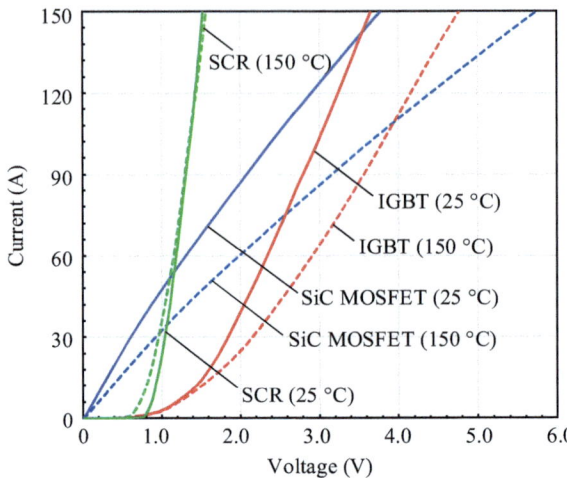

Figure 9.15   I–V curves of the SiC MOSFET [15], IGBT [16], and thyristor[17] at room temperature (25 °C) and high temperature (150 °C)

achieve the highest voltage and current ratings (up to 12 kV, 1.5 kA) in single package power devices, indicating a high commercial readiness and great potential for high voltage high power applications.

From a semiconductor current conduction mechanism point of view, the thyristor technology has the lowest forward conduction losses at high current due to its strong conductivity modulation, although the WBG power devices like the SiC MOSFETs demonstrate promising performances like the much lower specific on-state resistance, the high switching speed and blocking voltage capability compared to the Si counterpart device [14].

Figure 9.15 compares the I–V curves of the three selected power devices with same voltage rating (1,200 V) and similar current levels. It shows that the SiC MOSFET has better conduction characteristics at low current level (<30 A) due to its unipolar device characteristics. At high current level (>50 A), the thyristor demonstrates the lowest conduction voltage drop, thanks to the double-sided carrier injection and strong conductivity modulation. The thyristor's advantage of better conduction characteristics is more obvious at higher temperatures.

Another important advantage of the thyristor is its high pulse current (surge current) capability, about seven times better than SiC MOSFETs. The surge current capability is one important requirements for the power semiconductor devices in the SSCB application, as the power devices need to withstand the short circuit fault current which could rise to several times (e.g., five times) of the normal current.

Besides the advantages of low conduction voltage drop at high current level and the high surge current capability, the thyristors with symmetrical structure have bidirectional (forward and reverse) voltage blocking capability, which means no second device connected in anti-series is needed to block the reverse voltage in

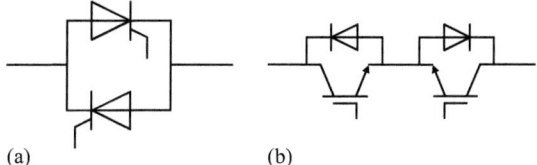

Figure 9.16 *(a) Two anti-parallel connected thyristors-based bidirectional SSCB topology and (b) two anti-series connected IGBTs-based bidirectional SSCB topology*

bidirectional SSCBs, as shown in Figure 9.16. This further reduces SSCB conduction losses with half number of devices in the conduction path.

The thyristor also demonstrates advantages in term of lower cost, appearing a more cost-effective solution due to the mature device fabrication technology and the relatively lower material (Si compared to SiC) cost.

In summary, the beneficial features of thyristors including (1) low conduction losses, (2) high surge current capability, (3) bidirectional voltage blocking capability, (4) high available voltage and current range, and (5) low prices, all make the thyristor technology a very attractive candidate for SSCB application.

Another essential requirement for the power semiconductor devices in the SSCB is the current interruption/turn-off capability. However, it's well known that the conventional thyristors SCRs are semi-controlled power devices, having no current interruption capability. To empower the thyristor with turn off current capability, two technologies are developed: (i) thyristor forced commutation technique [18] and (ii) modified thyristor structure with current turn-off capability. The thyristor forced commutation technique usually needs some auxiliary circuit (LC resonant circuit) connected in parallel with the thyristor.

For technology (i) (thyristor forced commutation technique), this chapter will investigate and evaluate some forced commutation thyristor-based SSCB topologies and summarize their advantages and limits in the applications of MVDC protection.

For the technology (ii) (modified thyristor structure), one good example is the gate turn-off thyristor (GTO) [19] or gate-commuted thyristor (GCT) technology [20,21]. To further increase the device's controllability and facilitate the gate driver design, the technologies like the IGCT [22,23], and ETO [24,25], etc., are invented. The IGCTs and ETOs have been investigated extensively in the past decade [20–26] and demonstrate great potentials in MVDC SSCB applications. This chapter reviews the technology status in the SSCB application and summarizes the advantages and disadvantages with these technologies.

The Z-source breaker [27–31] is also thyristor-based SSCB which can be categorized into the forced commutation thyristor-based SSCBs. However, due to its particularity in topologies and wide research interests, it was discussed and summarized in one separate section. This chapter systematically summarizes and compares the different types of Z-source breakers. Specially, the technical challenges limiting the wide application of the Z-source breaker are explained and summarized.

9.3.5.1   Silicon-controlled rectifier with active resonant circuit-based solid state circuit breakers

Figure 9.17 shows the topology of SCR with active resonant circuit-based SSCBs. The normal current conducts through the two anti-paralleled main SCRs (SCR1 and SCR2). Two pre-charged capacitors are needed to realize the bidirectional current interruption and there are two auxiliary SCRs (SCR3 and SCR4) in series with each pre-charged capacitor to excite the LC resonance. When there is a fault, the main SCRs (SCR1 and SCR2) gate signals are removed and the corresponding auxiliary SCR (SCR3 or SCR4, depending on the current direction) is turned on to excite the LC resonance. A zero current-crossing is created for the main SCR when the LC current resonates to the fault current level. Before the pre-charges capacitor is fully discharged, the capacitor applies a negative voltage to the anode of main SCR and help the main SCR turn-off by accelerating the minor carrier recommendation process. After the pre-charged capacitor is fully discharged, the fault current starts to charge the capacitor in reverse direction. The LC branch current starts to commutate to the MOV when the capacitor voltage reaches the MOV clamping voltage. The MOV absorbs the residual energy in the line inductances and the current decreases to zero. The auxiliary mechanical SCRs (SCR3 or SCR4) are turned off when the current goes to zero and the fault is isolated.

Two voltage charging circuits are needed for the capacitor C1 and C2, respectively. To reduce the number of pre-charged capacitors and the corresponding voltage charging circuits, Figure 9.18 shows an another topology by using a SCR bridge to realize bidirectional current interruption. Principles of operation of the topology in Figure 9.18 are similar as that in Figure 9.17.

Figure 9.19 shows a further improved topology by replacing the auxiliary SCRs with the active turn-off semiconductor switches like GTOs. With the GTO in the resonant circuit, it does not need to wait until the pre-charged capacitor is reversely charged to the MOV clamping voltage and rely on the MOV to interrupt the current. The GTO could turn off the fault current before the pre-charged capacitor is fully discharged, which could greatly reduce the total current

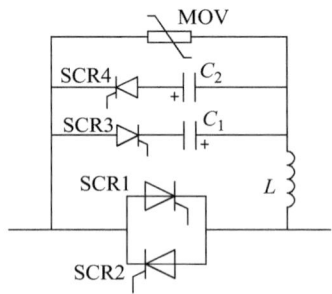

Figure 9.17   SSCB topology based on SCRs with active resonant circuits

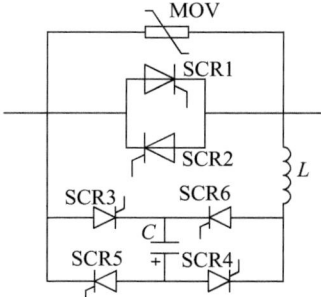

Figure 9.18 Another SSCB topology based on SCRs with active resonant circuits

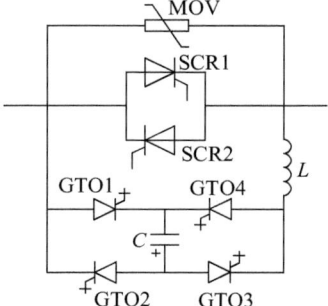

Figure 9.19 SSCB topology based on SCRs with active resonant circuits where the auxiliary switches are realized by GTOs

interruption time and the peak fault current level. Also the necessity of bipolar capacitors is avoided, which could save the circuit cost.

9.3.5.2 Silicon-controlled rectifier in series with commutation switch-based solid state circuit breakers

Figure 9.20 shows the SCR in series with commutation switch-based SSCB topology [33], which has two branches: the main current conduction branch and the current interruption branch. The main current conduction branch is composed of SCRs and low voltage fully controlled switches (e.g., IGBTs) in series. The current interruption branch is realized by series connection of fully controlled switches to achieve medium voltage level. During normal conduction, the SCRs and the commutation switches are turned on, and the current is conducting through the SCRs and the commutation switches. When there is a fault, the commutation switch is turned off first, commutating the current from the main conduction path to the current interruption branch. After all the current is commutated to the current interruption branch, a zero current crossing is created for the SCRs and some delay time is needed to complete the carrier recombination inside the SCRs. When the

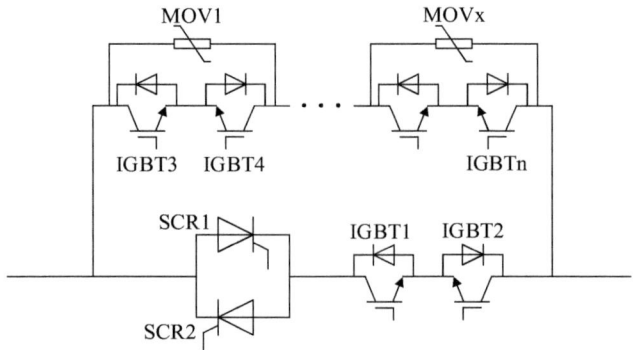

Figure 9.20   SCR-based SSCB with commutation switch in series with SCRs

SCRs can block the forward voltage, the IGBTs in the current interruption branch are turned off to interrupt the current.

With the current conducting through the SCRs and the low voltage commutation switch, there are relatively low conduction losses. However, this topology has relatively long current interruption time, as the SCRs need the reverse recovery time to be able to withstand the forward voltage. To speed up the SCR reverse recovery, it is preferred to creating a reverse-biased voltage for the SCRs by the commutation switch.

9.3.5.3   Silicon-controlled rectifier in series with commutation transformer-based solid state circuit breakers

Figure 9.21 shows the SCR in series with commutation transformer-based SSCB [34]. Replacing the commutation switch in Figure 9.20 with a transformer, the secondary side of the commutation transformer is connected in series with the SCRs and the primary side is connected with a pre-charged capacitor and auxiliary SCRs-based bridge. The topology has similar operation principles and features as that in Figure 9.20. When there is a fault, the corresponding auxiliary SCRs are turned on, and the capacitor discharge current is coupled at the secondary side of the transformer in opposite direction with the fault current. A current zero crossing is created for the main SCRs to realize their turn-off. Compared to the topology in Figure 9.20, the commutation-based topology can create a reverse biased voltage by the pre-charged capacitor for the main thyristors, which is beneficial during the main thyristor turn-off.

9.3.5.4   Z-source breakers

The Z-source breaker is also developed based on SCRs and has the advantage of autonomous current interruption (no current sensor needed), which eliminated the fault current sensing time, enabling the fast fault clearing speed when there is critical short circuit fault. The Z-source breaker is first proposed by Dr. Corzine [27] since 2010, which is named after the Z-source inverter [32], due to the similar impedance components, as shown in Figures 9.22–9.24. When there is a critical short circuit fault,

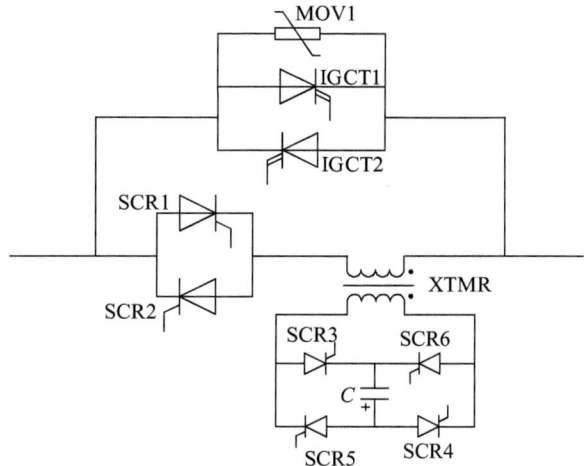

Figure 9.21 SCR-based SSCB with commutation transformer in series with SCRs [34]

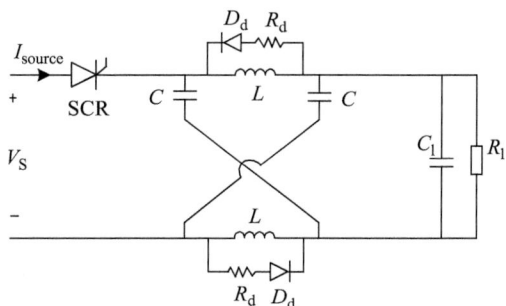

Figure 9.22 Crossed Z-source breaker topology

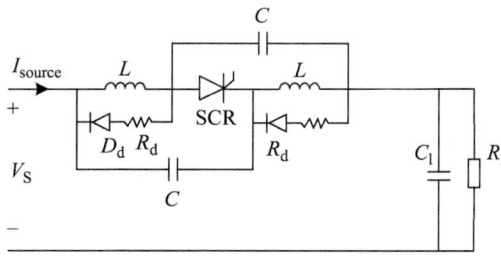

Figure 9.23 Parallel-connected Z-source breaker topology

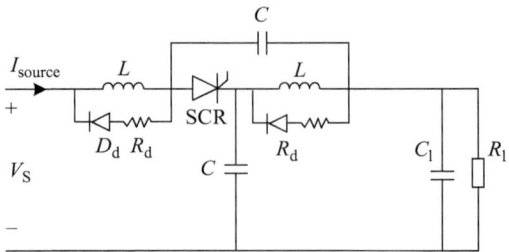

Figure 9.24   Series-connected Z-source breaker topology

the inductor current is assumed constant for short time period, and the capacitors provide transient current path, creating a current zero-crossing in the SCR. The SCR commutates off after current reaches zero and the short circuit fault is isolated from the power source. The commutation of the SCR does not need the OFF command from the control circuits, enabling a fast fault clearance speed.

Based on the locations of the LC resonant circuit, the Z-source breakers can be classified into three categories: crossed Z-source breaker (as shown in Figure 9.22), parallel-connected Z-source breaker (as shown in Figure 9.23) and the series connected Z-source breaker (as shown in Figure 9.24) [26]. The crossed Z-source breaker (as depicted in Figure 9.22), named according to the crossed LC circuit configuration, requires an inductor to be placed in the return path of the power source, which can be seen as a disadvantage in systems needing a common ground connection. The paralleled connected Z-source breaker (as depicted in Figure 9.23) is proposed with a common ground connection between the power source and load [29], which is named because the LC circuit is in parallel with the SCR after the SCR commutated off. One major issue with the paralleled connected Z-source breaker is the large source current reflected from the fault current after the SCR current commutates to zero. To further improve the circuit, the series connected Z-source breaker (in Figure 9.24) is proposed in Ref. [28]. The new topology is termed as series connected Z-source breaker because the LC circuit are connected in series with the SCR after it commutates off. The series connected Z-source breaker provides part of the fault current from the capacitors instead of all from the power source. Hence, the reflected current to the source during current interruption is greatly reduced.

The Z-source breaker has the advantages of low conduction losses and autonomous current interruption, while there are also some limitations with the Z-source breakers. The first limitation is incompetence of overload protections. The crossed Z-Source breaker and the parallel-connected Z-source breaker can only protect the most critical fault with high fault current ramp rates (e.g., the short circuit fault), which is only a subset of faults in practical power system. Although the series-connected Z-source breaker can protect the overload faults in theory, the overload current tripping level is extremely high (>10 times of the rated current) [28], making the overload protection not so practical and meaningful. Secondly, the Z-source breakers in Figures 9.22–9.24 have only unidirectional current conduction capability. To realize bidirectional current conduction and interruption, more components (SCRs, capacitors, inductors) are needed [30,31], causing higher cost and complicated circuit. Another limitation is from the necessity of the output capacitors (C_l) in all the Z-source breakers in Figures 9.22–9.24 [29]. In absence of the load capacitors, the load step change also introduces a high di/dt, and the Z-source breakers could misinterpret load change as a short circuit fault, mistakenly turning off the load current.

9.3.5.5 Integrated gate-commutated thyristor-based solid state circuit breakers

The IGCT technology has found wide applications since its invention in medium voltage drives, wind power conversion, STACOMs, etc. thanks to its low conduction losses and convenient ON/OFF control with the integrated gate unit. In the past 10 years, the IGCT technology has experienced continuous optimization and customization to better suite different applications. Recently, the IGCT devices become a more attractive option for SSCB application with the further optimization of the device structure to drive down the conduction losses.

The asymmetrical IGCT (A-IGCT) can only block forward bias voltage needs a diode in series connection to build a bidirectional SSCB [Figure 9.25 (a)]. Similarly, the reverse conduction IGCT (RC-IGCT), which can only block forward biased voltage, needs to be connected in anti-series with another RB-IGCT [Figure 9.25(b)]. Reverse-blocking IGCT technology [20,21,39], shows promising prospect due to its very low conduction losses and compatible with a circuit topology with a single device in the conduction path for bidirectional SSCBs.

Reference [37] demonstrates a 1 kV, 1 kA bidirectional SSCB prototype based on 2.5 kV RB-IGCTs, which has on-state voltage drop as low as 0.9 V at 1 kA, at 125 °C and fast fault current interruption speed (2.6 kA current interruption within 200 μs at 1 kV system voltage), validating the suitableness of the RB-IGBT technology for the SSCB application [35,36].

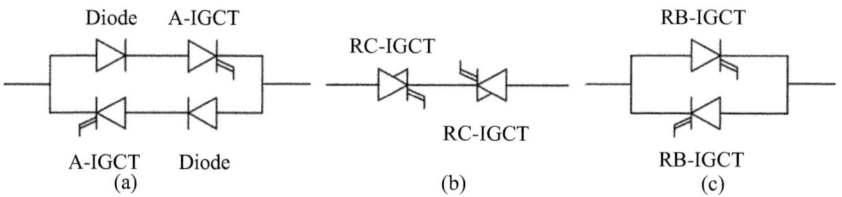

Figure 9.25   *IGCT-based SSCB topologies. (a) A-IGCT-based SSCB topology; (b) RC-IGCT-based SSCB topology; and (c) RB-IGCT-based SSCB topology*

9.3.5.6   Emitter turn-off thyristor-based solid state circuit breakers

GTO devices add current interruption capability to the thyristor's typical advantages, like low conduction voltage drop, high surge-current capability, and bidirectional voltage-blocking capability with symmetrical device structure. However, one major drawback of the GTO device is the complicated and slow transient-response gate driver circuit, resulting in a very long turn-off time. Besides IGCT technology, the ETO technology is another major effort made to overcome these drawbacks and improve the GTO concept.

The ETO technology was developed from mature GTO and power-MOSFET technologies and combines the advantages of both. In particular, GTOs' high voltage/current ratings and low forward-voltage drop, and power-MOSFET's simple interface for gate driving, similar to IGBT's. In addition, ETOs also exhibit wide reverse-biased safe operation area (RBSOA), and relatively high switching speed, making them good candidates for SSCBs.

Figure 9.26 shows a bidirectional SSCB based on two ETOs connected in anti-series [38,39]. Applying the ETO concept to the SiC, the SiC ETO presents a unique opportunity for medium voltage SSCB applications: the ultra-high blocking voltages, large RBSOA, low forward drop and potential for high-temperature operation. Reference [39] shows a MV SSCB developed based on the SiC ETOs, where two 15 kV SiC ETOs and two 15 kV PiN diodes are connected in anti-series. Thanks to the SiC devices' high blocking voltage capability, no devices series connection is needed to reach about 7 kV operation voltage, which greatly simplified the circuit and increases the power density as well as the circuit reliability.

9.4   Hybrid circuit breaker

9.4.1   Introduction to hybrid circuit breaker

The HCB is consisted of both MSs and power semiconductor switches and combines the advantages of both MCBs and SSCBs. During current normal conduction, the load current mainly conducts through the MSs, and the conduction losses are greatly reduced. When there is a fault, the current commutates from the MS to the

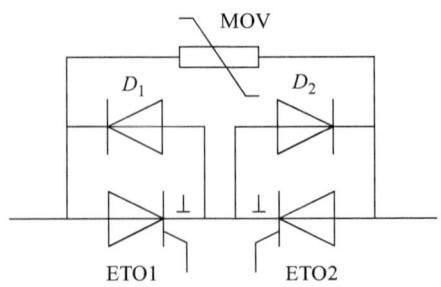

Figure 9.26   ETO-based SSCB topology

power semiconductor switch, and the power semiconductor switch interrupts the fault current very fast and arc-free. HCBs can usually interrupt the current in a millisecond or a few milliseconds. Reference [40] shows a MS opening time of 70 μs was achieved.

The MSs in many HCBs can be realized by a fast disconnect switch (contacts open at zero current) which does not need to interrupt the current with arcing, while some HCB topologies need the MS arcing voltage to commutate the current to the power semiconductor switches. The current interruption speed of HCBs is highly dependent on the MS contacts separation speed. The power semiconductor switch needs to wait until the MS contact distance is large enough to withstand the transient interruption voltage.

The power semiconductor switches in HCBs only need to conduct for a short time (usually within a few milliseconds) during a fault current interruption, so the conduction losses are usually not a big concern due to the short conduction time. However, lower conduction voltage drops are still preferred to facilitate the current commutation from the MS to the power semiconductor switch. Same with SSCBs, the current surge capability and current turn-off capability are required for the power semiconductor switches in HCBs.

Compared to MCBs and SSCBs, HCBs obviously have the most complicated circuits, which also lead to a relatively large form factor, although heat sinks or other bulky cooling systems are not necessary for the power semiconductor switches due to short conduction time. But the MSs in HCBs usually require some high current driving circuits to realize fast contacts separation speed, which increases the total HCB volume.

9.4.2 Classification of hybrid circuit breakers

There are different classification methods for HCBs, e.g., based on types of power semiconductor switches used, types of MSs used, or current commutation method, etc.

Figures 9.27–9.29 show three types of HCBs classified based on the current commutation method. Figure 9.27 shows HCBs using the MS arcing voltage for current commutation, and Figures 9.28 and 9.29 show HCBs with additional current commutation units (CCUs) which are located in the MS branch and current

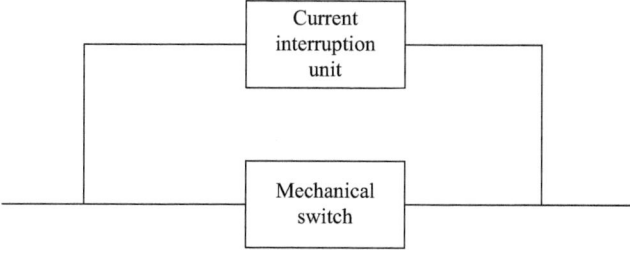

Figure 9.27 HCBs using the MS arcing voltage for current commutation

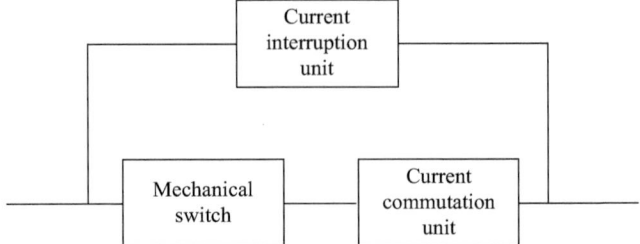

Figure 9.28   HCB with CCU in series with the MS

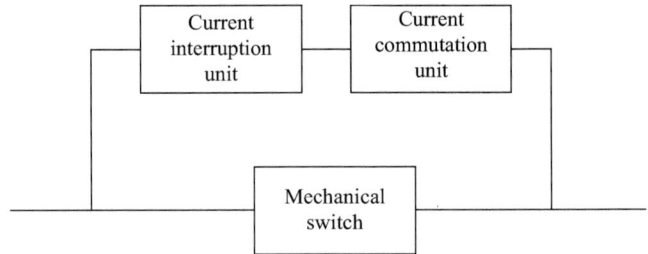

Figure 9.29   HCB with CCU in parallel with the MS

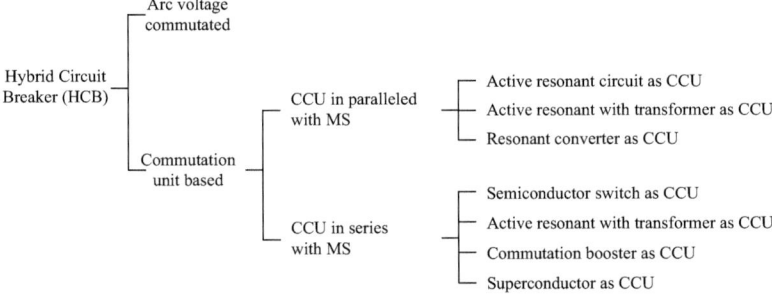

Figure 9.30   HCB topology classification in categorization tree

interruption unit (CIU) branch, respectively. With the help of CCU, zero current crossing can be created for the MS, which can realize arc-free opening and extend the MS's lifetime.

Figure 9.30 further classified HCBs based on how the CCU is realized, e.g., the commutation unit can be realized by some active resonant circuits, semiconductor switches, transformer coupled resonant circuits, or even superconductors, etc. In the following section, each type of HCBs will be introduced with basic principles of operation as well as their advantages and disadvantages.

9.4.3 Arc voltage commutated hybrid circuit breaker

Figure 9.31 shows a typical topology of the arc commutated HCB [41]. During normal current conduction phase, the MS is closed to conduct the nominal current, which has very low conduction losses. When there is a fault, the power semiconductor switches are turned on and the MS is opened simultaneously, generating an arc voltage, and commutating the current from the MS path to the power semiconductor switch path. After the current is fully commutated to the power semiconductor switch path and the gap between the contacts of the MS is large enough to withstand the transient interruption voltage, the power semiconductor switch is turned off, and the current is commutated to MOVs, which drives the current to zero and absorbs the residual energy stored in the system loop inductances.

The power electronics switches can be fulfilled by fully controllable power semiconductor devices, like IGBTs, IGCTs, etc. Series connection of power semiconductor devices may be needed to achieve the system voltage requirements. SCRs in series with a capacitor [33] as shown in Figure 9.32 can also be used as the power semiconductor switch to interrupt the fault current.

Arc commutated HCBs have the advantages of low conduction losses and relatively simple circuits to realize. The MSs need to be carefully designed or selected, which require fast contact separation speed and high enough arc voltage (higher than the power semiconductor switch conduction voltage drops) to guarantee the current can be successfully commutated from the MS.

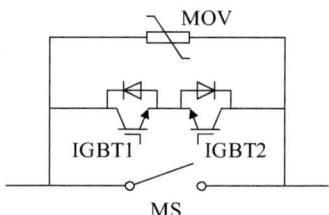

Figure 9.31   Arc voltage commutated HCB using IGBTs for current interruption

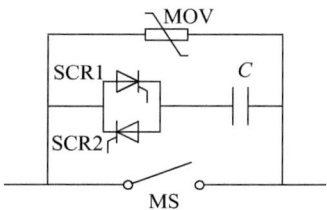

Figure 9.32   Arc voltage commutated HCB using SCRs for current interruption

9.4.4   Current commutation unit in parallel with mechanical switch

9.4.4.1   Active resonant hybrid circuit breaker

Figure 9.33 shows a typical active resonant HCB topology [34], which is very similarly to active resonant DC MCB as shown in Figure 9.4. The operation principles are also the same with active DC MCBs with the only difference that auxiliary MSs are replaced by SCRs (SCR1 and SCR2).

In Figure 9.33, two voltage charging circuits are needed for capacitors C1 and C2 which increase the circuit complexity and cost. To reduce the number of pre-charged capacitors and the corresponding voltage charging circuit, Figure 9.34 shows an alternative topology by using a SCR bridge to realize bidirectional current interruption [34,42]. The principles of operation and features of the topology in Figure 9.34 are same with that in Figure 9.33.

Compared to the arc voltage commutated HCB, the active resonant HCB's current commutation speed is much faster with the help of the voltage across the pre-charged capacitors. The requirement on the MS's arc voltage is relaxed which gives more flexibility in the MS selection and design. However, the circuit is more complexed compared to the arc voltage commutated HCB, and LC resonant branch as well as the charging circuits are needed. The right control logic and timing is needed for the MS and the auxiliary SCRs.

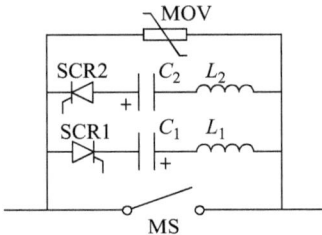

Figure 9.33   An HCB topology based on active resonant circuits

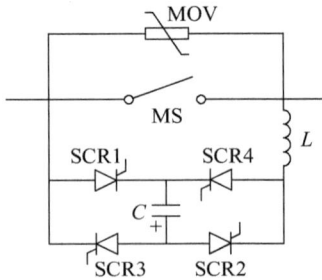

Figure 9.34   Another HCB topology based on active resonant circuits

9.4.4.2 Active resonant with transformer-based hybrid circuit breaker

Figure 9.35 shows a topology where the capacitor-based resonant circuit is coupled with the MS by a transformer [43,44]. The control logic is the same with that in Figure 9.34. When there is a fault, the corresponding auxiliary SCRs are turned on, and the capacitor discharge current is coupled at the secondary side of the transformer to commutate the current from the MS to transformer. A current zero crossing is created for the MS to extinguish the arc. Usually, the transformer primary side has more turns to generate higher current in the transformer secondary side to speed up the current commutation. With the help of the transformer, the resonant circuit is isolated with the high voltage main power circuit, and low voltage capacitors could be use. Also, with different turn ratios between the transformer primary side and secondary side, the current commutation capability is improved. A transformer with high coupling coefficient between the primary side and the secondary side is preferred.

Figure 9.36 shows another transformer-based CCU located in the current interruption branch [45]. When there is a fault, the IGBTs and the corresponding thyristors are turned on to initiate the current commutation. The MS is opened when there is a current zero crossing. When the MS contact gap is big enough, the IGBTs are turned off to interrupt the fault current.

Compared with the topology in Figure 9.33, the capacitor C1 is replaced by series connected IGBTs, which enable faster fault isolation speed by avoiding the capacitor charging time. Obviously, the cost is increased as the IGBTs are much more expensive than a capacitor.

9.4.4.3 Resonant converter-based hybrid circuit breaker

Figure 9.37 shows the voltage source converter-based HCB [46,47]. The voltage source converter-based HCB has three branches: the MS branch for normal current

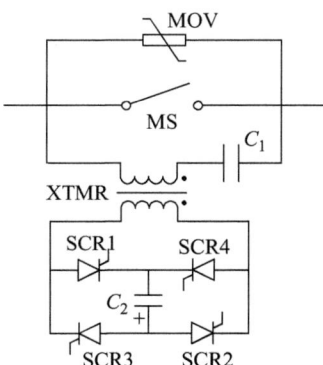

Figure 9.35   An HCB topology based on transformer coupled active resonant circuits

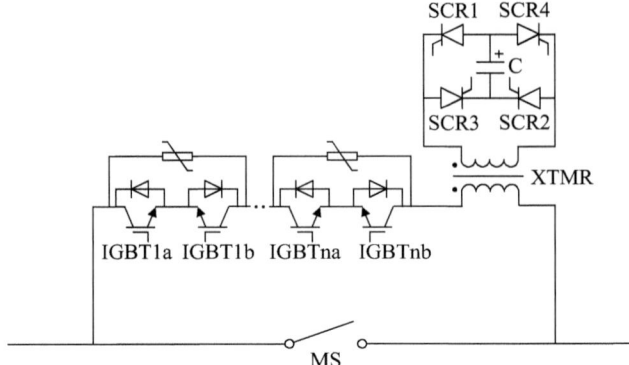

Figure 9.36   Another HCB topology based on transformer coupled active resonant circuits [45]

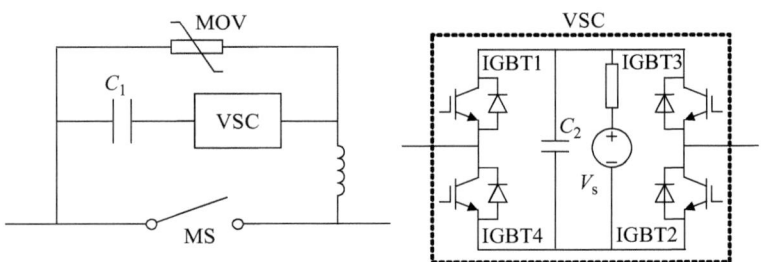

Figure 9.37   An HCB topology based on resonant converter for current commutation

conduction, the voltage source converter (VSC) in series with the resonant capacitor and inductor for current commutation, and the MOV branch for current interruption and residual energy absorbing. When there is a fault, the MS actuator is triggered instantaneously (contacts start to separate) and the voltage source converter starts to excite the current resonance between the capacitor and inductor. With the resonant current amplitude increasing gradually, the MS contact gap is also expanding. When the LC resonant current reaches the magnitude of the fault current, creating a current zero crossing for the MS, the arc in the MS is extinguished and the MS is totally opened. Note that at the current zero crossing point, the MS contact gap needs to be wide enough to withstand the turn-off voltage. After the MS is opened, the capacitor voltage across C1 increases linearly until it reaches the MOV clamping voltage and the current starts to commutate to MOVs, which force the fault current to zero.

In this topology, the DC bus voltage Vs of the VSC can be much lower than the system voltage, which means low voltage rating IGBTs can be used to save cost. With the LC resonance, the current amplitude keeps increasing and eventually creates a current zero crossing for the MS when the resonant current is equal to the fault current. This means the same resonant converter-based HCB design can adapt to and interrupt different fault current levels. However, the resonant converter-based HCB has relatively slow current interruption speed, as it usually takes multiple resonant cycles for the amplitude of the resonant current to reach the fault current level. Precise timing control is needed to excite the LC resonance, which is relatively more complicated compared to other HCBs.

9.4.5   Current commutation unit in series with mechanical switch

9.4.5.1   Semiconductor switch as current commutation unit

The CCU can also be realized by power semiconductor switches, which can quickly turn off the conduction current and create a zero current crossing for the MS. The MS can realize arc-free opening and remove the arc quenching unit.

The voltage rating of power semiconductor switches in the CCU can be lower than the system voltage, which can be realized by low voltage Si MOSFETs, IGBTs, etc. To realize bidirectional current conduction and interruption, anti-series connection of two semiconductor switches or diode bridge may be needed.

The topology could be as simple as Figure 9.38 [33], where the CCU is realized by two anti-series connected IGBTs and the current interrupter is realized by a capacitor. When this is a fault, the IGBTs are turned off and the current is commutated to the capacitor. After the current is fully commutated to the capacitor, the MS is opened at zero current, the fault current starts MS is opened at zero current, and the fault current starts to charge the capacitor until the MOV clamping voltage is reached. Then the fault current is forced to zero by the MOV clamping voltage and the fault energy is consumed by the MOV.

In this topology, large capacitances for C are needed to make sure the current can be fully commutated to the capacitor C when IGBTs are turned off. And the capacitor voltage should increase slowly to leave enough time for the MS contacts separating to withstand the MOV clamping voltage. During current interruption, there may be some resonance between the system loop inductance and the capacitor C, which extends current interruption time. Also, during the HCB OFF state, the

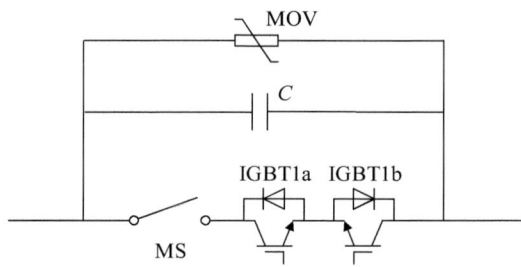

Figure 9.38   Simple HCB topology using semiconductor switches as current commutation unit

capacitor C is charged to the system voltage, and caused large discharge current during ON state.

To reduce the LC resonance, thyristors SCR1 and SCR2 are connected in series with the capacitor C as shown in Figure 9.39 [33]. With the help of the thyristors, a faster current interruption speed can be achieved. When the capacitor voltage is charged to the MOV clamping voltage, the thyristor is naturally turned off with a negative biased voltage from MOV clamping voltage. Also, the capacitor C is not charged during the HCB OFF state, avoiding the high discharging current during ON state.

Different fully controlled power semiconductor devices can be adopted for the CIU and CCU. In this section, the IGBT is take as an example to explain the different circuit topologies, which can be replaced by IGCTs, SiC MOSFETs, ETOs, etc.

Figure 9.40 shows the HCB topology proposed by ABB [48–50], where the two anti-series connected IGBTs are used to realize the CIU and CCU. When this is a fault, IGBT1 and IGBT2 are turned off, commutating the current from the MS to IGBT3-IGBTn. The MS is open under zero current condition without arcing concerns. When the contacts inside the MS is wide enough, IGBT3-IGBTn are turned off and the fault current is interrupted with the help of MOVs. Thanks to the IGBT's fast current turn-off speed, the fault current interruption time is greatly reduced compared to the topologies in Figures 9.38 and 9.39.

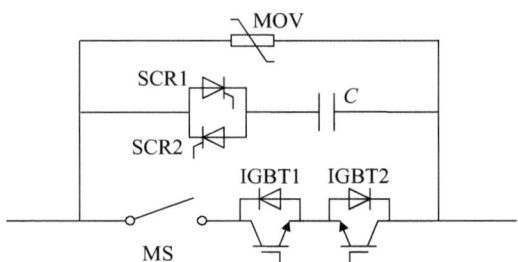

Figure 9.39   Another HCB topology using semiconductor switches as current commutation unit

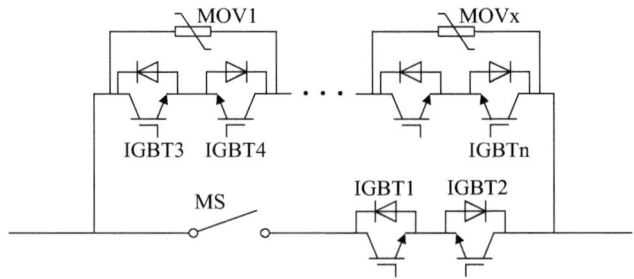

Figure 9.40   An HCB topology using anti-series connected IGBTs as CCU and CIU

Figure 9.41 shows the HCB topology proposed by KIT [51,52], where the CIU is realized by a diode bridge with series connected IGBTs. This HCB has similar principles of operation with that in Figure 9.40, but the cost is lower due to less number of IGBTs.

Figure 9.42 shows the HCB topology proposed by SGCC [53,54]. The CIU and CCU are realized by the IGBT bridges with a snubber capacitor which improves the IGBT's turn-off capability by zero voltage switching. During current conduction, two parallel current paths are formed in the IGBT bridges to reduce the conduction losses. But the cost is increased accordingly as more IGBTs are used. During the IGBT turn-on, the snubber capacitors could also generate large discharge current through the IGBTs, causing thermal stress concerns for the IGBTs.

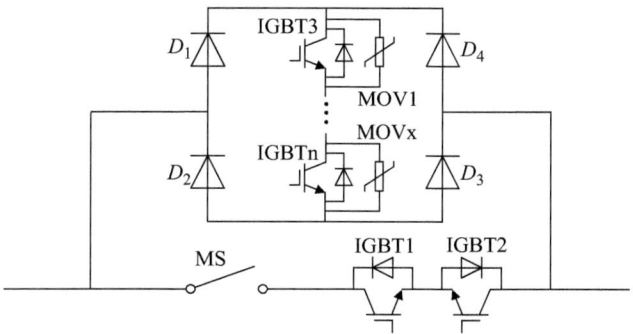

Figure 9.41   An HCB topology using anti-series connected IGBTs as CCU and diode bridge as CIU

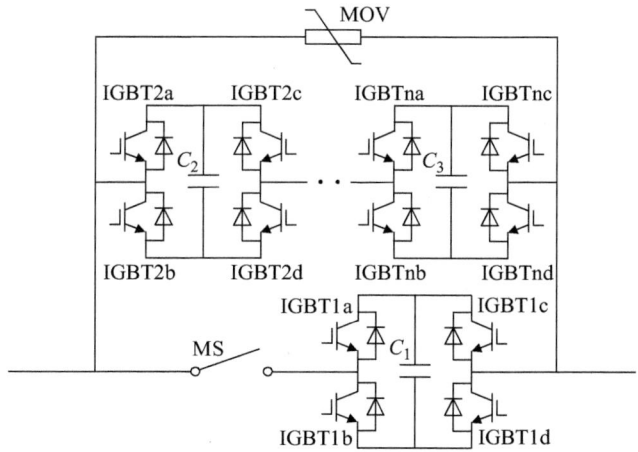

Figure 9.42   An HCB topology using IGBT bridge as CCU and CIU

9.4.5.2   Resonant circuit with transformer as current commutation unit

The semiconductor-based CCU could be coupled with the MS through a transformer [55], as shown in Figure 9.43. The transformer primary side is connected to the pre-charged capacitor while the secondary side is connected in series with the MS. When there is a fault, the corresponding SCRs at the transformer primary side are fired to excite the capacitor discharging, and the current is coupled to the transformer secondary side, which is opposite to the fault current direction, to create a current zero crossing point for the MS.

One benefit of the adoption of transformer is lower conduction losses compared to the power semiconductor switch-based CCU. Like previous topologies, the current interrupter switches could be realized by IGBTs, IGCTs or any other power semiconductor switches that have current turn-off capability. Series connection of multiple devices may be needed to reach the desired voltage level.

9.4.5.3   Commutation booster-based hybrid circuit breaker

Another interesting HCB topology [56,57] is shown in Figure 9.44, where the CCU is called commutation booster. The commutation booster-based HCB has three

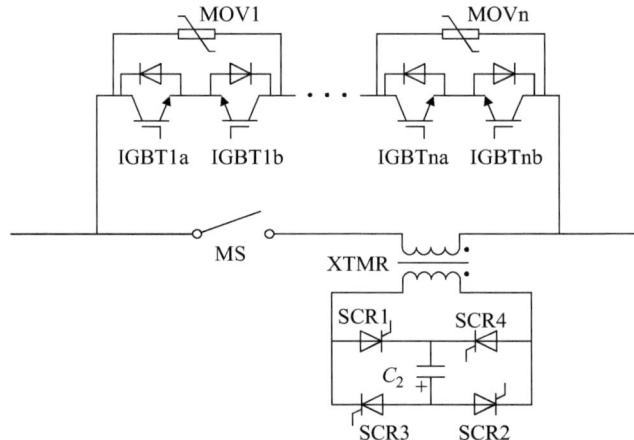

Figure 9.43   An HCB topology using transformer coupled resonant circuit as CCU [55]

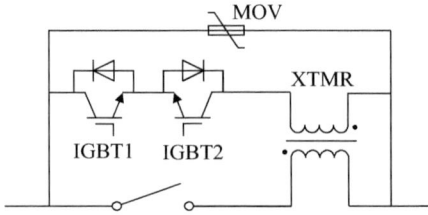

Figure 9.44   An HCB topology using commutation booster as CCU [55]

branches: the MS branch in series with the commutation booster primary side for normal current conduction; the power semiconductor switch branch (IGBTs as an example) in series with the commutation booster secondary side for current interruption; and the MOV branch for voltage clamping and residual energy absorbing. The primary side of commutation booster has higher inductances compared to the secondary side. When there is a fault, the rising fault current generates a voltage drop across the commutation booster primary side, which commutates the current from the MS to the IGBTs. When all the current is commutated to the IGBT branch, a current zero crossing is created for the MS and the MS opens without arcing or with reduced arcing voltages. The power semiconductor switch will turn off the fault current when the MS contact gap is large enough to withstand the turn-off voltage.

In the commutation booster based HCB, the current commutation speed between the MS and the power semiconductor switch is dependent on the fault current rising rate: the higher the fault current rising rate is, the faster the current is commutated. In that sense, when there is slow fault current rising, such as the overload conditions, the current commutation would be slow and difficult.

9.4.5.4 Superconductor as current commutation unit

Figure 9.45 shows the superconductor commutation-based HCB [3,58] which has two branches: the MS branch in series with the superconductor coil for normal current conduction and the power semiconductor switch branch (use IGBTs as an example) with MOVs in parallel for fault current interruption. When the fault current rises to the critical current level of the superconductor coils, the superconductor coils conduction voltage increases due to increased coil resistances. The increased superconductor voltage drop forces the current commutates from the MS to the power semiconductor switch branch and the MS would open at zero current crossing or low current levels. The power semiconductor switch will turn off the fault current when the MS contact gap is large enough to withstand the transient interruption voltage from MOVs.

The superconductor commutation-based HCB does not need current sensors to sense the fault current, which relies on the superconductor coil critical current level. This could offer faster fault current detection and more robust and reliable fault current sensing. In addition, the conduction losses are relatively low compared

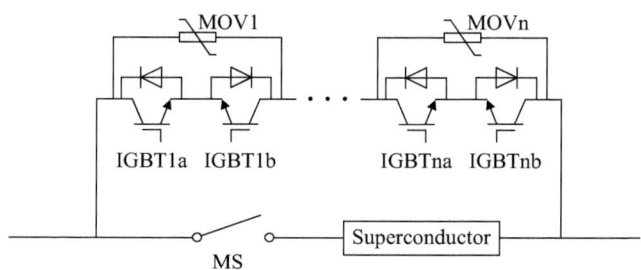

Figure 9.45   An HCB topology using superconductor as CCU [3,58]

to HCBs using semiconductor switches as CCU. However, the cryogenic cooling system are needed for the superconductor coils, which may be bulky and limit its wide application.

9.5 Conclusion

This chapter reviewed circuit breaker technologies for MVDC system protection, including MCBs, SSCBs and HCBs. The fundamentals, benefits, and design considerations of the three MVDC circuit breaker technologies are explained and discussed. A further classification of each type of circuit breakers is provided with the circuit topology and brief description of operating principles. Advantages and disadvantages of different topologies are also discussed and compared.

MCBs have low conduction losses, but the long fault tripping time causes high fault current level and limits the maximum operational voltage level. The high arcing energy causes wear or damage to the circuit breaker contacts during each short circuit fault current interruption, reducing its lifetime and requiring frequent maintenance service.

SSCBs demonstrate ultra-fast current interruption speed and have low fault-stress on application systems. The zero arc energy exposure during current interruption enables less fire hazard and arc flash mitigation for people and equipment. However, the high conduction losses propose great challenges in the cooling system design, especially when the series connection of multiple power semiconductor devices is needed to reach the targeted voltage rating. WBG power semiconductor devices (SiC or GaN power devices) show superior device performances like low conduction resistances and high blocking voltage capability, making the SSCB technology more feasible and promising.

By integrating the MCB and SSCB, HCBs combine both advantages of low conduction losses and fast fault current interruption speed, showing significant potential for MVDC applications. The current interruption time is highly determined by the MS contact separation speed, so the design and optimization of a MS with fast contact opening speed is one important research direction. The current commutation time between the MS and power semiconductor switches is another key factor on current interruption time, and different techniques to achieve fast current commutation are proposed and need further investigation.

References

[1]   L. Liljestrand, M. Backman, L. Jonsson, E. Dullni, and M. Riva, "Medium Voltage DC Vacuum Circuit Breaker," in 2015 3rd International Conference on Electric Power Equipment – Switching Technology (ICEPE-ST), 2015, pp. 495–500, doi: 10.1109/ICEPE-ST.2015.7368340.

[2]   L. Liljestrand, L. Jonsson, M. Backman, and M. Riva, "A New Hybrid Medium Voltage Breaker for DC Interruption or AC Fault Current

Limitation," 2016 18th European Conference on Power Electronics and Applications (EPE'16 ECCE Europe), 2016, pp. 1–10, doi: 10.1109/EPE.2016.7695562.

[3]   X. Pei, O. Cwikowski, D. S. Vilchis-Rodriguez, M. Barnes, A. C. Smith, and R. Shuttleworth, "A Review of Technologies for MVDC Circuit Breakers," *IECON 2016 – 42nd Annual Conference of the IEEE Industrial Electronics Society*, 2016, pp. 3799–3805, doi: 10.1109/IECON.2016.7793492.

[4]   T. Erikksson, M. Backman, and S. Halen, "A Low Loss Mechanical HVDC Breaker for HVDC Grid Applications", CIGRE, Paris, 24–29 Aug. 2014.

[5]   Y. Ren, X. Yang, F. Zhang, F. Wang, L. M. Tolbert, and Y. Pei, "A Single Gate Driver Based Solid-State Circuit Breaker Using Series Connected SiC MOSFETs," *IEEE Transactions on Power Electronics*, vol. 34, no. 3, pp. 2002–2006, 2019, doi:10.1109/TPEL.2018.2861920.

[6]   L. Zhang, S. Sen, Z. Guo, X. Zhao, A. Q. Huang, and X. Song, "7.2-kV/60-A Austin SuperMOS: An Enabling SiC Switch Technology for Medium Voltage Applications," in *2019 IEEE Electric Ship Technologies Symposium (ESTS)*, 2019, pp. 523–529, doi: 10.1109/ESTS.2019.8847863.

[7]   A. M. Roshandeh, Z. Miao, Z. A. Danyial, Y. Feng, and Z. J. Shen, "Cascaded Operation of SiC JFETs in Medium Voltage Solid State Circuit Breakers," *2016 IEEE Energy Conversion Congress and Exposition (ECCE)*, 2016, pp. 1–6, doi: 10.1109/ECCE.2016.7854907.

[8]   Z. Miao, G. Sabui, A. Moradkhani Roshandeh, and Z. J. Shen, "Design and Analysis of DC Solid-State Circuit Breakers Using SiC JFETs," *IEEE Journal of Emerging and Selected Topics in Power Electronics*, vol. 4, no. 3, pp. 863–873, 2016, doi:10.1109/JESTPE.2016.2558448.

[9]   X. Song, A. Q. Huang, S. Sen, L. Zhang, P. Liu, and X. Ni, "15-kV/40-A FREEDM Supercascode: A Cost-Effective SiC High-Voltage and High-Frequency Power Switch," *IEEE Transactions on Industry Applications*, vol. 53, no. 6, pp. 5715–5727, 2017, doi: 10.1109/TIA.2017.2737627.

[10] M. Kempkes, I. Roth, and M. Gaudreau, "Solid-state Circuit Breakers for Medium Voltage DC Power," in *2011 IEEE Electric Ship Technologies Symposium*, 2011, pp. 254–257, doi: 10.1109/ESTS.2011.5770877.

[11] O. Vodyakho, C. Widener, M. Steurer *et al.*, "Development of Solid-state Fault Isolation Devices for Future Power Electronics-based Distribution Systems," in *2011 Twenty-Sixth Annual IEEE Applied Power Electronics Conference and Exposition (APEC)*, 2011, pp. 113–118, doi: 10.1109/APEC.2011.5744584.

[12] O. Vodyakho, M. Steurer, D. Neumayr, C. Edrington, G. Karady, and S. Bhattacharya, "Solid-State Fault Isolation Devices: Application to Future Power Electronics-Based Distribution Systems", *Electric Power Applications IET*, vol. 5, no. 6, pp. 521–528, 2011.

[13] A. Q. Huang, "Power Semiconductor Devices for Smart Grid and Renewable Energy Systems," *Proceedings of the IEEE*, vol. 105, no. 11, pp. 2019–2047, 2017.

[14] X. Song, A. Q. Huang, M. Lee, and C. Peng, "Theoretical and Experimental Study of 22 kV SiC Emitter Turn-OFF (ETO) Thyristor," *IEEE Transactions on Power Electronics*, vol. 32, no. 8, pp. 6381–6393, 2017.

[15] Wolfspeed, 1200 V SiC MOSFET C3M0021120K datasheet, https://assets.wolfspeed.com/uploads/2020/12/C3M0021120K.pdf.

[16] Littelfuse, 1200 V IGBT IXYR100N120C3 datasheet,https://m.littelfuse.com/~/media/electronics/datasheets/discrete_igbts/littelfuse_discrete_igbts_xpt_ixyr100n120c3_datasheet.pdf.pdf.

[17] Littelfuse, 1200 V single thyristor CS60-12io1 datasheet, https://ixapps.ixys.com/DataSheet/CS60-12io1.pdf.

[18] G. K. Dubet, "Classification of Thyristor Commutation Methods," *IEEE Transactions on Industry Applications*, vol. IA-19, no. 4, pp. 600–606, 1983.

[19] J. A. Deacon, J. D. van Wyk, and J. J. Schoeman, "An Evaluation of Resonant Snubbers Applied to GTO Converters," *IEEE Transactions on Industry Applications*, vol. 25, no. 2, pp. 292–297, 1989.

[20] K. Kurachi, K. Taguchi, and G. Majumdar, "GCT Technologies and Their Applications," in *2019 10th International Conference on Power Electronics and ECCE Asia (ICPE 2019-ECCE Asia)*, Busan, Korea (South), 2019, pp. 2158–2165.

[21] K. Satoh, T. Nakagawa, M. Yamamoto, K. Morishita, and A. Kawakami, "6 kV/4 kA Gate Commutated Turn-off Thyristor with Operation DC Voltage of 3.6 kV," in *Proceedings of the 10th International Symposium on Power Semiconductor Devices and ICs. ISPSD'98 (IEEE Cat. No. 98CH36212)*, Kyoto, Japan, 1998, pp. 205–208.

[22] I. Nistor, T. Wikstrom, and M. Scheinert, "IGCTs: High-Power Technology for Power Electronics Applications," in *2009 International Semiconductor Conference*, Sinaia, 2009, pp. 65–73.

[23] Hitachi-ABB, Asymmetric and reverse conducting IGCTs, https://www.hitachiabb-powergrids.com/offering/product-and-system/semiconductors/integrated-gate-commutated-thyristors-igct/asymmetric-and-reverse-conducting.

[24] Y. Li, A. Q. Huang, and K. Motto, "Experimental and Numerical Study of the Emitter Turn-off Thyristor (ETO)," *IEEE Transactions on Power Electronics*, vol. 15, no. 3, pp. 561–574, 2000.

[25] B. Chen, A. Q. Huang, S. Atcitty, A. Edris, and M. Ingram, "Emitter Turn-off (ETO) Thyristor: An Emerging, Lower Cost Power Semiconductor Switch with Improved Performance for Converter-based Transmission Controllers," in 31st Annual Conference of IEEE Industrial Electronics Society, 2005. IECON 2005., Raleigh, NC, 2005, 6 pp.

[26] A. Qawasmi, J. Teichrib, N. Venkatesh, and R. W. De Doncker, "A New Thyristor-Based Power Electronic Device for DC Circuit Breakers in Medium-Voltage Applications," in *2018 9th IEEE International Symposium on Power Electronics for Distributed Generation Systems (PEDG)*, Charlotte, NC, 2018, pp. 1–6.

[27] K. A. Corzine and R. W. Ashton, "A New Z-Source DC Circuit Breaker," *IEEE Transactions on Power Electronics*, vol. 27, no. 6, pp. 2796–2804, 2012.

[28] A. H. Chang, B. R. Sennett, A. Avestruz, S. B. Leeb, and J. L. Kirtley, "Analysis and Design of DC System Protection Using Z-Source Circuit Breaker," *IEEE Transactions on Power Electronics*, vol. 31, no. 2, pp. 1036–1049, 2016.

[29] A. Maqsood, A. Overstreet, and K. A. Corzine, "Modified Z-Source DC Circuit Breaker Topologies," *IEEE Transactions on Power Electronics*, vol. 31, no. 10, pp. 7394–7403, 2016.

[30] S. G. Savaliya and B. G. Fernandes, "Analysis and Experimental Validation of Bidirectional Z-Source DC Circuit Breakers," *IEEE Transactions on Industrial Electronics*, vol. 67, no. 6, pp. 4613–4622, 2020.

[31] D. Keshavarzi, T. Ghanbari, and E. Farjah, "A Z-Source-Based Bidirectional DC Circuit Breaker With Fault Current Limitation and Interruption Capabilities," *IEEE Transactions on Power Electronics*, vol. 32, no. 9, pp. 6813–6822, 2017.

[32] F. Z. Peng, "Z-Source Inverter," *IEEE Transactions on Industry Applications*, vol. 39, no. 2, pp. 504–510, 2003.

[33] Gu, C., Wheeler, P., C. Alberto, W. Alan, and E. Francis. (2017). Semiconductor Devices in Solid-State/Hybrid Circuit Breakers: Current Status and Future Trends. Energies. doi: 10. 495. 10.3390/en10040495.

[34] C. Meyer and R. W. De Doncker, "Solid-State Circuit Breaker Based on Active Thyristor Topologies," *IEEE Transactions on Power Electronics*, vol. 21, no. 2, pp. 450–458, 2006.

[35] P. Cairoli, L. Qi, C. Tschida, V. R. R. Ramanan, L. Raciti, and A. Antoniazzi, "High Current Solid State Circuit Breaker for DC Shipboard Power Systems", in *IEEE Electric Ship Technologies Symposium (ESTS) 2017*, Washington, DC, 14–16 Aug. 2019.

[36] P. Cairoli, R. Rodrigues, U. Raheja, Y. Zhang, L. Raciti, and A. Antoniazzi, "High Current Solid-State Circuit Breaker for Safe, High Efficiency DC Systems in Marine Applications," in *IEEE Transportation Electrification Conference & Expo (ITEC 2020)*, Chicago, IL, July 17, 2020.

[37] F. Agostini, U. Vemulapati, D. Torresin *et al.*, "1MW Bi-directional DC Solid State Circuit Breaker Based on Air Cooled Reverse Blocking-IGCT," in *2015 IEEE Electric Ship Technologies Symposium (ESTS)*, 2015, pp. 287–292, doi: 10.1109/ESTS.2015.7157906.

[38] Z. Xu, B. Zhang, S. Sirisukprasert, X. Zhou, and A. Q. Huang, "The Emitter Turn-off Thyristor-based DC Circuit Breaker," in *2002 IEEE Power Engineering Society Winter Meeting. Conference Proceedings (Cat. No.02CH37309)*, New York, NY, USA, 2002, pp. 288–293 vol. 1.

[39] M. A. Rezaei, G. Wang, A. Q. Huang, L. Cheng, and C. Scozzie, "Static and Dynamic Characterization of a >13kV SiC p-ETO Device," *2014 IEEE 26th International Symposium on Power Semiconductor Devices & IC's (ISPSD)*, Waikoloa, HI, 2014, pp. 354–357.

[40] J. Zyborski, T. Lipski, J. Czucha, and S. Hasan, "Hybrid Arcless Low-Voltage AC/DC Current Limiting Interrupting Device," *IEEE Transactions on Power Delivery*, vol. 15, no. 4, pp. 1182–1187, 2000.

[41] Y. Feng, X. Zhou, S. Krstic, Y. Zhou, and Z. J. Shen, "Molded Case Electronically Assisted Circuit Breaker for DC Power Distribution Systems," *IEEE Transactions on Power Electronics*, vol. 36, no. 6, pp. 6586–6595, 2021, doi: 10.1109/TPEL.2020.3037477.

[42] Y. Wu, Y. Wu, F. Yang, M. Rong, and Y. Hu, "Bidirectional Current Injection MVDC Circuit Breaker: Principle and Analysis," *IEEE Journal of Emerging and Selected Topics in Power Electronics*, vol. 8, no. 2, pp. 1536–1546, 2020, doi: 10.1109/JESTPE.2018.2888590.

[43] Y. Wu, Y. Hu, Y. Wu, M. Rong, and Q. Yi, "Investigation of an Active Current Injection DC Circuit Breaker Based on a Magnetic Induction Current Commutation Module," *IEEE Transactions on Power Delivery*, vol. 33, no. 4, pp. 1809-1817, 2018, doi: 10.1109/TPWRD.2018.2813139.

[44] Y. Wu, Y. Wu, F. Yang, M. Rong, and Y. Hu, "A Novel Current Injection DC Circuit Breaker Integrating Current Commutation and Energy Dissipation," *IEEE Journal of Emerging and Selected Topics in Power Electronics*, vol. 8, no. 3, pp. 2861–2869, 2020, doi: 10.1109/JESTPE.2019.2911103.

[45] W. Wen, Y. Huang, B. Li, Y. Wang, and T. Cheng, "Technical Assessment of Hybrid DCCB With Improved Current Commutation Drive Circuit," *IEEE Transactions on Industry Applications*, vol. 54, no. 5, pp. 5456–5464, 2018, doi: 10.1109/TIA.2018.2791404.

[46] L. Ängquist, A. Baudoin, T. Modeer, S. Nee, and S. Norrga, "VARC – A Cost-Effective Ultrafast DC Circuit Breaker Concept," 2018 IEEE Power & Energy Society General Meeting (PESGM), 2018, pp. 1–5, doi: 10.1109/PESGM.2018.8586299.

[47] S. S. Mirhosseini, S. Liu, J. C. Muro, Z. Liu, S. Jamali, and M. Popov, "Modeling a Voltage Source Converter Assisted Resonant Current DC Breaker for Real Time Studies," *International Journal of Electrical Power & Energy Systems*, Vol. 117, 2020, doi: 105678, ISSN 0142-0615, https://doi.org/10.1016/j.ijepes.2019.105678.

[48] M. Callavik, A. Blomberg, J. Hafner, and B. Jacobson, "The Hybrid HVDC Breaker," in ABB Grid System, Technical paper, Nov, 2012.

[49] R. Derakhshanfar, T. Jonsson, U. Steiger, and M. Habert, "Hybrid HVDC breaker - Technology and applications in point-to-point connections and DC grids," in CIGRE Session, 2014, pp. 1–11.

[50] P. Sellier, R. Besrest, and C. Zimmermann, "Hybrid Circuit Breaker Device," US Patent 7,508,636, Mar. 24, 2009.

[51] R. Sander and T. Leibfried, "Considerations on Energy Absorption of HVDC Circuit Breakers," in Power Engineering Conference (UPEC), 2014 49th International Universities, IEEE, 2014, pp. 1–6.

[52] R. Sander, M. Suriyah, and T. Leibfried, "A Novel Current-Injection Based Design for HVDC Circuit Breakers," in PCIM Europe 2015; International

Exhibition and Conference for Power Electronics, Intelligent Motion, Renewable Energy and Energy Management; Proceedings of VDE, 2015, pp. 527–533.

[53] G. Tang, C. Gao, X. Luo, X. Wei, Y. Qiu, and J. Cao, "Direct-Current Circuit Breaker and Implementation Method Therefor," WO Patent 2 014 131 298 A1, 2014.

[54] W. Zhou, X. Wei, S. Zhang, G. Tang, Z. He, J. Zheng, Y. Dan, and C. Gao, "Development and Test of a 200kV Full-bridge based Hybrid HVDC Breaker," in Power Electronics and Applications (EPE'15 ECCE-Europe), 2015 17th European Conference on IEEE, 2015, pp. 1–7.

[55] Wen, Weijie, Huang, Yulong, Cheng, Tiehan, Gao, Shutong, Chen, Zhengyu, Zhang, Xiangyu, Yu, Zhanqing, Zeng, Rong, and Liu, Weidong. "Research on a Current Commutation Drive Circuit for Hybrid DC Circuit Breaker and Its Optimization Design," *IET Generation, Transmission & Distribution*, 2016, doi: 10.10.1049/iet-gtd.2015.0840.

[56] Magnusson, J., R. Saers, L. Liljestrand, and A. Ab. "The Commutation Booster, a New Concept to Aid Commutation in Hybrid DC-Breakers." 2015.

[57] J. Magnusson, L. Liljestrand, and R. Saers, "Apparatus Arranged to Break an Electrical Current," Patent WO2014032692, Mar. 6, 2014.

[58] O. Cwikowski, R. Shuttleworth, and M. Barnes, "Apparatus and Method for Controlling a DC Current," Patent WO 2014/177874 A2, Nov. 2014.

Chapter 10

DC marine vessel electric system design with case studies

Li "Lisa" Qi[1] and John O. Lindtjørn[2]

With more and more stringent environmental laws and policies developed and enforced worldwide, the shipping industry requires higher efficiency and less emission. Marine vessels tend to change from full mechanical propulsion to electrical propulsion. With the development of high-voltage and high-current semiconductor devices, DC electrical systems on marine vessels become economically feasible and competitive to conventional AC systems. Low voltage DC (LVDC) has been applied to small commercial marine [1–3] and naval vessels [4,5]. Medium voltage DC (MVDC) may be implemented on large commercial and naval vessels requiring high power.

FOR utility power systems, the DC transmission and distribution systems have the main advantages of high power transfer capability over long distance, low loss, and ease of energy storage integration. A marine vessel electric system is a finite inertia electric power system. Due to limited space, the connecting lines on vessels are short because of the short distance between equipments. The system is prone to dynamics since the generation capacity is comparable to the load demand. The advantages of DC marine vessel electric systems over AC systems could be different from the DC utility power systems.

DC marine electric systems is gaining attraction in a wide range of vessel types for different reasons. Offshore support vessels choose it primarily for the heightened fault tolerance, variable speed generators and ease of energy storage integration, while some icebreakers need a way to reduce the power plant footprint so that it will fit within the confines of the hull. Ferries choose DC because it is the most cost efficient and functional platform for integrating energy storage, making hybrid and fully electric operation a reality. The examples include two retrofit ferry projects Aurora and Tycho Brahe [6], whose AC power plants have been upgraded into a predominantly DC power plant to achieve maximum benefits out of their new plant. Shuttle-tankers will choose DC for simple and functional integration of variable speed shaft generators and, right around the corner, expedition cruises for

[1]Research Center, ABB Inc., Raleigh, NC, USA
[2]Marine and Ports, ABB AS, Fornebu, NO, USA

its suitability to integrate batteries and fuel cells for extended zero-emissions operation in sensitive areas. Implementing DC marine electric systems will also enjoy the possibility of distributing main power at higher DC voltages compared to AC, representing savings of up 40% or more on cabling.

This chapter discusses the applications of DC power distribution systems for marine vessels. In March 2013, ASEA Brown Boveri (ABB) delivered the first vessel DC electric power system – Onboard DC Grid^{TM} system – on a DP class offshore support vessel Dina Star owned by Myklebusthaug Management [1,2]. It is the first International Marine Organization vessel in the world powered by a modern primary DC power system. Different aspects in the Onboard DC Grid^{TM} design, including system topology, protection and safety, stability analysis and control, energy storage integration, as well as vessel control, are studied. Different case studies are presented targeting the different design aspects and applicability to different types of marine vessels. The benefits and advantages of DC marine electric systems are discussed. The lessons learned from its one-year field operation of the Onboard DC Grid^{TM} on Dina Star are also summarized and described.

10.1   Marine electric systems

The transformation from an LV/MVAC to a LV/MVDC marine electric distribution system is shown in Figure 10.1. A MVAC system has MVAC and LVAC subsystems interconnected. Transformers provide voltage transformation and power quality at common AC buses. In DC distribution, AC to DC, DC to AC, and DC to DC converters are used widely to convert power at different stages. Fully distributed systems are used on commercial vessels. The distributed systems provide simplificity and enough redundancy and thus have relatively low cost and sufficient reliability. Multiple distributed buses can be connected by bus tie breakers. High requirements on reliability and survivability are drivers of the ring configurations and zonal distribution for the commercial and naval vessels. Main buses are connected in a ring configuration in order to provide high fault tolerance and thus high system reliability during serious faults or emergency conditions.

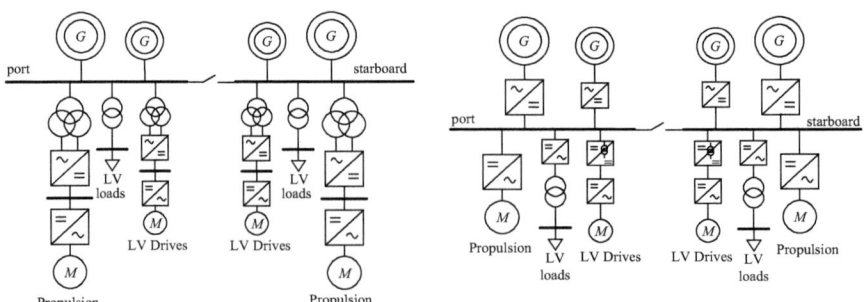

Figure 10.1   Transformation from AC to DC shipboard distribution system

This shift from AC to DC power is primarily driven by high efficiency and low emission requirements in the shipping industry. The power rating of a LV vessel electric power system is up to tens of megawatts and the power rating of a MV vessel electric power system can be up to well above 100 MW. Both AC and DC marine electric systems are still finite inertia power systems. However, the DC operation can have the following features and advantages.

- Variable speed operations with less conversion stages. Variable speed generators and motors are allowed with optimum machine operation at different loading conditions, which results in reduced fuel consumption, maintenance and operation cost.
- Easy integration of energy storage and other DC based sources, such as fuel cells. Most energy storage units are DC sources. The direct or indirect connection of the DC sources permits less energy conversion devices and smaller footprint.
- Footprint reduction. The removal of AC switchboards and transformers in the LV/MVDC systems has resulted in a reduced footprint compared with the AC systems. In addition, two poles in DC instead of three phases in AC can also save a large amount of cables.
- Improved power quality. The harmonics at AC buses due to AC/DC rectifiers are no longer a problem.
- Increased electric system efficiency. From AC to DC distribution, the electricity efficiency is improved up to 1% with reduced energy conversion stages and no reactive loss in DC distribution.

10.2 DC marine system concepts

In the following sections, two representative LV/MVDC system concepts are given. They are generated from a generalized DC system configuration diagram. The challenges and design concerns of the representative DC system concepts are also discussed.

10.2.1 A generalized DC network configuration

Due to the symmetric structure, only the starboard or port side of a marine electric system is required to be considered in its network configuration analysis. A generalized configuration diagram for both AC and DC systems is developed to fully include different variations. Electric power is provided to four types of loads (ACLV, ACMV, DCLV, and DCMV loads); three voltage transformation equipment (generator, transformer, and converter) can be used to change voltages between MV and low voltage LV levels; four connecting equipment link different components at the same voltage level. The aforementioned loads, voltage transformation equipment, and connection equipment are summarized in Table 10.1. The generalized network configuration diagram is illustrated in Figure 10.2. For simplification, DC sources are not included in the figure.

Table 10.1   Components used in the generalized network configuration

Load types	MVAC, MVDC, LVAC, LVDC
Voltage transformation equipment	1. Generators (multi-stator-winding, doubly-fed, open-winding, etc.)
	2. Transformers
	3. DC/DC converters
Connecting equipment	1. AC/DC rectifiers
	2. DC/AC inverters
	3. AC/AC drives
	4. Cables or bus ducts

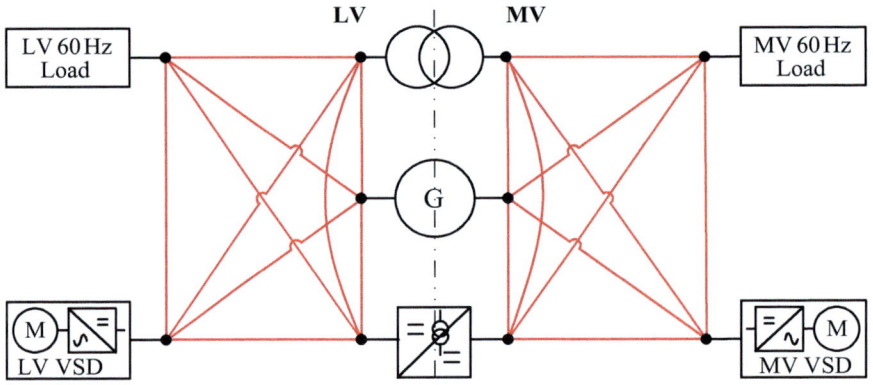

Figure 10.2   A generalized network configuration diagram for AC and DC shipboard power systems [7]

Using a variable speed generator as a source, a generalized DC network configuration diagram is derived as Figure 10.3. Three different electric paths, 60 Hz AC, variable speed AC, and DC, exist in this diagram. A generator, a transformer, and a DC/DC converter connect MV and LV subsystems. Converters and cables connect different AC and DC components. The specific DC system shown in Figure 10.1 is obtained from Figure 10.3. Further, by assigning various voltage transformation and connecting components, different DC system configurations can be generated from the derived generalized DC configuration diagram.

10.2.2   LVDC system concept and Onboard DC Grid^{TM}

Figure 10.4(a) shows a LVDC network configuration diagram from Figure 10.3. In this configuration, a common LVDC bus distributes power from the source to LV AC and DC loads. This LVDC system concept is illustrated in Figure 10.4(b). Compared to the LVAC system demonstrated in Figure 10.1, the transformers in front of the LV drives for AC power quality are removed from the LVDC system. However, AC/DC and DC/AC conversion stages are added to the auxiliary LVAC

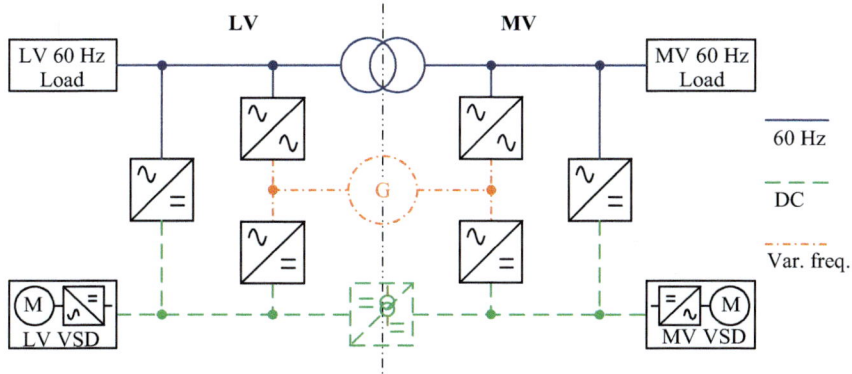

Figure 10.3 A generalized DC system configuration diagram [7]

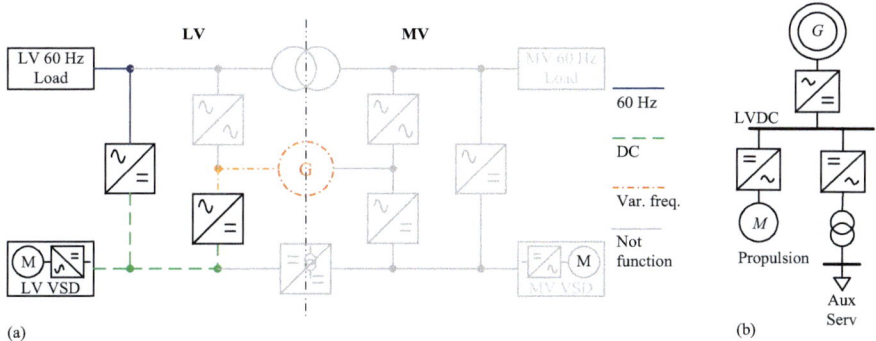

(a) (b)

Figure 10.4 A LVDC (a) configuration diagram and (b) system concept [7]

loads. Therefore, this LVDC system concept is only beneficial to the marine vessels with heavy propulsion and drive loads installed. If the auxiliary loads are dominant, the added DC/DC converters would significantly increase system overall cost.

Electric propulsion became a preferred choice for some types of vessels from the late 1980s because of applications of variable speed drives. ABB has completed the development of a new LVDC based distribution system. The primary focus has been on LV distribution systems and using offshore support vessels as one reference example. The LVDC distribution system is targeting all vessels types with total installed power up to several tens of megawatts.

On Dina Star delivered by ABB, the main focus of the DC electric power system are variable speed generators, space savings, and dynamic performance [8,9]. The Onboard DC Grid^{TM} system is a new way of distributing energy for the LV electrical systems on ships. It can be used for any electrical ship application operating at a nominal voltage of 1,000 V DC. Figure 5(a) and (b) shows the layout of the machinery, electric distribution, and propulsion system for a typical Platform

Supply Vessel (PSV) with the traditional AC distribution system and the new Onboard DC Grid[TM], respectively.

On commercial marine vessels, a DC power distribution can be arranged with all cabinets in either a centralized or distributed line-up throughout the vessel (an example shown with short-circuit proof DC bus bars) as demonstrated in Figure 10.6. In the centralized configuration, all converters are located in one or multiple line-ups as in a multi-drive system and occupies the same space as today's main AC switchboard. In the centralized DC bus configuration, the need for various ambient conditions (temperature and humidity) and cleanliness is reduced in thruster rooms, as an example. This is particularly useful during construction and commissioning.

Although some DC systems are highly centralized, the Onboard DC Grid[TM] platform also supports fully distributed systems using cables or bus-ducts. In the distributed Onboard DC Grid[TM], each converter can be placed freely on the vessel and close to the corresponding power source or load. The rectifiers can be even integrated into the synchronous generators. The integrated generator and rectifier units directly feed power into a common DC bus that distributes the electrical power to different loads. Each main load can be fed by a separate inverter unit. The sensitive AC loads, such as low voltage hotel loads, are fed using island inverters, which are specially developed to feed clean power to the sensitive loads. Further, energy storage devices, such as batteries or supercapacitors, can be integrated into the system and used for a wide variety of functions, including load leveling, peak shaving, and zero-emission operations.

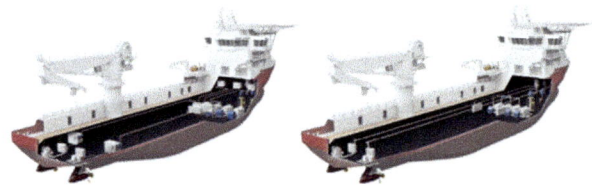

Figure 10.5   PSV with (a) traditional AC system and (b) Onboard DC Grid[TM] [8,9]

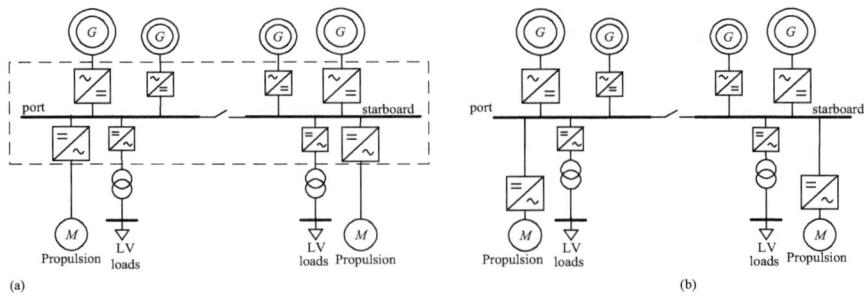

Figure 10.6   Vessel DC electric power system with (a) centralized line-up and (b) distributed line-up

10.2.3   MVDC system concept

Figure 10.7(a) shows a voltage transformation DC/DC converter based MVDC network configuration diagram obtained from Figure 10.3. In this configuration, a common MVDC bus distributes power from the generator to all four types of MV and LV loads. The DC/DC converter converts DC voltage from MV to LV to supply LV variable speed drives and other LVDC loads. This DC/DC converter based MVDC system concept is illustrated in Figure 10.7(b). Compared to the MVAC system demonstrated in Figure 10.1, the conversion stages of the MV propulsion loads in this MVDC system require one stage less by removing the transformers. However, the conversion stages are increased for all other loads by adding converters. In certain vessel types, if the loads other than propulsion loads are dominant, the increased conversion stages may result in overall reduced efficiency, reliability, and increased cost.

The MV generator rectifiers, MV inverters, DC/DC converters, and MVDC protection are the key components of the MVDC system design in Figure 10.7(b). The generator rectifiers can be either current source converters (CSCs) or voltage source converters (VSCs). Among different CSC technologies, a thyristor rectifier with fault current limiting capability can be implemented. Allowing a 20% variable speed operation range, the variable speed generator needs to supply the nominal voltage even at 80% of its nominal speed. With a six-pulse thyristor rectifier directly connected to the MV generator, the AC generator must be substantially oversized for high current harmonics and over-excitation at its low operation speed boundary. Different measures, such as harmonic filters and multiphase generators, may be adopted to reduce current harmonics. VSCs can reduce current harmonics and maintain rated voltage output even at low operation speed. Limited generator oversizing is thus required if VSCs are implemented as the generator rectifiers. Some generator oversizing is still required since the extra insulation by dv/dt of VSCs should be considered. A modular multilevel converter (MMC) is thus preferred to a conventional two-level VSC since its dv/dt imposes little extra insulation requirement to the MV generator.

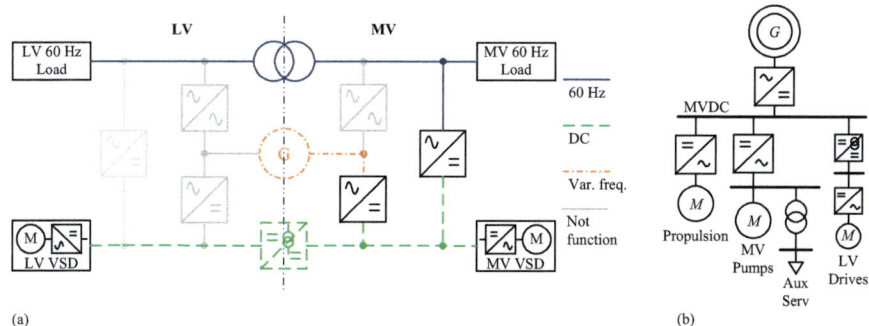

(a)   (b)

Figure 10.7   A DC/DC converter based MVDC (a) configuration diagram and (b) system concept [7]

MV DC/AC inverters interfaces the MV propulsion systems and fixed frequency AC loads in Figure 10.7(b). Similarly, because of little extra insulation on the MV motors, MMCs are better options than a conventional two-level VSC. The required voltage ratio for a DC/DC converter between MVDC and LVDC subsystems ranges from 5:1 to 20:1. In order to be comparable to a variable speed LV drive train in AC, the efficiency of the DC/DC converter should be equal to or better than a conventional transformer. For fixed speed AC loads at MV and LV levels, such as service and hotel loads, AC power supply through regular transformers is still desired for its high efficiency, safety using galvanic isolation, and low common-mode EMI.

Since all loads are connected to the MVDC bus, a feasible MVDC system protection solution is necessary. Different protection speeds are required in different system designs. Ultrafast solid-state circuit breakers (SSCBs), fast hybrid circuit breakers, fast mechanical circuit breakers, and converters can all be used as MVDC protection devices depending on different protection speed requirements. Considering fast discharging currents capacitors and energy storage in some MVDC system designs, fast DC circuit breakers (DCCBs) are a desirable MVDC protection option due to fast fault interruption speeds. Without fast increasing DC fault currents, converters can be adopted by MVDC protection. However, most existing low-cost commercially available converters do not have fault current limiting or interruption capability. One main challenge of the converter based MVDC protection is how to modify the existing converter topologies and controls in a cost-effective way to allow sufficient DC fault limiting or interruption capability.

10.3 Benefits and advantages

The drivers for DC in the marine industry have been multifaceted and vessel type specific, however there are a few key features of DC that are worth highlighting

- Optimized operation of combustion engines,
- Efficient integration of DC based power sources, and
- Potential for very fault tolerant power systems.

Optimized operation of combustion engines. Today almost all power onboard electric ships are derived from combustion engines, most operating on liquid oil (HFO/MDO), some on gas (from LNG mainly), and even some with Dual Fuel capability (liquid fuel or gas). For standard AC generator sets, the specific fuel oil consumption (SFOC) of the engines is typically lowest at a narrow operating window, often around 85% of rated load. If engines are operated far away from this optimum point, SFOC can deteriorate significantly. When introducing DC distribution, the generators are no longer required to operate at the speed dictated by a system electrical frequency but are free to adjust the engine rpm as a function of load to optimize SFOC and other parameters such as emissions, noise and vibrations and maintenance intervals.

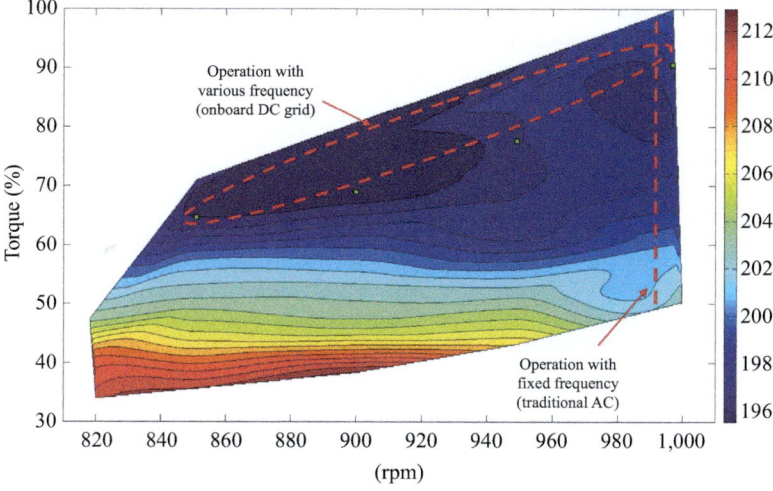

Figure 10.8   SFOC as a function of RPM and torque [8,9]

For a test engine, the SFOC was measured at various speeds and torques of a diesel engine and is shown in Figure 10.8. The dashed circle indicates the variable speed operation range of the engine while the dashed line is the fixed speed operation. The dark regions of the figure indicate where the SFOC is at the lowest. The test results clearly confirm that the engine can run at lower powers in combination with lower RPM and thereby lower and widen the SFOC optimal setpoint. As can be seen from the figure, the biggest impact of this is operating at lower loads.

Improved SFOC at lower loads is particularly attractive to dynamically positioned vessels in the offshore industry. Many of these vessels spend extended periods of time in critical operations where propeller thrust is used to keep vessel position. The criticality of the operation often calls for multiple redundant engines to be online simultaneously resulting in very low loads on individual engines. In these modes, variable speed generator sets can reduce fuel consumption by 30%–40%. In addition, noise from generators will be drastically reduced and maintenance intervals lengthened.

Efficient integration of DC power sources. The DC backbone of Onboard DC Grid[TM] makes integration of DC based consumers and loads more functional and cost efficient than in an equivalent AC system. Most notably this includes most energy storage types such as batteries and super capacitors, fuel cells and variable speed shaft generators. In a system with an AC backbone, all of these would need additional (or oversized) power electronics and (in many cases) a transformer to interface with the AC bus.

Potential for a very fault tolerant power system. Maintaining propulsion power is often critical to a vessel operation. A good example of this is when a vessel is dynamically positioned in close proximity to another installation. A DC power system has improved upon an AC equivalent system in multiple ways;

- Solid state protection circuitry and controlled voltage ramp-up following a short-circuit means that single points of failure have minimal (if any) impact on healthy system sections.
- DC infrastructure combined with generator rectifiers eliminate reactive power-flow between parallel generators and reverse power from the system to the genset. This practically means that AC common mode failures of governor and AVR faults, where a fault on a single generator can take out the entire system, have been eliminated.
- No need for synchronization when connecting new energy sources means that start-up from a black-out is very low risk and can be done much more quickly than an equivalent AC power system.
- Cost efficient and highly functional integration of energy storage means that more can afford this additional barrier against black-out or sudden load-changes.

The effect of these characteristics is that an operator can have more faith in their power plant which in turn often leads to lower incidence rates, more economic use of the power plant, both drivers for lower operational costs. In addition to the aforementioned benefits:

- The classic issue with ramp limits on thruster RPM and power will also be different in Onboard DC Grid^{TM}. In traditional AC power systems with variable speed thrusters, the rate of change of power is strictly limited by the control system based on available and usually low engine ramps. The engine ramps are restricted by the allowable frequency and voltage variations in the 50 or 60 Hz AC network. With Onboard DC Grid^{TM} these restrictions can be relaxed due to tighter control of sources and loads, and higher ramp rates may be utilized in the thruster control.
- **Footprint reduction of up to 30%.** Taking power directly from the common DC bus, the transformers and main AC switchboards are no longer needed. This can result in a reduced footprint of up to 30% as compared with a traditional AC system. Further, the power distribution on 1,000 Vdc instead of 690 Vac can reduce the weight and volume of cables by as much as ~40%.
- **Improved power quality.** The harmonics common in AC systems with frequency converters is no longer a problem. The costly and bulky harmonic filters or 12/18/24-pulse transformers associated with the high power frequency converters are no longer needed.
- **Increased distribution efficiency.** The electrical efficiency from the generators to the loads will be improved with less installed components, such as no transformers. In the process of going from AC to DC distribution the electrical system has improved by 0.5%–1%.
- **Easy ship-to-shore connection.** For the 50 or 60 Hz AC systems, ship-to-shore connection systems are required in order for any vessels to be connected

in different ports of different countries. The DC side vessels can be more easily connected to shore connections because the frequency is no longer an issue. Also, by Onboard DC Grid^{TM}, there are no starting currents from motors and transformers drawn from the shore-connection and thus low-power feeders are allowed in shore-connection in ports.

- **A DC power system is highly digitalized due to the high converter content in the system.** This offers many opportunities with respect to data collection and system control that has previously not been possible with standard AC systems or mechanical systems.

10.4 Protection and safety

In a DC electric distribution system, the fault sources are capacitors, DC sources, AC sources, such as generators and motors. The DC fault current is a summation of all fault currents from all contributing sources. The system topology is reflected by different weighting factors associated with different sources. The fault currents in DC shipboard distribution normally have high derivatives due to low inductance of DC connection lines, including cables, bus ducts, and bus bars. Fault currents of different sources can have different time constants and thus different DC fault developing speeds. Detailed analysis of different DC fault currents can be found in [10–12]. Normally, capacitor discharging currents have rising times from a few microseconds to hundreds of microseconds; AC source fault currents have rising times determined by their own 50 or 60 Hz frequencies; and DC source fault currents have rising times ranging from a few milliseconds to several milliseconds depending on circuit parameters. Therefore, DC fault current derivatives can vary from a few to a few hundreds of mega-amperes per second. Depending on the rising speeds of the DC fault currents, protective devices having different response speeds should be applied.

According to key protective devices, DC protection solutions can be categorized into

- DCCB (DC Circuit Breaker)-based protection;
- Fuse-based protection;
- Converter-based protection;
- Fault current limiter-based protection; and
- Hybrid (of different protection devices) protection;

Compared to the traditional fuse-based DC protection, DCCB based protection is more reliable because of repetitive operations of breakers. For the fault current limiter-based protection, both converters and fault current limiters can be applied. However, considering the wide implementation of converters in DC and the need of extra passive elements for fault current limiting in fault current limiters, converters are better choices than solid state fault current limiters in terms of convenience for DC protection. However, not all converters are capable of fault current limiting. Different DC protection methods, including overcurrent protection, differential

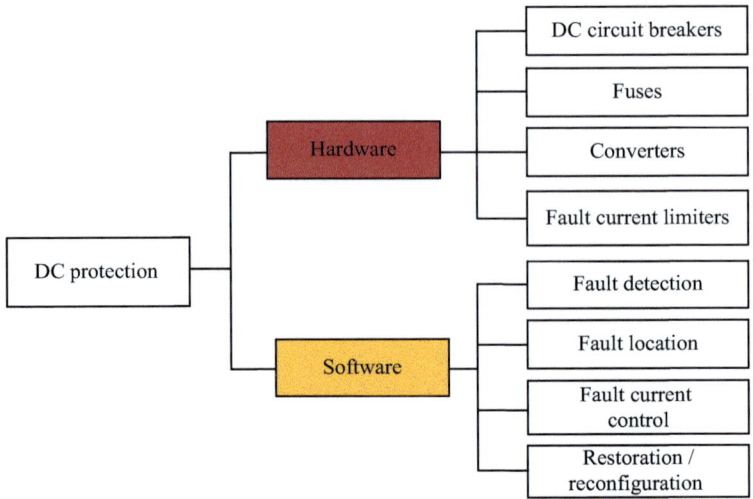

Figure 10.9   Hardware and software required in DC protection [10]

protection, distance protection, and current derivative, can be used for DC protection. Depending on whether communication is critical or not, a DC fault identification and protection coordination method can also be characterized as

- a local measurement-based method; or
- a communication-based method.

10.4.1   Protection of Onboard DC Grid^{TM}

In developing the new Onboard DC Grid^{TM}, the entire system has been designed and optimized as opposed to merely optimizing on a component level. However, some important protection principles have been carried over from the traditional AC system and have formed the framework for the new DC system protection philosophy:

- Equipment and personnel shall be protected in case of failures.
- Proper selectivity shall be ensured in such a way that safe operation is maintained after any single failure.

Since the main AC switchboard is removed, including its AC generator and feeder circuit breakers, a new protection system should be developed. It is essential to design a new DC protection philosophy that fulfills class requirements for equipment and personnel protection and protection selectivity. Proper protection of the Onboard DC Grid^{TM} can be achieved by a design as shown in Figure 10.10 with a combination of fuses, isolator switches, and controlled turn-off of semiconductor power electronics devices [8,9]. This can be realized in a distributed (a) and centralized (b) topology.

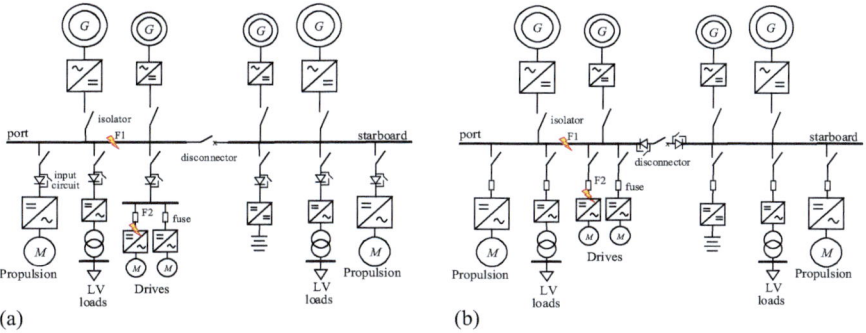

Figure 10.10 *Illustration of the breaker-less protection scheme for Onboard DC Grid^{TM}*

In the new protection scheme, fuses are used to protect and isolate inverter modules in case of serious module faults. This is the same as the existing fuse protection of LV multi-drives. This is a proven, compact and cost-efficient way or realizing protection of smaller feeders. In addition, input circuits of controlled switches separate the inverter modules from the main DC bus and provide full control of reverse power, both in fault and normal conditions (as for example in propeller braking mode). This means that faults on a single consumer will not affect other consumers on the main DC distribution system. In the event of severe faults on the distributed DC bus, the system is protected from all generators by turning off the controllable thyristor rectifiers. Once all fault contributions from generators become low, the load disconnector between the main DC buses are turned off to isolate the faulty part from the rest of the system. The isolators are installed in each circuit branch and can also galvanically isolate faulty sections from the healthy sections. For added simplicity and fault tolerance, but at higher cost and footprint, isolators can be replaced with air circuit breakers (bus-tie is always implemented this way) – thereby avoiding the need for coordination with generator rectifiers or other sources.

For a major fault F1 in the distributed system (a) at or close to either the port or starboard DC bus, all generators contribute fault currents to the fault location. The capacitor discharging of other loads are blocked by the input circuits. After fault detection, the bus-tie opens and the thyristor rectifiers on the faulty side start to turn off and interrupt the fault contributions. After the fault currents are reduced on faulty side there is limited current at the disconnector switch. This disconnector switch opens to isolate the faulty bus from the healthy bus and voltage is quickly restored on the bus. The foldback protection principle employed by thyristor rectifiers requires the consideration to the temporary voltage outage during major fault clearing and ride-through capabilities of all downstream equipment. If there is a fault F2 (a) at the inverter of the multi-drives, the capacitors of the multi-drive system contribute high fault currents to the fault location. The fuse closest to the fault location melt quickly and isolate the fault from the rest of the system. In the centralized system (b), the opposing input circuits placed on opposite sides of the bus-tie form a solid-state circuit breaker and interrupt fault current from starboard side within a few microseconds thereby preventing capacitor discharge and ensuring un-interrupted supply on the healthy side of the system. A fault in location F2 can be handled in the same way as for F2 location in (a).

For Onboard DC Grid^{TM}, this protection design has resulted in:

- Closed bus operation in DP2 without additional equipment because common mode faults like governor and AVR failures is handled more effectively; and
- Clearing of major short-circuit currents in a "soft" way so that the voltage of the capacitors can be maintained and the converters can restart quickly. Therefore, the system recovers quickly and predictably.

Even when a rare major DC fault occurs, the system recovers quickly and reliably. This often translates into significantly more economical operation of the vessel because the operator is not worried about operating with a closed bus-tie and can reduce the required need for generators online. In non-critical operations the operator may even operate with only a single generator connected, knowing that in the event of a diesel trip, a new one will start-up and connect very quickly.

10.4.2 DCCB-based protection

DCCB based protection is selected for high reliability for fast fault interruption speed and selective fault isolation. At a DC fault, with low inductance of the short lines between electrical components on vessels, DC fault currents from DC sources can quickly rise to very high magnitudes. Fast DC fault interruption is thus required to prevent damages to equipment and devices due to high DC fault currents. With quickly decreased DC voltage, undervoltage of converters and other equipment may trip and results in a disconnection of many loads. Fast DC protection is also required to prevent loss of components and system. Certain converters with fault handling capability can provide fast fault interruption or limiting. However, if the source converters are blocked, this may cause system wide loss of power supply and is not acceptable by the commercial and naval vessels requiring high reliability and survivability. In comparison, DCCBs can be placed system wide and coordinately operated to isolate only faulted system and thus restrict fault impacts as minimum as possible.

Depending on the specific DCCB implemented, DCCB-based protection can be further subcategorized by different circuit breaker technologies. In recent years, three DCCB technologies, mechanical, hybrid, and solid-state, have been developed. Because of non-zero-crossing for DC voltages and currents and rapidly increasing DC fault currents, main technical challenges of DCCB technologies for DC shipboard distribution protection are

- the creation of artificial zero-crossing; and
- fast turning-off speeds for fast fault interruption.

If the same fault identification technology is used, solid-state circuit breakers (SSCBs) have the fastest turning-off speeds as much as less than one millisecond;

while hybrid DCCBs are capable of turning-off within a few milliseconds, and mechanical DCCBs need a few to several milliseconds. One example of the SSCB protection for shipboard power system is to use SSCBs as tie breakers to separate faulted from healthy parts and then use generators to limit fault currents [3].

The design of SSCB based shipboard system protection, including the SSCB design and its corresponding protection algorithms can be found in [12]. As demonstrated below, the application of SSCBs system wide can achieve full DC protection by appropriate thresholds and ultrafast turning-off speeds of SSCBs together with proper coordination between upstream and downstream SSCBs. Figure 10.11(a) indicates a ring configuration system with distributed loads L1~L3. Capacitors are installed at the converter inputs. Figure 10.11(b) is the fault equivalent circuit at the port side DC bus. The protection of this port side DC subsystem can be expanded into the rest of the DC shipboard system. The capacitors C1~C3 are the equivalent capacitors of the three zonal loads. The capacitor C4 is the equivalent line capacitor connecting the port and starboard DC bus. Seven solid-state DCCBs, CB1–7, are placed in the system as shown in Figure 10.11. Figure 10.12 illustrates the performance of the DCCB based overcurrent protection.

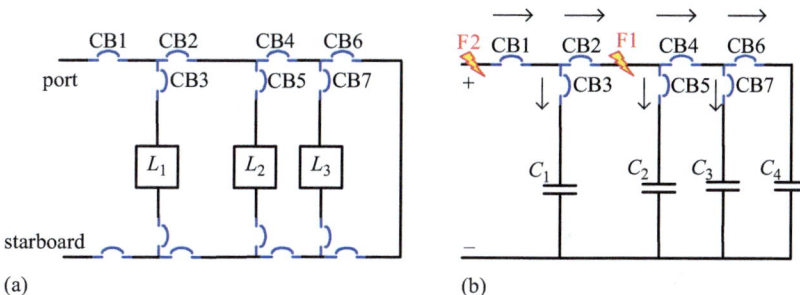

Figure 10.11   A ring configuration DC shipboard power system and its fault equivalent circuit of the port DC bus [10]

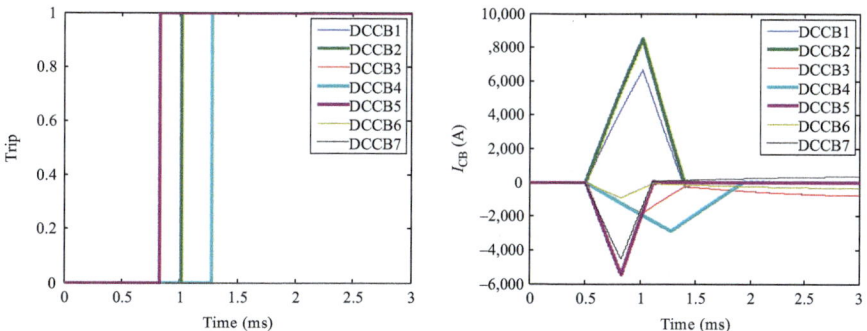

Figure 10.12   Simulated trip signals and current of DCC1-7s at F1 [10]

The simulated trip signals and currents of each DCCB at F1 are shown in the figure. At the fault F1, CB2, CB4, and CB5 are the closest breakers and thus react to open due to high DC fault currents in Figure 10.12. These trip signals were sent to their corresponding breakers and commanded them to open. After the opening of the breakers, the currents CB2, CB4, and CB5 become zero while there are still currents flowing through other DCCBs.

Overcurrent protection is selected to be used with solid-state DCCBs because of its fast speed resulting from its simplicity and minimum communication. The fast fault detection, location, and protection coordination is especially important to solid-state DCCBs since rapidly increasing DC fault currents should be interrupted before reaching the overcurrent and thermal limits of solid-state switches. The overcurrent coordination among different DCCBs at different scenarios are investigated as shown in Figure 10.13. Figure 10.13(a) and (b) illustrates the coordination schemes among CB1, 2, and 3 at F1 and F2 given in Figure 10.11. The same coordination principles can be expanded to all other DCCBs in Figure 10.11.

The fault currents, i_{CB1}, i_{CB2}, i_{CB3}, and thresholds, I_{th1}, I_{th2} and I_{th3}, of CB1, CB2, and CB3 are indicated in Figure 10.13. Since the solid-state DCCBs have ultrafast response speed of less than one millisecond, the fault current can be linearized as straight lines. The positions of downstream and upstream DCCBs is relative and depends on fault locations. The upstream and downstream relations between different breakers could be reversed at different fault locations. For example, at F1 in Figure 10.11, CB1 is upstream of CB2; while, at F2, CB1 is downstream of CB2. Since correct protection selectivity only allows the breakers closest to the fault locations to open, the current thresholds should also be adaptive to different fault locations in order to ensure selectivity. Figure 10.13, at fault F1, I_{th1} is higher than I_{th2} and I_{CB2} is higher than I_{CB1}. Lower threshold and higher DC fault current of CB2 guarantees the faster fault interruption of CB2 than CB1 at F1. I_{th1} and I_{th2} are adjusted for fault F2. CB1 is opened before CB2 since I_{th2} is higher than I_{th1} and I_{CB1} is higher than I_{CB2}. This adaptive threshold setting can be realized by sensing different fault current directions flowing through the same DCCB at different fault conditions.

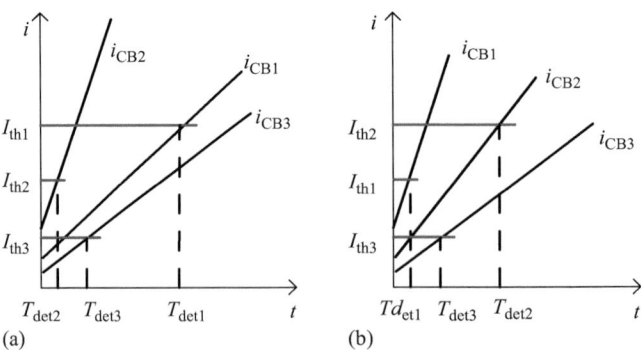

Figure 10.13   Overcurrent protection coordination of CB1,2,3 at fault F1, 2 [10]

10.5 Stability assessment and control

Stability is an important aspect in DC marine electric system design. Stability should be ensured all the time during system operation. In this section, the stability issues caused by network resonance is studied. DC marine electric systems are converter-based power systems. The network resonance is induced by the harmonic amplification and propagation due to the interactions between different converters directly connected to the DC buses.

10.5.1 Stability issues by integration of motor drives

Figure 10.14 presents a generic MVDC shipboard system under the stability study. The system has two generator-rectifier units, which allow high surge current and limit DC-side short-circuit fault currents. The load sharing among generators is achieved by the DC voltage droop control. Two 6 MW MV motor drives adopt neutral point clamped (NPC) three-level converter topology. These high-power motor drives are controlled by direct torque control (DTC) for fast dynamic response, with an average switching frequency equal to 500 Hz. DC capacitors are assumed to be distributed at each motor drive instead of centralized DC capacitor banks. Such a distributed DC capacitor configuration allows the use of diodes at motor drives to reduce the peak fault current stress at a major DC fault.

One technical challenge in implementing a marine MVDC electrical distribution system is the integration of voltage source inverter motor drives. A transformer and a diode rectifier provide important roles in the harmonic attenuation and DC-link dynamic stability in a conventional AC distribution system. The transformer

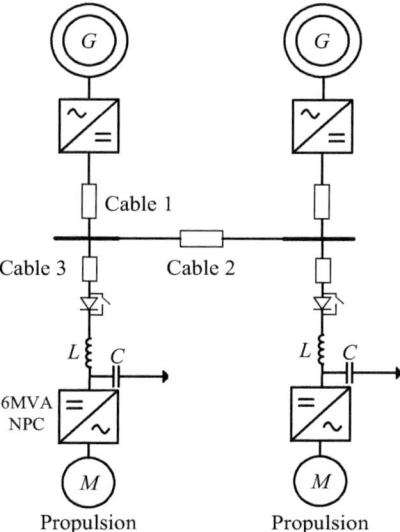

Figure 10.14 A generic MVDC shipboard power system for stability study

leakage inductance provides effective filtering for differential mode harmonics, while the transformer delta connection provides effective galvanic isolation for common mode noise. The diode rectifier, due to its commutation voltage drop, provides effective damping to improve DC-link voltage stability. In a marine MVDC electrical distribution system, since the drive inverters are directly coupled to the common DC bus, the small-signal DC system instability, common mode noise, and differential mode harmonic instability, all may occur during system operation.

10.5.2   Impedance-based stability analysis

The DC impedance analysis can be used to analyze and assess the system stability. A DC shipboard power system can be divided into the load and source subsystems. Its small-signal representation is demonstrated in Figure 10.15. $Z_s(s)$ represents the DC source impedance, including the generator-rectifier source impedance, cable impedance and LC filters impedance. $Z_l(s)$ represents the DC load impedance, for example the motor drive impedance. From the superimposition of different resources in a linear system, V_l and I_l can be represented by equations (10.1) and (10.2):

$$V_l(s) = \frac{V_s(s)Z_l(s)}{Z_s(s)+Z_l(s)} + I_s(s)\frac{Z_l(s)Z_s(s)}{Z_l(s)+Z_s(s)} = V_s(s)\frac{1}{1+\frac{Z_s(s)}{Z_l(s)}} + I_s(s)Z_s(s)\frac{1}{1+\frac{Z_s(s)}{Z_l(s)}}$$

$$= [V_s(s)+I_s(s)Z_s(s)]\frac{1}{1+\frac{Z_s(s)}{Z_l(s)}}$$

(10.1)

$$I_l(s) = \frac{V_s(s)}{Z_s(s)+Z_l(s)} - I_s(s)\frac{Z_l(s)}{Z_l(s)+Z_s(s)} = \frac{V_s(s)}{Z_l(s)}\frac{1}{1+\frac{Z_s(s)}{Z_l(s)}} - I_s(s)\frac{1}{1+\frac{Z_s(s)}{Z_l(s)}}$$

$$= \left[\frac{V_s(s)}{Z_l(s)} - I_s(s)\right]\frac{1}{1+\frac{Z_s(s)}{Z_l(s)}}$$

(10.2)

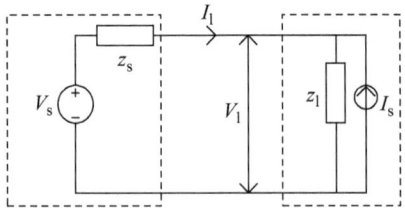

Figure 10.15   *The small-signal representation of a DC system with a source and a load*

From (10.1) and (10.2), the studied DC system is stable if (1) each individual component, including the voltage and current source, is stable; and (2) the impedance ratio $Z_s(s)/Z_l(s)$ satisfies the Nyquist stability criterion. According to the conservative Middlebrook stability criterion [13], a system design should satisfy (10.3) in order to guarantee system stability:

$$\|Z_s(S)\| < \|Z_l(S)\| \tag{10.3}$$

10.5.3 Impedance of generator-thyristor system

The equivalent DC impedance of a generator-thyristor unit is critical to understand the small-signal DC stability and harmonic stability issues. For simplification purpose, the generator is assumed to work in unsaturated area and constant speed mode. In synchronous generators, the positive and negative sequence impedances are coupled due to the effects of salient rotors. The asymmetrical d–q axis parameters of salient rotors also introduce the time-variance into the sequence impedance.

Introducing a small perturbance in rotor current, two perturbation currents at the frequencies of $f_m + f_1$ and $f_m - f_1$ are simultaneously imposed upon the fundamental component in stator current. With f_m as the perturbation frequency and f_1 as the fundamental frequency, the resulted stator perturbation currents at the frequency of $f_m + f_1$ and $f_m - f_1$ are of the positive sequence and negative sequence, respectively.

Considering the positive and negative perturbations in stator current, the three-phase stator current can be written as (10.4):

$$\begin{bmatrix} i_a(t) \\ i_b(t) \\ i_c(t) \end{bmatrix} = \begin{bmatrix} I_1\cos(\omega_1 t + \theta_i) + I_p\cos(\omega_p t + \theta_{pi}) + I_n\cos(\omega_n t + \theta_{ni}) \\ I_1\cos(\omega_1 t + \theta_i - 2\pi/3) + I_p\cos(\omega_p t + \theta_{pi} - 2\pi/3) + I_n\cos(\omega_n t + \theta_{ni} + 2\pi/3) \\ I_1\cos(\omega_1 t + \theta_i + 2\pi/3) + I_p\cos(\omega_p t + \theta_{pi} + 2\pi/3) + I_n\cos(\omega_n t + \theta_{ni} - 2\pi/3) \end{bmatrix} \tag{10.4}$$

where $f_p = f_m + f_1, f_n = f_m - f_1$ Subscripts "p" and "n" denote the variables associated with positive and negative sequence, respectively. I_1 is the fundamental current magnitude f, I_p is the positive sequence current perturbation magnitude and I_n is the negative sequence current perturbation magnitude. θ_i, θ_{pi}, and θ_{ni} are the phase angles of the fundamental, the positive, and the negative sequence perturbation currents, respectively Similarly, the three-phase armature voltage in time domain is given by (10.5):

$$\begin{bmatrix} u_a(t) \\ u_b(t) \\ u_c(t) \end{bmatrix} = \begin{bmatrix} V_1\cos(\omega_1 t + \theta_v) \\ V_1\cos(\omega_1 t + \theta_v - 2\pi/3) \\ V_1\cos(\omega_1 t + \theta_v + 2\pi/3) \end{bmatrix} + \begin{bmatrix} V_p\cos(\omega_p t + \theta_{pv}) \\ V_p\cos(\omega_p t + \theta_{pv} - 2\pi/3) \\ V_p\cos(\omega_p t + \theta_{pv} + 2\pi/3) \end{bmatrix}$$
$$+ \begin{bmatrix} V_n\cos(\omega_n t + \theta_{nv}) \\ V_n\cos(\omega_n t + \theta_{nv} + 2\pi/3) \\ V_n\cos(\omega_n t + \theta_{nv} - 2\pi/3) \end{bmatrix} \tag{10.5}$$

where V_1, V_p, and V_n are the amplitudes of the fundamental, the positive, and the negative sequence perturbation voltages, respectively. θ_v, θ_{pv}, and θ_{nv} are the phase angles of the fundamental, the positive, and the negative sequence perturbation voltages, respectively.

In the frequency domain, the perturbation stator voltage, the impedance matrix, and the perturbation stator current can be described by (10.6). The impedance matrix can be represented as Z_{sim} as in (10.7). $L_{dr}(s)$ and $L_{qr}(s)$ are the rotor d and q axis equivalent inductance with the damping impedances included. The detailed derivation of these equations can be found in [14]:

$$\begin{bmatrix} Vp(s+j\omega_1) \\ Vn(s+j\omega_1) \end{bmatrix} = \begin{bmatrix} Z_{pp}(s) & Z_{pn}(s) \\ Z_{np}(s) & Z_{nn}(s) \end{bmatrix} \begin{bmatrix} I_p(s+j\omega_1) \\ I_n(s-j\omega_1) \end{bmatrix} \tag{10.6}$$

$$Z_{sim}(s) = \begin{bmatrix} R_s + \frac{1}{2}(s+j\omega_1)[L_{ds}+L_{qs}-L_{dr}(s)-L_{qr}(s)] & -\frac{1}{2}(s+j\omega_1)[L_{ds}-L_{qs}-L_{dr}(s)+L_{qr}(s)]e^{2j\theta_r} \\ -\frac{1}{2}(s-j\omega_1)[L_{ds}-L_{qs}-L_{dr}(s)+L_{qr}(s)]e^{-2j\theta_r} & R_s + \frac{1}{2}(s-j\omega_1)[L_{ds}+L_{qs}-L_{dr}(s)-L_{qr}(s)] \end{bmatrix} \tag{10.7}$$

$$L_{qr}(s) = \frac{sL_{mq}^2}{Z_Q(s)} \qquad L_{dr}(s) = \frac{sL_{md}^2[Z_D(s)+Z_f(s)-2sL_{md}]}{Z_D(s)Z_f(s)-s^2L_{md}^2} \tag{10.8}$$

Here, θ_r is the angle between rotor d-axis and phase A armature winding. I_p $(s+j\omega_1)$ and $I_n(s-j\omega_1)$ are the positive and negative sequence perturbations in stator current. $V_p(s+j\omega_1)$ and $V_n(s-j\omega_1)$ are the positive and negative sequence perturbation in stator voltage. The matrix $Z_{sim}(s)$ is the sequence impedance matrix of the synchronous generator. The non-zero off-diagonal elements indicate there is cross coupling between the positive and negative sequence components.

The parameters of a synchronous generator to be verified are listed in Table 10.2. The calculated values from the analytical model (10.8) are compared with its numerical simulation results in Figures 10.16 and 10.17.

Figure 10.16 compares the magnitude and phase angle of $Z_{pp}(s)$ and $Z_{pn}(s)$ from the simulations and the analytical model (8). The amplitudes decrease as frequency increases up to about 8 Hz. This is due to the characteristics of rotor

Table 10.2   Parameters of a 7MVA synchronous generator [14]

Symbol	Value	Unit	Symbol	Value	Unit
L_{ds}	16.3	mH	L_{md}	16.2	mH
L_{qs}	9.5	mH	L_{mq}	9.4	mH
R_D	16.1	mΩ	L_D	16.745	mH
R_Q	2.145	mΩ	L_Q	9.452	mH
R_f	8.05	mΩ	L_f	16.857	mH
R_s	3	mΩ	f_1	60	Hz
θ_r	30	degree			

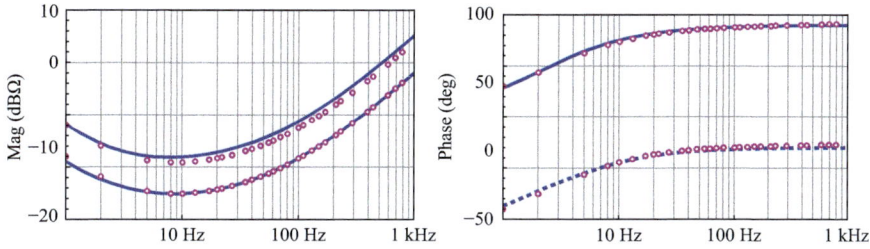

Figure 10.16 *Impedance response of $Z_{pp}(s)$ and $Z_{pn}(s)$. Solid line: $Z_{pp}(s)$; dashed line: $Z_{pn}(s)$ [14]*

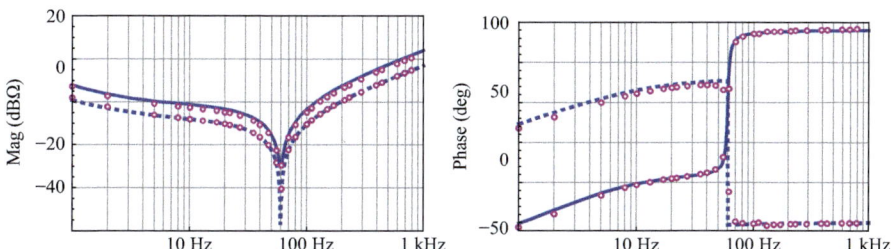

Figure 10.17 *Impedance response of $Z_{nn}(s)$ and $Z_{np}(s)$. Solid line: $Z_{nn}(s)$, dashed line: $Z_{np}(s)$ [14]*

equivalent inductance. Equation (10.8) can be rewritten as (10.9):

$$L_{qr}(s) = \frac{L_{mq}^2}{\frac{R_Q}{s} + L_Q} \qquad L_{dr}(s) = \frac{L_{md}^2\left[L_{1D} + L_{1f} + \frac{R_D}{s^2} + \frac{R_f}{s^2}\right]}{\frac{R_D R_f}{s^2} + L_f \frac{R_D}{s} + L_D \frac{R_f}{s} + L_f L_D - L_{md}^2} \tag{10.9}$$

where L_{1D} and L_{1f} are the leakage inductance of the d-axis damping winding and the excitation winding. At the low frequency range, below 8 Hz in this case, the resistance of the damping windings is dominant, which makes the amplitude of the $Z_{pp}(s)$ and $Z_{pn}(s)$ decrease as the frequency increases. Beyond this frequency range, the impedances are dominated by the inductances. The phase angle of $Z_{pp}(s)$ is close to 90° and $Z_{pn}(s)$ is $2\theta_r - 90°$, which is affected by the rotor angle position due to the asymmetry between the rotor d- and q-axis parameters. As the rotor angle θ_r is changing over time, the sequence coupling is time-variant. Figure 10.17 depicts $Z_{np}(s)$ and $Z_{nn}(s)$, which have the same behaviors as those of $Z_{pp}(s)$ and $Z_{pn}(s)$. In addition, there are singular points at the fundamental frequency.

The AC impedance of the generator-rectifier unit can be mapped to DC as (10.10). R_{gdc} is determined by both the stator winding resistance and commutation

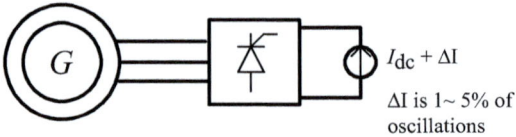

Figure 10.18   Frequency sweeping of the generator-rectifier unit

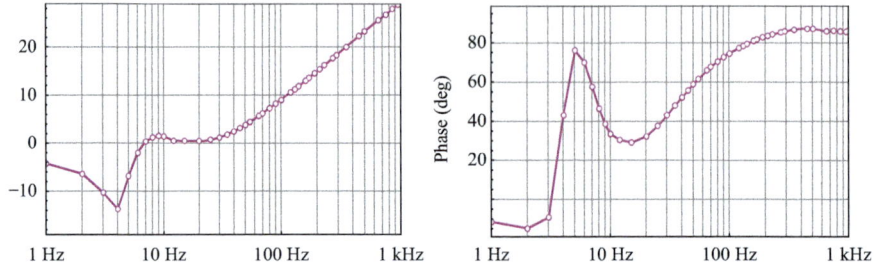

*Figure 10.19   Bode plots of simulated DC-side impedance response of a
generator-rectifier unit [14]*

of the thyristor rectifier:

$$Z_{gdc}(s) = j\omega \times L_{gdc}(s) + R_{gdc} \approx a \times j\omega \times L_{pp}(s) + R_{gdc} \tag{10.10}$$

The constant a can be estimated from the DC side small-signal frequency sweeping implemented on a detailed simulation model of the generator-rectifier unit as shown in Figure 10.18. The DC-side impedance response from the frequency sweeping is shown in Figure 10.19. Between 20 Hz and tens of kilohertz, the generator-rectifier unit behaves very close to a constant RL impedance. The frequency range below 20 Hz can be neglected since it is far below typical DTC inverter switching harmonics frequencies. The constant a can be estimated from comparing the results from the frequency sweeping and the analytical model (10.10). This impedance modeling is developed for three-phase generators but can be extended to six-phase generators, which are more widely applied in MVDC electrical distribution systems in order to supply sufficiently high voltage.

10.5.4   System stability assessment

Figure 10.20 shows the impedance network of the example MVDC system in Figure 10.14. The two generator-rectifier units are represented by their DC impedance $Z_{gdc}(s)$. Z_{cable1}, Z_{cable2}, and Z_{cable3} are the impedances of the cables with length of 25 m, 45 m, and 200 m, respectively. The cables can be simplified into series resistance and inductance, ignoring the cable capacitance due to the short length and the less than 10 kHz harmonics frequency range of high power DTC motor drive.

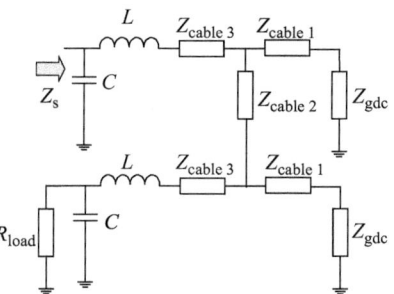

Figure 10.20   Impedance model of the example MVDC network

There are two DTC motor drive systems. In small signal analysis, one is represented by its negative impedance R_{load} and the other is represented by a harmonics current source I_T. For the motor drive represented by its negative impedance, with an average DTC switching frequency of 500 Hz, the DC terminals of the motor drives have negative resistance behavior up to several hundreds of Hz. The negative resistance amplitude is determined by the output power level as given by (10.11). At the terminal of each three-level NPC DTC motor drive, there is a *LC* filter. The filter design and selection of the capacitance and inductance should consider the small-signal DC stability and harmonics stability.

$$R_{load} = -\frac{V_{dc}^2}{P_{load}} \tag{10.11}$$

A typical DTC motor drive system has a wide spectrum of current harmonics. Figure 10.21 presents the harmonics spectrum of the simulated DTC motor drive current I_T. With an average DTC switching frequency of 500 Hz, the DC current harmonics spread from one hundred Hz up to several kHz. In addition, the dominant current harmonics frequencies may vary with different operating speeds. As analyzed before, these current harmonics may be amplified and propagated due to the resonance. The widely distributed DTC harmonics could excite different resonance modes in the DC network and thus cause serious overcurrent issues. Figure 10.22 demonstrates the cable current harmonics being amplified at around 450 Hz. At 1 pu speed, the NPC drive has low current harmonics at 450 Hz, as shown in Figure 10.22(a), however, the 450 Hz harmonics is significantly amplified in the cable as shown by the spectrum of the cable current I_c in Figure 10.22(b).

Figure 10.23 shows the bode plot of the impedance $Z_s(s)$ of the example MVDC network seen by the NPC motor drive. It can be observed that there are two resonance peaks at frequency f_1 and f_2, which are corresponding to the following two oscillation modes.

- mode 1: oscillations between generator and all DC capacitor banks.
- mode 2: oscillations between the cable and distributed DC capacitor banks.

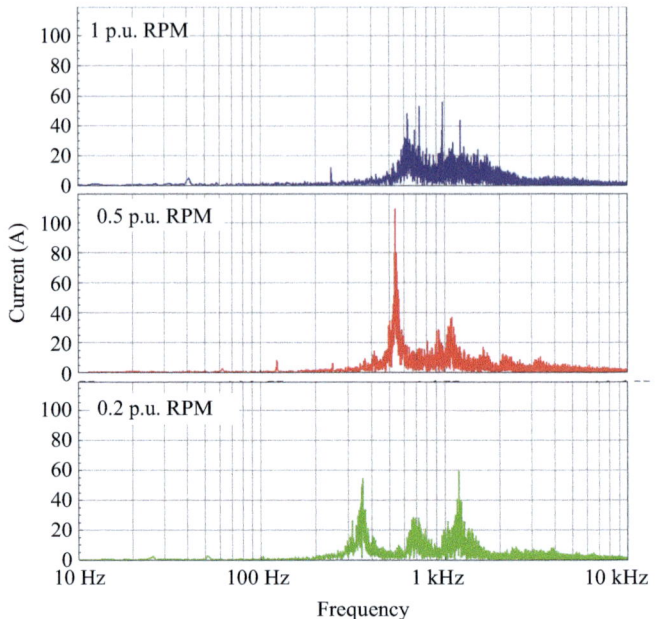

Figure 10.21 The current harmonics spectrum of an NPC DTC motor drive at different speeds [14]

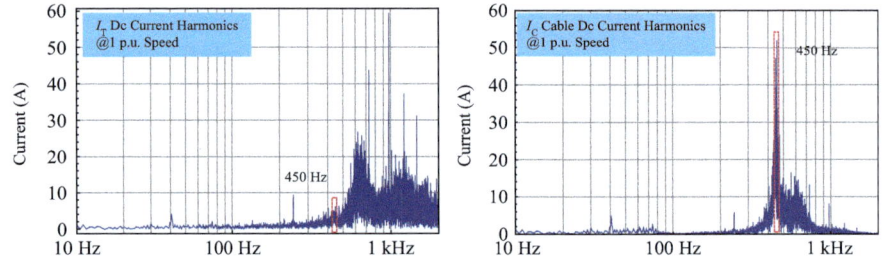

Figure 10.22 The current spectrum of (a) motor drive inverter, and (b) cable [14]

At mode 1, the resonance is caused by the oscillations between the generator-rectifier impedance Z_{gdc} and the motor drive capacitor C. The cable impedance and the filter inductance may be negligible at f_1. The simplified equivalent circuit at mode 1 is shown in Figure 10.24(a). The resonance frequency f_1 can be approximated by (10.12) and then (10.12) is substituted back into $Z_s(s)$. The amplitude of the first resonance peak can be approximated by (10.13):

$$f_1 = \frac{1}{2\pi\sqrt{L_{gdc}C}} \tag{10.12}$$

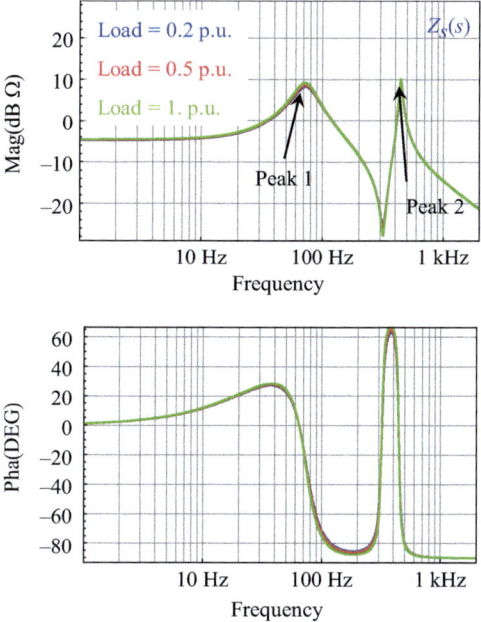

Figure 10.23   Bode plot of source impedance Zs(s) [14]

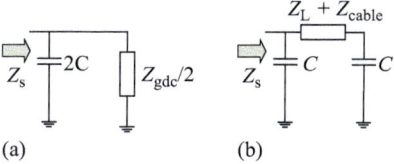

(a)   (b)

Figure 10.24   Equivalent circuit at (a) mode 1 and (b) mode 2

$$\|Z_s(j2\pi f_1)\| = \sqrt{\frac{L_{gdc}^2 + CL_{gdc}R_{gdc}^2}{C^2 R_{gdc}^2}} \tag{10.13}$$

where L_{gdc} and R_{gdc} are the inductance and resistance of the DC impedance of the generation unit, and C is the filter capacitance.

At mode 2, the resonance is induced by the oscillation between the DC cable impedance Z_{cable} and the motor drive filter capacitor C. As a result, the simplified equivalent circuit at mode 2 is shown in Figure 10.24(b). The resonance frequency f_1 can be approximated by (10.14). The amplitude of the second peak can be approximated by (10.15):

$$f_2 = \frac{1}{2\pi\sqrt{(L_{cable} + L)\frac{C}{2}}} \tag{10.14}$$

$$\|Z_s(j2\pi f_2)\| = \sqrt{\frac{(L_{cable} + L)(L_{cable} + L + 2C^2 R_{cable}^2)}{4C^2 R_{cable}^2}} \qquad (10.15)$$

Given the constant power operation of the motors, if the two resonance peaks derived from (10.13) and (10.15) are lower than the negative resistance of the DTC motor drive derived from (10.11), then the system is stable at modes 1 and 2. The aforementioned conditions are listed as (10.16) and (10.17):

$$\|Z_s(j2\pi f_1)\| = \sqrt{\frac{L_{gdc}^2 + CL_{gdc} R_{gdc}^2}{C^2 R_{gdc}^2}} < \frac{V_{DC}^2}{P_{load}} \qquad (10.16)$$

$$\|Z_s(j2\pi f_2)\| = \sqrt{\frac{(L_{cable} + L)(L_{cable} + L + 2C^2 R_{cable}^2)}{4C^2 R_{cable}^2}} < \frac{V_{DC}^2}{P_{load}} \qquad (10.17)$$

10.5.5   Stability control and simulation verification

As mentioned earlier, the LC filter at the terminal of the motor drive as in Figure 10.20 can be designed to attenuate the current harmonics and ensure the small-signal stability. Due to the wide spectrum harmonics emission of the DTC, the cut-off frequency of the *LC* filter should be set much lower than the average switching frequency. With the average switching frequency equal to 100 Hz, it can be represented by (10.18):

$$\frac{1}{2\pi\sqrt{LC}} < 100 \text{ HZ} \qquad (10.18)$$

Equations (10.16), (10.17), and (10.18) together define the design criteria for the *LC* filters considering the small-signal stability and harmonics filtering. The design criteria (10.16), (10.17) and (10.18) are illustrated as the stability requirement 1 line, stability requirement 2 line, and harmonic requirement line in Figure 10.25 and the shaded area surrounded by the three lines is the final feasible parameter design range of the *LC* filter.

Without any filter, the MVDC cable current is shown in Figure 10.26. The current harmonics is 35% of the DC current and the high current harmonics result in substantial cable derating and losses. Four sample DC filter designs A, B, C, and D are selected from Figure 10.25 to verify the harmonic attenuation and stability. Detailed time domain simulation results of the four sample filter designs are shown in Figures 10.27–10.30. The simulation results match the analytical analysis results quite well. The design B results in stable voltages and currents with the least amount of harmonics. The designs A and C result in instability of the DC bus and amplified harmonics in voltages and currents; although the system is stable at the design D, the harmonics are more than two times those of the design point B. It should be mentioned that the design of the *LC* filters is constrained by the DC fault tolerance as well as weight and volume limitations on vessels.

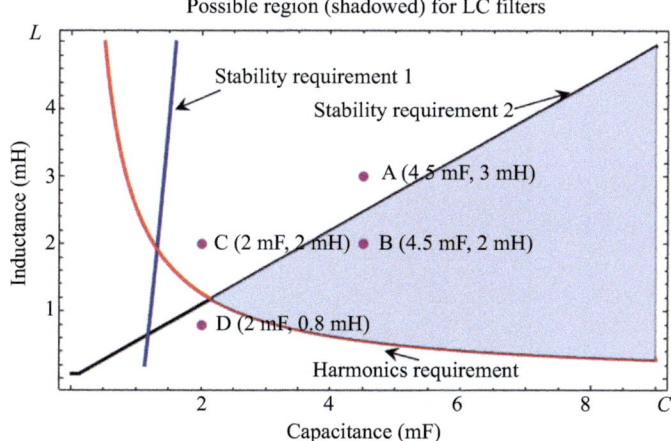

Figure 10.25 LC filter design analysis [14]

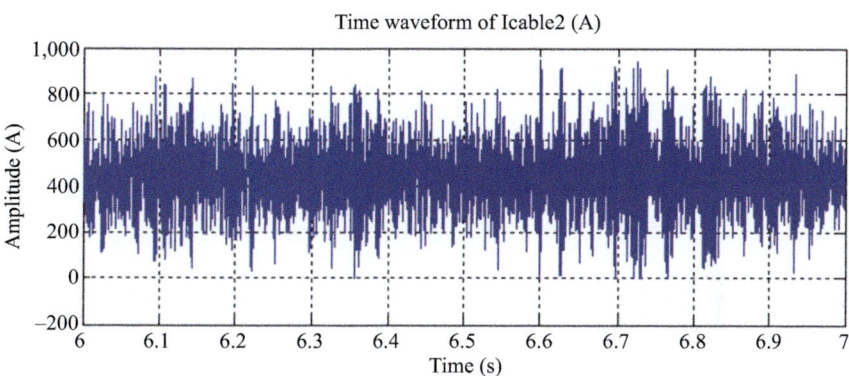

Figure 10.26 MVDC cable current before inserting LC filter inductor [14]

10.6 Integration of energy storage

10.6.1 Functions and benefits

Different types of energy storage technologies have received a dramatic increase in attention in recent years, not least in the maritime industry. Whilst this can be attributed to a number of different factors, what is certain is that it has the potential to improve safety, efficiency and performance of future vessels. Energy storage can be used in a wide range of ways onboard a vessel and most applications can be broken down into the few basic functions (or combinations thereof) described in Table 10.3.

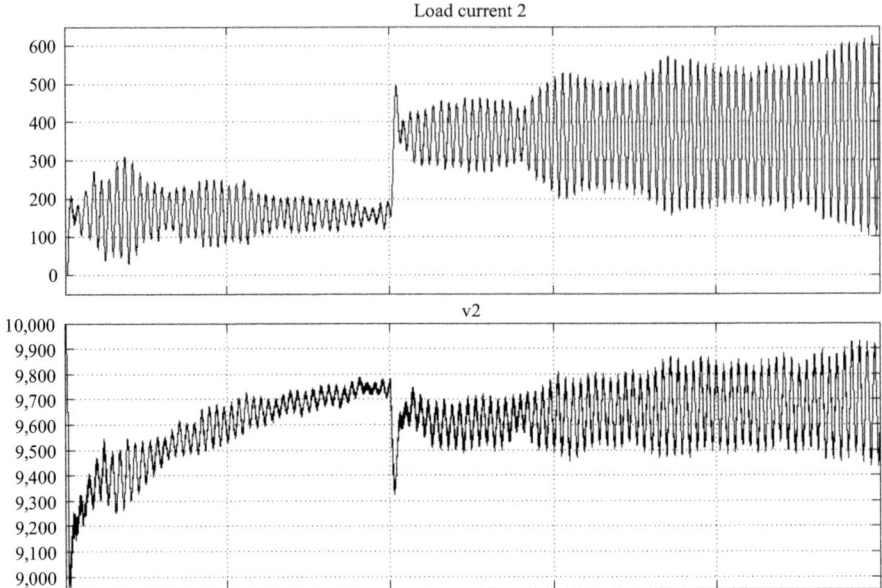

Figure 10.27   Cable current and DC link voltage resulting from inserting DC filter design A [14]

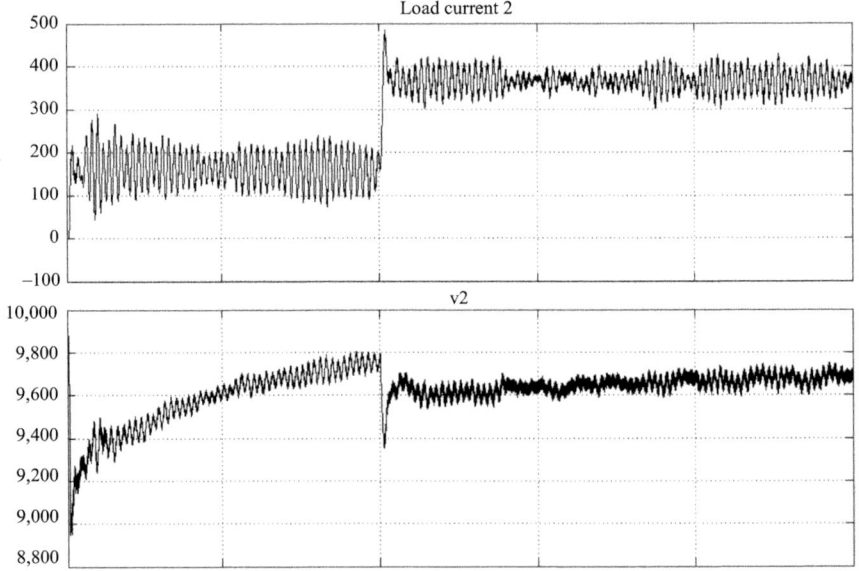

Figure 10.28   Cable current and DC link voltage resulting from inserting DC filter design B [14]

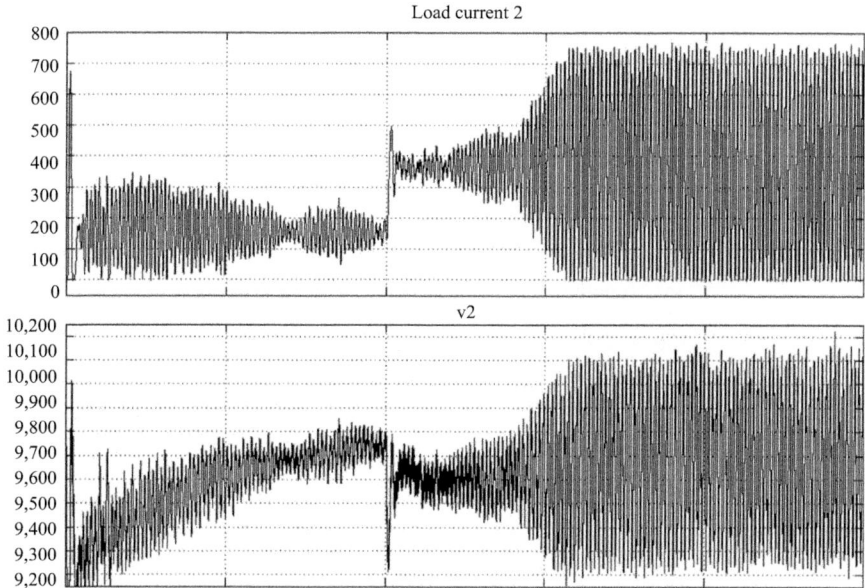

Figure 10.29   Cable current and DC link voltage resulting from inserting DC filter design C [14]

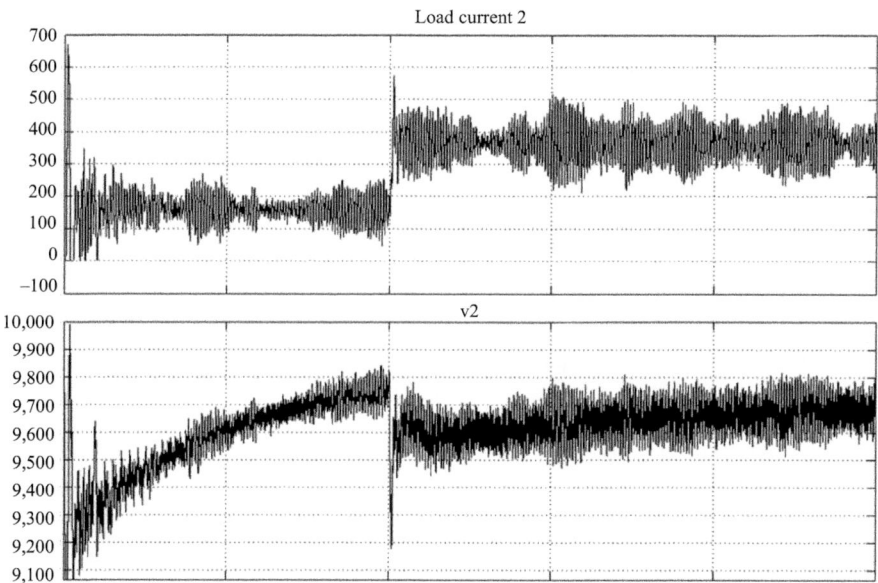

Figure 10.30   Cable current and DC link voltage resulting from inserting DC filter design D [14]

Table 10.3   ES functions and services

Name	Description	Purpose
Spinning reserve	Unit is connected and running but not charging or discharging energy into the system. On loss of generating capacity it steps in to take the load for a predefined period of time.	Backup for running gensets; Fewer engines needed online; Improved fuel efficiency through higher partial load;
	If other functions are activated simultaneously, this function ensures that sufficient energy is left in battery	Reduced engine running hours;
Enhanced ride through	Same as spinning reserve, but on a local level in a sub-system like a thruster or drilling drive	Energy storage solutions can give UPS like functionality for all or portions of power system; New ways of achieving higher environmental regularity number; Higher power system availability;
Peak shaving	Unit absorbs load variations in the network so that engines only see the average system power	Level the power seen by engines; Offset the need to start new engine; Improved fuel efficiency; Reduced engine running hours;
Enhanced dynamic performance	Unit absorbs sudden load changes and then ramps the change over on running engines. If peak shaving is used, then this function is automatically included	Instant power in support of running gensets; Enable use of "slower" engines, LNG/Dual fuel engines, fuel Cells;
Strategic loading	Unit charges and discharges to optimize the operational point of running engines, ensuring that energy is produced at the lowest cost, taking the efficiency of the ES system into account	Charging and discharging ES media in such a way that it optimizes the operating point of the gensets; Power is produced at peak efficiency;
Zero emissions operation	Unit powers the system so that engines can be turned off	Zero emissions in harbour and on sea; Quiet engine room;

Since most energy storage media are DC sources, the DC integration requires less equipment than AC in general as in Figure 10.31. The DC energy storage may be directly connected to a DC system without any converters or transformers. The DC/DC converter, if applicable, is significantly more compact than its AC counterpart since less semiconductors are used. For basic functions where no selectivity or starting scenarios are considered, the DC/AC converter is almost twice the length of the DC equivalent. If the selectivity, overvoltage and starting scenarios is considered, then this ratio becomes even higher. Generally, the integration of DC energy storage into a DC distribution system becomes simpler and more functional for less added cost than doing the same into an AC distribution system.

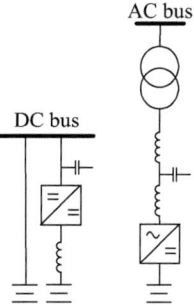

Figure 10.31   Integration of energy storage into DC and AC systems

The option of connecting the energy storage directly to the DC link can offer a slight reduction in length and improved efficiency compared to the DC/DC converter option. However, this is at the expense of controllability of the current in and out of the energy storage and system voltage level. From a control perspective, this direct connection option means that energy storage power flow is determined by the sum of the actions of all other sources and loads in the system. This method of direct connection is only suited to a limited number of applications, typically systems of low complexity where batteries represent a dominant power source. In addition, from a system voltage perspective, this option of direct connection means that the system voltage is defined by the energy storage and its state of charge. The DC bus thus becomes a variable DC bus. The variations can be significant and may therefore require the rest of the system to be over-dimensioned.

For these reasons, direct online solution is often chosen when efficiency is more important than controllability. An example of this implementation are ferries that operate in zero-emissions mode where a large amount of the consumed energy passes through the batteries on its way to the propellers. The DC/DC converter solution is preferred in applications where controllability and fault tolerance are of higher importance than the efficiency of the energy storage system. A good example of this is a dynamically positioned vessel where the battery is used to support the power system by means of functions like peak shaving, enhanced dynamic support and spinning reserve. In these cases, energy storage efficiency does not have a significant impact on overall electrical system efficiency because the battery is primarily used as an energy buffer and relatively little energy is passed through it during normal operation.

As discussed earlier, one major benefit of the Onboard DC Grid™ is the fuel optimization. In a system with variable speed engines where energy storage is not included, the engine needs to be operated in such a way that it always has enough reserves to be able to absorb fast load changes. The power margins in reserve means that some optimization potential in fuel efficiency is left untouched. The synergy of energy storage and variable speed engines offers full optimization at variable speed operation. When a system is equipped with energy storage and the

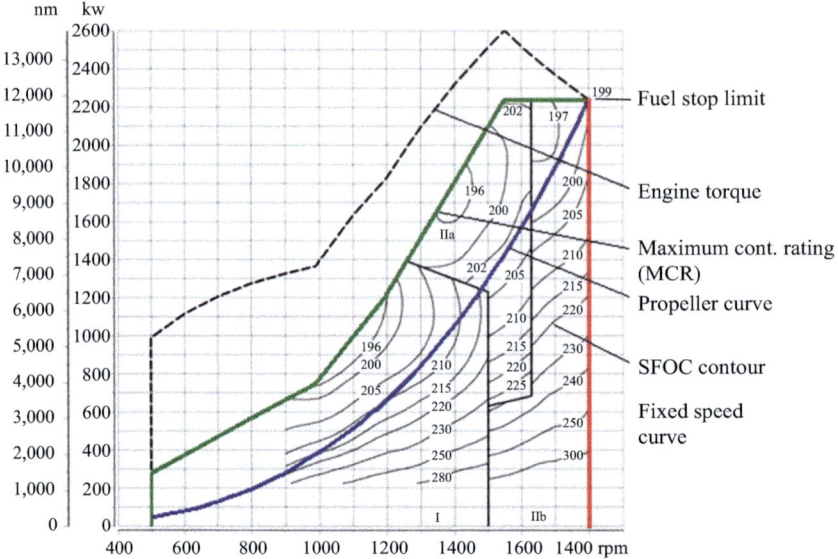

Figure 10.32   SFOC as a function of RPM and torque

enhanced dynamic support function is activated, the energy storage can take the role of absorbing quick load changes and the engine optimisation at variable speed operation has one constraint less and can achieve even further fuel reduction. Figure 10.32 demonstrates the fuel saving with energy storage added into the Onboard DC Grid™. Going from fixed speed to variable speed operation, the speed vs. load path is moved from the vertical 1,800 rpm axis (vertical line) to the propeller curve (solid parabolic line). When energy storage is added, then this path can be moved even further to the left, possibly all the way to the Maximum Continuous Rating (MCR) curve (piecewise solid linear line segment).

Optimum energy storage size can be determined according to load profile, the energy storage, and other available sources in the system. The selected energy storage is controlled to achieve targeted control objectives. The following sections gives examples of the optimum sizing and control of energy storage for ferries and drilling vessels.

10.6.2   Planning optimization of energy storage for ferries

The energy storage planning optimization designed for a ferry ship is demonstrated in Figure 10.33. The planning optimization consists of two optimization processes, which can be solved by two loops. The inner loop determines the optimal fuel saving operation for a single ferry trip. Based on the single-trip fuel saving information, the outer loop evaluates the total fuel saving and energy storage cost over its lifetime. In this planning optimization, the inputs are ferry operation parameters, including long term load profiles and equipment parameters. The constraints are

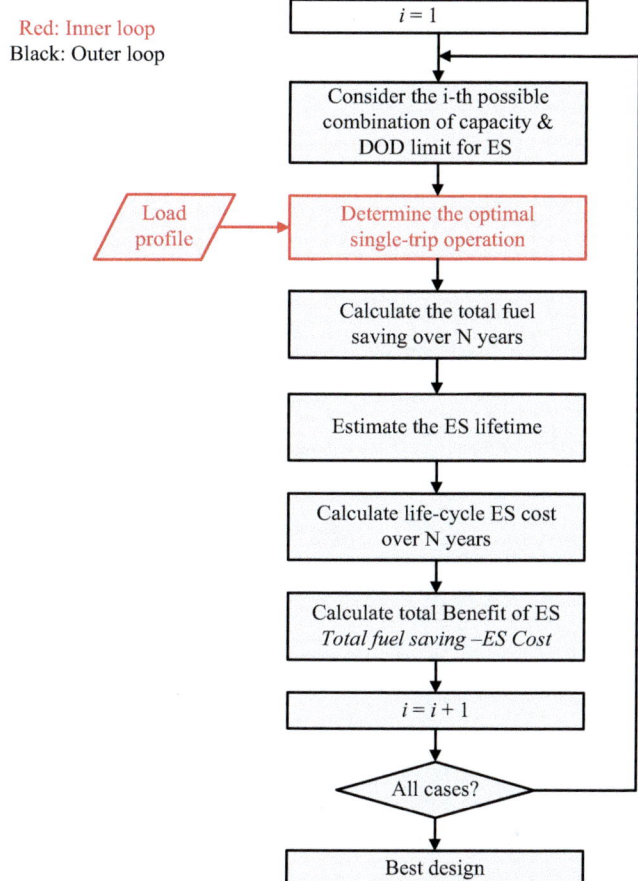

Figure 10.33   Flowchart of the overall energy storage planning optimization process for a ferry ship

operation limitations and the objective is to minimize the fuel cost. The outcome includes the long-term economic evaluation of single or multiple energy storage units. The final results determine the best energy storage technology and capacity design for the ferry ship.

For the single ferry trip optimization, assuming all diesel units are online, for a marine system with N_G diesel gensets and N_{ES} energy storage units, the single trip fuel cost optimization has $(N_G + N_{ES})$ N variables, where N is the number of samples in the long-term load profile. $P_i(n)$ and $P_{ESi}(n)$ are the output power of the i_{th} generator and the i_{th} energy storage unit at time step n. $E_i(0)$ is initial charge of the i_{th} energy storage unit at port. Considering the initial charges of the energy storage units to be additional variables, the decision vector is given by (10.19) and the optimization problem is to minimize the total fuel cost over the operation period as in (10.20):

$$
x = \begin{bmatrix} \overline{P_1} \\ \vdots \\ \overline{P_{N_G}} \\ \overline{P_{ES1}} \\ \vdots \\ \overline{P_{ESN_G}} \\ E_1(0) \\ \vdots \\ E_{N_G}(0) \end{bmatrix}, \quad \overline{P_i} = \begin{bmatrix} P_i(1) \\ \vdots \\ P_i(N) \end{bmatrix}, \quad \overline{P_{ESi}} = \begin{bmatrix} P_{ESi}(1) \\ \vdots \\ P_{ESi}(N) \end{bmatrix} \tag{10.19}
$$

$$
\underset{\substack{P_i(n) \\ P_{ESi}(n)}}{Min}\ f = \sum_{i=1}^{N_{ES}} Q \times (E_i(0) - E_i^{min}) + \sum_{n=1}^{N}\sum_{i=1}^{N_G} F_i(P_i(n), P_i(n-1)) \times \Delta t
$$

$$
F_i(P_i(n), P_i(n-1)) = (P_i(n) \times P_i^{rated}) \times \left[a_i(P_i(n))^2 + b_i(P_i(n)) + c_i \right]
$$

$$
\times \left[1 + K_i \left(\frac{P_i(n) - P_i(n-1)}{\Delta t} \right)^2 \right]
$$

$$
s.t. \sum_{i=1}^{N_G} P_i(n) \times P_i^{rated} + \sum_{i=1}^{N_{ES}} P_{ESi}(n) = P_L(n)
$$

$$
\left| \frac{P_i(n) - P_i(n-1)}{\Delta t} \right| \leq R_i, i = 1, 2, ..., N_G
$$

$$
P_i^{min} \leq P_i(n) \leq P_i^{max}, i = 1, 2, ..., N_G
$$

$$
P_{ESi}^{min} \leq P_{ESi}(n) \leq P_{ESi}^{max}, i = 1, 2, ..., N_{ES}
$$

$$
E_i^{min} \leq E_i(n) \leq E_i^{max}, E_i(n) = E_i(n-1) - P_{ESi}(n) \times \Delta t, i = 1, 2, ..., N_{ES}
$$

$$
\tag{10.20}
$$

where Q is the electricity price at the harbor, $F_i(P_i(n), P_i(n-1))$ is the generation cost of the i_{th} generator at time step n, a_i, b_i and c_i are cost coefficients for the i_{th} generator, P_i^{rated} is the rated power of the i_{th} generator, K_i is the transient penalty factor for the i_{th} generator, Δt is the time step, $P_L(n)$ is the forecasted load at time step n, R_i is the percentage ramping limit of the i_{th} generator, $E_i(n)$ is the stored energy of the i_{th} energy storage unit at time step n, P_i^{min} and P_i^{max} are the minimum and maximum output power of the i_{th} generator, P_{ESi}^{min} and P_{ESi}^{max} are the minimum and maximum output power of the i_{th} energy storage unit, E_i^{min} and E_i^{max} are the minimum and maximum energy of the i_{th} energy storage unit.

The first term in the objective function is related to the cost of the energy storage initial charge, which is priced at the utility electricity price at a port. The second term is the diesel fuel cost over the operation period. It uses a quadratic function to approximate the steady-state specific fuel cost ($/kW output) and a transient penalty factor K_i to approximate the additional fuel consumption due to the ramping of the diesel generator. The constraints include the power balance, the diesel generator ramping limits, the diesel generator output power limits, and the energy storage output power limits, and the energy storage stored energy limits, or SOC. Once a depth of discharge (DoD) limit is set from the output loop, the minimum and maximum energy limits can be set to constrain the DoD of an energy storage unit. This optimization problem is a linearly-constrained differentiable problem and thus can be solved using available optimization tools, such as the fmincon function in the Matlab Optimization Toolbox. More details on the problem formulation and optimization algorithms of the inner loop optimization can be found in [16].

The objective of the outer loop is to determine the optimal energy technology and capacity combination by maximizing the benefit of the energy storage system over the life cycle (e.g., an N year period). The life-cycle ES benefit is given by (10.21). The total life-cycle fuel saving in present value from the energy storage system can be calculated according to (10.22):

$$Benefit(N) = S_{TOT}(N) - C_{TOT}(N) \tag{10.21}$$

$$S_{TOT}(N) = \sum_{n=1}^{N} \frac{NTY \times SFT \times FP1 \times FCR^{n-1}}{(1 + IR)^n} \tag{10.22}$$

$C_{TOT}(N)$ is the total cost of the energy storage system over N years in present value, and $S_{TOT}(N)$ is the total fuel saving from the energy storage system over N years in present value. NTY is the number of trips per year. SFT is the saved fuel per trip. $FP1$ is the fuel price at year 1. FCR is the fuel price increase rate per year. IR is the yearly interest rate for determining the present value of expenses occurring in the future. The total life-cycle cost C_{TOT} of an energy storage system is a sum of the capital and operation cost of all equipment and devices, including the balance of plant.

A simplified power system model is considered for an electric ferry ship. As shown in Figure 10.34, all components are connected to a common DC bus. This system is composed of several diesel generators of various sizes, several energy storage units, and one lumped load which represents all system loads plus losses. The two diesel generators are 6 MVA each with Pmax = 1 pu, Pmin = 0.1 pu, R = 0.3 pu/min.

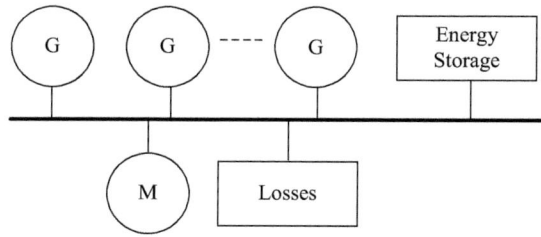

Figure 10.34   A Ferry ship model

With different load profiles assumed, optimization results of a single trip using a hybrid of energy storage solution, including one battery and one super capacitor, are shown in Figure 10.35. Based on the single-trip ferry optimization results, the total fuel saving and the hybrid energy storage system cost over a 10-year period are evaluated. Figure 10.36 illustrates the total fuel saving and the energy storage system cost over a 10-year period for different batteries and super capacitors combinations with 10% variations in the load profile. Assuming the fuel cost of

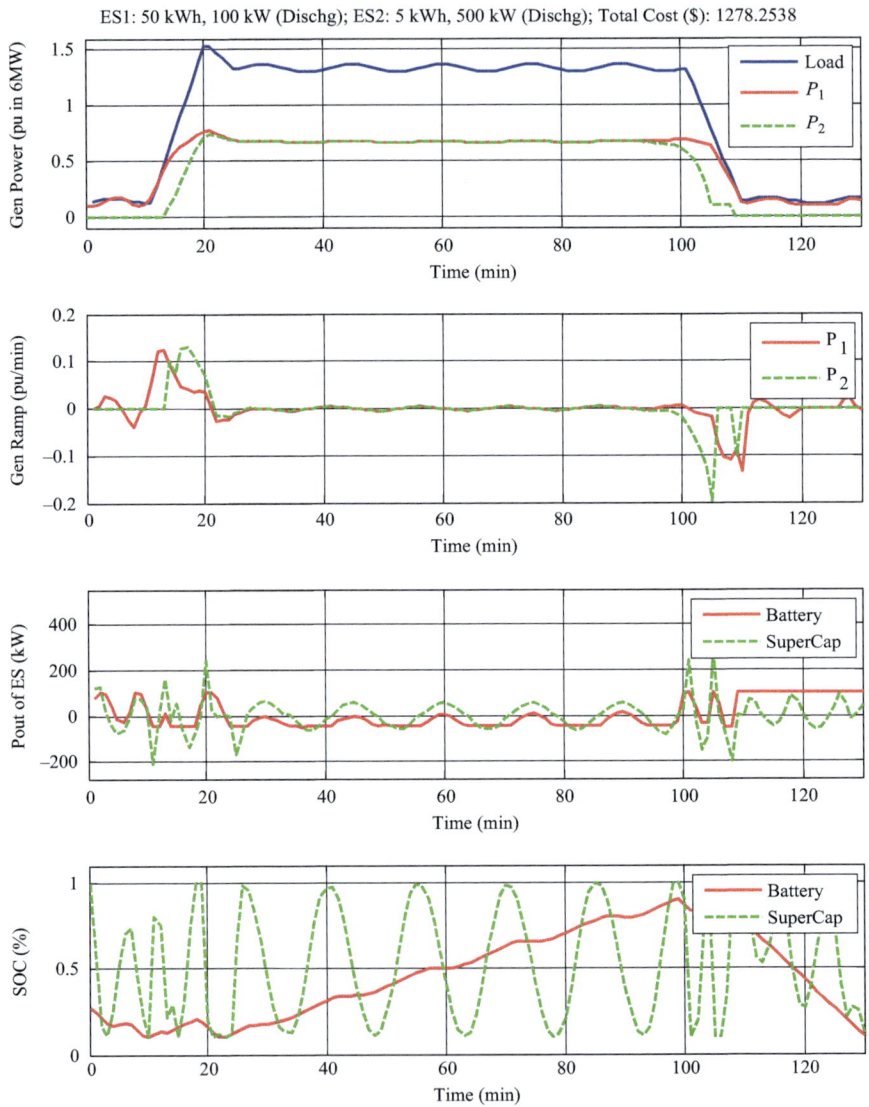

Figure 10.35   One single-trip optimization using a hybrid ES

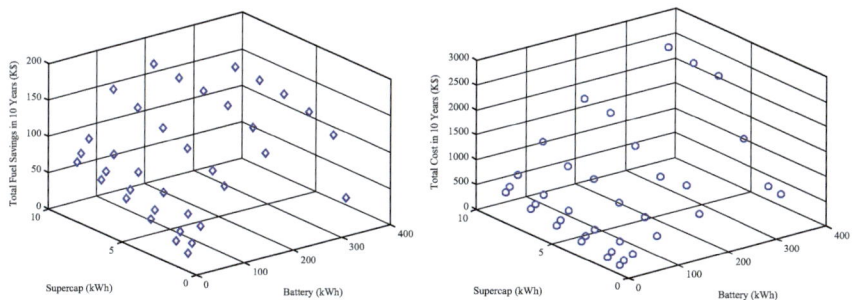

Figure 10.36   Total fuel saving and total cost of ES (excluding W&V cost) in 10 years – ferry case

$0.5/kg, it is found that the combinations of a super capacitor of 5–10 kWh and a battery of 150–200 kWh are better selections than other choices since they have higher fuel savings and lower costs. These combinations therefore provide better options for the investment payback.

10.6.3   Voltage control of energy storage for drilling vessels

The voltage control of a reduced DC shipboard power system on a drilling vessel is illustrated in Figure 10.37. Since the net load has both fast and slow variations, a hybrid energy storage system consisting of both short and long discharge energy storage is potentially the most economical solution for load smoothing to increase diesel fuel efficiency, reduce emission, and improve load factor. The hybrid energy storage system should be coordinately controlled to be fully utilized, but not violate, both power and energy limitations of individual energy storage units. One control method is to directly measure the load power, separate fast and slow variations by filters, and then use the correspondingly fast and slow energy storage units to compensate for the fast and slow load variations. However, communication and processing delays in measurement and power calculation causes the unsatisfactory performance of this power-measurement-based method in real systems. Alternatively, a frequency-dependent DC voltage droop control can be used to coordinate the power sharing between different power sources with different specific power and energy ratings.

From a control standpoint, the major distinguishing factor for different energy storage units and diesel generators is the discharge time or response speed. If a super capacitor has a short discharge time of tens of seconds, then a diesel genset has practically an "infinite" discharge time. The control coordination should fully exploit the capability of different sources while ensuring compliance with their power and energy limits. The conventional droop control is basically a proportional coordination strategy with constant droop gains for different sources for the entire frequency bandwidth. To utilize the droop concept for sources with different response speeds, a frequency-dependent response is needed for each type of source.

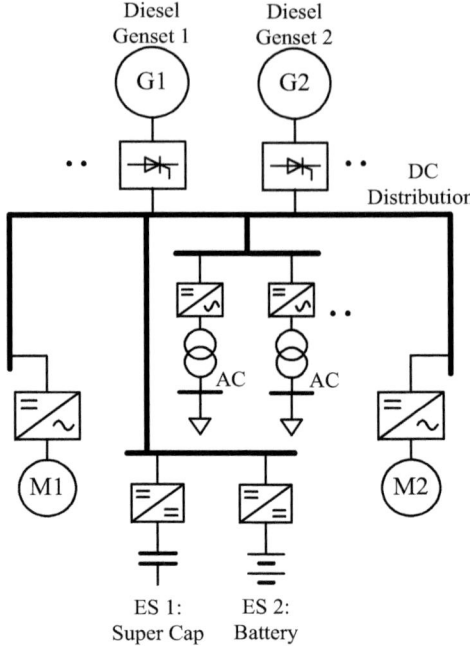

*Figure 10.37   A reduced DC shipboard power system with hybrid energy storage
[15]*

This would allow the coordination in both magnitude (power) and frequency
(discharge time).

Figure 10.38 illustrates the diagram of a frequency-dependent DC voltage
droop control for various types of sources with different response speeds. The DC
bus voltage variation signal passes through a set of filters that attenuate or amplify
this variation signal based on its frequency range. For example, for a very-low-
frequency load variation, diesel generators are the main responders; while, for a
high-frequency load variation, super capacitors can be designed to provide most
supply. The droop gains of different types of sources are designed separately for
their different frequency ranges. Since this frequency-dependent DC voltage droop
control only uses local voltage measurements, it has fast response and immune
from communication delay or failure.

For the frequency-dependent voltage droop control, the ratios between the
droop gains will determine the power sharing between different energy sources.
The droop gain ratios, instead of the absolute gains, determine the coordination
between sources. Simplified input filters for the frequency-dependent droop control
is shown in Figure 10.39, which consists of a first-order high pass filter and a first-
order low-pass filter. The low pass filter reduces the gains at high frequency and
contributes to closed-loop stability. This input filter design can be used in the droop
controller for both super capacitors and battery energy storage systems. The

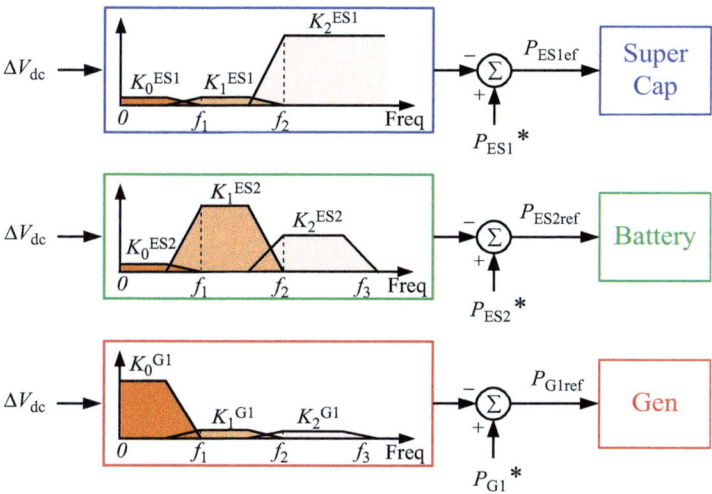

Figure 10.38   Multi-frequency DC voltage droop control ($\Delta Vdc = Vdc - Vdcref$) [15]

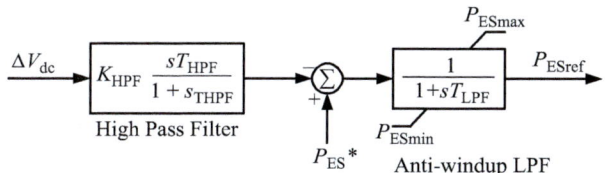

Figure 10.39   Implementation using first order filters [15]

linearized system model of the example system in Figure 10.37 with the frequency-dependent droop control is shown in Figure 10.40.

The generator response to DC bus variation is given by (10.23). N_G is the number of online generators, R_G is the generator droop gain, and T_G is the generator voltage control constant (0.01 s–0.1 s). The generator excitation control has slower dynamics compared to super capacitors or batteries, which naturally limits the generator response to low-frequency variations. The battery response to the DC bus variation is given by (10.24). K_{BA}, T_{HPFBA}, T_{LPF} are filter parameters to be designed, and T_{BA} is the battery unit time constant (~0.001 s). The super capacitor response to the DC bus variation is given by (10.25). K_{SC} and T_{HPFSC} are the filter parameters to be designed:

$$G_G(s) = \frac{P_G(s)}{\Delta V_{dc}(s)} = \frac{N_G R_G}{1 + s T_G} \qquad (10.23)$$

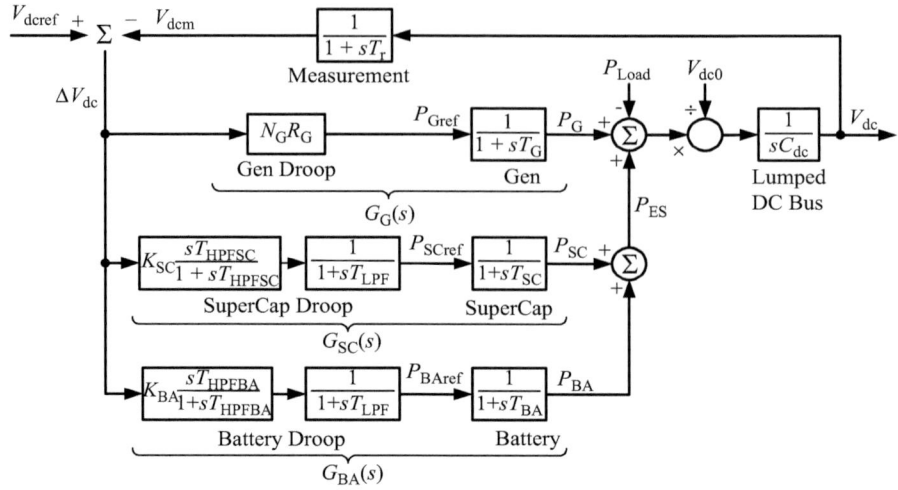

Figure 10.40   Linearized system model with frequency-dependent droop control for energy storage units [15]

$$G_{BA}(s) = \frac{P_{BA}(s)}{\Delta V_{dc}(s)} = K_{BA} \frac{sT_{HPFBA}}{1 + sT_{HPFBA}} \cdot \frac{1}{1 + sT_{LPF}} \cdot \frac{1}{1 + sT_{BA}} \qquad (10.24)$$

$$G_{SC}(s) = \frac{P_{SC}(s)}{\Delta V_{dc}(s)} = K_{SC} \frac{sT_{HPFSC}}{1 + sT_{HPFSC}} \cdot \frac{1}{1 + sT_{LPF}} \cdot \frac{1}{1 + sT_{SC}} \qquad (10.25)$$

The open and close loops of the DC voltage control are given by (10.26) and (10.27). T_r is the time constant for DC voltage measurement (~0.001 s). A stable system requires the open loop control function to have a phase margin greater than zero. All energy storage control parameters thus affect the overall system stability. However, the phase margin of the open loop control is difficult to find analytically. A parametric scanning is thus performed on all energy storage control parameters to identify the stability limits as Figure 10.41. From the figure, a few important observations are

- The phase margin primarily depends on the low pass filter (LPF) time constant, T_{LPF} and the sum of all energy storage control gains, $K = K_{BA} + K_{SC}$. The high pass filter (HPF) time constants have negligible effects on the phase margin of $G_0(s)$.
- A certain phase margin corresponds to a close-to-linear relationship between K and T_{LPF}, which is $K \leq A * T_{LPF} + B$.
- For a certain phase margin, the stability constraints A and B depend on the genset voltage droop, DC voltage measurement delay, and DC bus capacitance:

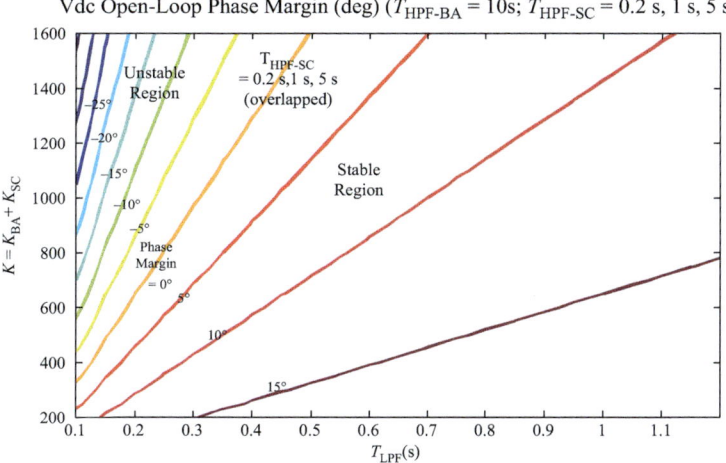

Figure 10.41 *Stability constraints on gains and time constants of the frequency-dependent droop [15]*

$$G_o(s) = \frac{1}{1 + sT_r} \frac{1}{sC_{dc}V_{dc0}} [G_G(s) + G_{SC}(s) + G_{BA}(s)] \qquad (10.26)$$

$$\frac{V_{dc}(s)}{V_{dcref}(s)} = \frac{\frac{1}{sC_{dc}V_{dc0}}[G_G(s) + G_{SC}(s) + G_{BA}(s)]}{1 + G_0(s)} \qquad (10.27)$$

The design variables to be optimized are the filter parameters in Figure 10.40 and are defined as x in (10.28). Given a load profile, the optimization objective is to minimize the load variations of the diesel generator as (10.29), where $P_G(n)$, $P_{BA}(n)$, and $P_{SC}(n)$ are obtained by analytically modeling and simulating the example system shown in Figure 10.37. Penalty is added when constraints are violated. Two optimization algorithms are used sequentially. First, a global search using Particle Swarm Optimization (population-based stochastic search) is performed. Based on the results, a local search using Nelder–Mead Simplex Method (heuristic search) is performed to find the optimal filter parameters. The constraints to the optimization include the parameter boundaries, the energy storage energy capacity limits, and the stability limit $K \leq A*T_{LPF} + B$ from parametric scanning:

$$x = \begin{bmatrix} K_{SC} & T_{HPFSC} & K_{BA} & T_{HPFBA} & T_{LPF} \end{bmatrix}' \qquad (10.28)$$

$$\min_x \sum_{n=1}^{N} \{[P_G(n) - P_G(n-1)]^2 + 0.001 \times [P_{BA}(n) - P_{BA}(n-1)]^2$$

$$+ 0.001 \times [P_{SC}(n) - P_{SC}(n-1)]^2\} + penalty \qquad (10.29)$$

In the example system as shown in Figure 10.37, the total generation capacity is 5 MW. The sizes of the two energy storage units are shown in Table 10.4. Two

Table 10.4   Sizes of energy storages [15]

	P (kW)	E (kWh)
Super capacitor	−788 to 788	4.75
Battery	−320 to 639	312

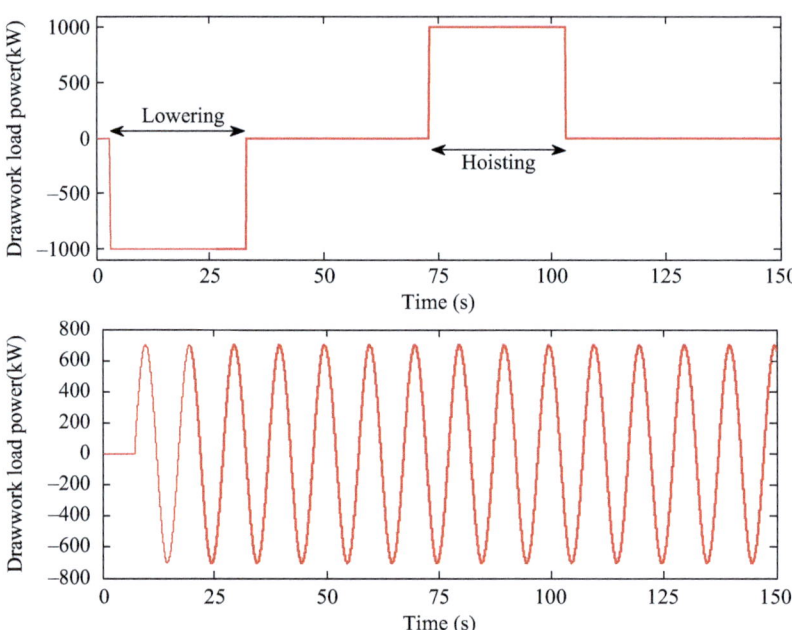

Figure 10.42   Two load patterns for optimizing ES droop control parameters [15]

different load patterns shown in Figure 10.42 are used to optimize the filter parameters in the droop control design. In both patterns, the load variations are predicted to be less than $+/-1$ MW, which means that the combined power from both
energy storage units is sufficient to compensate almost all load variations. The
optimized parameters are then used to control the super capacitor and battery
energy storage units in Figure 10.37. The actual load pattern for testing the control
performance is shown by the solid line in the second graph of Figure 10.43. It has
some randomness but is mainly composed by the two load patterns in Figure 10.42.
With the hybrid battery and super capacitor to compensate almost all load variations, the diesel generator output power is much smoother than the load variation
and the DC bus variation is within 3% of its nominal value. The high frequency
load variations are mainly picked up by the super capacitors. If the super capacitors
are close to its power limit, the battery picks up a part of the high frequency load
variations. The square-wave load variations at the beginning and end of the load
cycle are mainly compensated by the battery as expected since the super capacitor

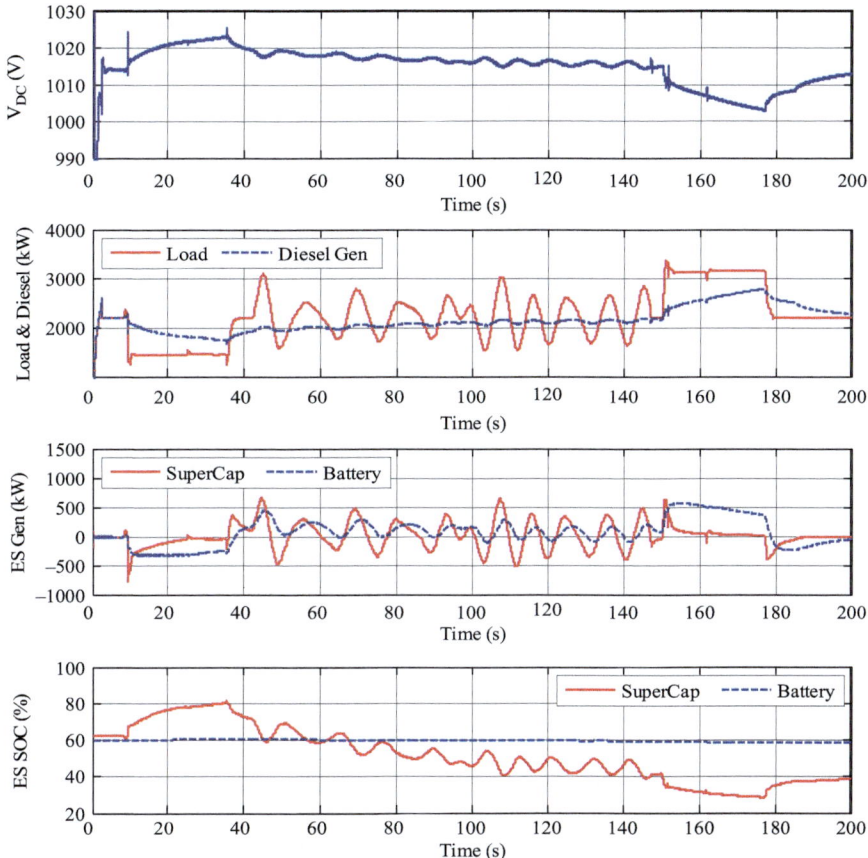

Figure 10.43   Actual load pattern, DC bus voltage, and the performance of the proposed frequency-dependent droop control [15]

does not have enough energy to supply or absorb the load variations in the long-term.

10.7   Vessel control system

Onboard DC Grid™ uses ABB's automations platform 800xA to implement system control functions, including power energy management system (PEMS) and vessel management system (VMS). The system integration and control are designed in such a way that it plays to the strengths of the various energy sources in the system and keeps tight control on consumers. ABB has adopted a new approach to power and energy management by using a PEMS. PEMS manages both the balance of the traditional power and the newly added stored energy in the shipboard power system. The latter becomes important when adding sources such as batteries and super capacitors with very finite amounts of energy available. The balance of power also takes on new dimensions in a DC Grid system when sources of variable speed generators, shaft generators and batteries operate in parallel.

To achieve the maximum benefits of a system, some of the PEMS functionality is implemented at lower levels, closer to the load and source converters – normally functions requiring fast response such as standard load sharing and overload protections. This can be accomplished autonomously by the different energy sources. Other functions have been implemented at a higher level such as the traditional power management system (PMS) domain – normally functions that require a high level of coordination between different sources. Optimal functionality and performance is achieved through tight horizontal integration between power sources and consumers, as well as tight vertical integration between fast embedded control of converters and generators and the system level application.

Onboard DC Grid^{TM} has a harmonized control and communication infrastructure that allows for a transparent and lightning-fast flow of information between system components. This ensures a holistic approach to the task of coaching the best performance, be it for safety or efficiency, out of a power system. The high level of integration also means that high quality information is available to an operator or remote support service engineer should he or she need it. Figure 10.44 shows an example ECR HMI for a three-split DC Grid system with a battery connected to the port-side section. The same section also has a generator rectifier that also serves as shore-connection rectifier.

For vessels with automatic charging from shore, the PEMS coordinates the process of connecting, charging and disconnecting from the charging station. The PEMS is structured so that each energy source forms an autonomous subsystem. This increases the fault tolerance of the system controls by reducing interdependence between energy sources. Sub-system functionality is realized as far as

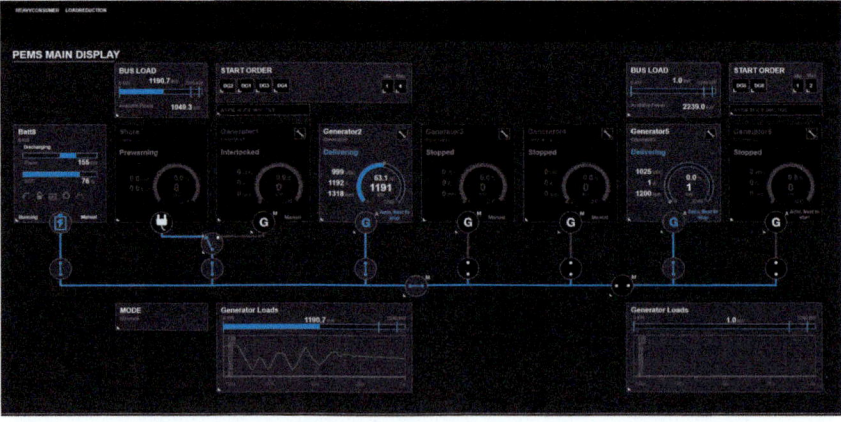

Figure 10.44   Illustration of a PEMS main display

practicable on a sub-system level, only involving the wider system when it becomes necessary. This also means that operation of the vessel remains intuitive and simple even when done so from local control since the majority of sub-system functionality remains intact.

10.8 Summary and lessons from field operation

The Onboard DC Grid^{TM} system is a new way of distributing energy for LV installations in a ship. Onboard DC Grid^{TM} combines the best of both AC and DC components/systems available, it is fully compliant with rules and regulations and provides advantages including low emissions and low fuel consumption. It can be used for any electrical ship application up to at least 20 MW and operates at a nominal voltage of 1,000 V DC. The power distribution can be arranged with all cabinets in a single line up (multidrive approach) or distributed throughout the vessel with cables or DC bus ducts.

The most immediate benefit of this change is reduced fuel consumption. The benefits of the variable speed operation on the engines are summarized below:

- Reduced Specific fuel consumption by up ~20% and ~40% for medium and high speed engines respectively for partial load operation.
- Cleaner combustion process with less build-up of soot when operating at partial loads.
- Reduced GHG emissions due to lower fuel consumption and reduced particle emissions due to cleaner combustion.
- Increased temperature of exhaust gases at lower loads means that SCRs can be fully operational at all load levels, reducing both NOx emissions and urea consumption.
- Potential reduction in audible noise level by more than 6dB.
- Reduced maintenance costs due to reduced wear and tear on the engine by up to 30%.

For the ship-owners, the following main benefits are expected:

- Up to 20% fuel saving if taking full advantage of all features including energy storage and variable speed engines.
- Reduced maintenance of engines by more efficient operation.
- Improved dynamic response by use of energy storage, which may give a better dynamic positioning performance with lower fuel consumption or more accurate positioning.
- Increased space for payload through lower footprint of electrical plant and more flexible placement of electrical components.
- More functional vessel layout through more flexible placement of electrical components.
- A system platform that affords simple "plug and play" retrofitting possibilities to adapt to future energy sources.

After one year's operation, in 2014, a third-party company Pon Power Scandinavia released its verification of the following benefits of the Onboard DC Grid^{TM} installed on "Dina Star":

- Identified reduction of specific fuel oil consumption of up to 27%.
- Tests showed 30% noise reduction, contributing to improved working conditions aboard the vessel.

References

[1]   ABB, "ABB Delivers First Onboard DC Grid System," 2013. [online] http://www.abb.com/cawp/seitp202/7199db5e8cd3924e85257b3b00491f07.aspx.

[2]   ABB, "Test Confirm up to 27% Fuel Savings on Ships from Onboard DC Grid," 2014. [online] http://www.abb.com/cawp/seitp202/6f0d5472c16d3fc4c1257cf9002661ed.aspx.

[3]   E. Haugan, H. Rygg, A. Skjellnes, and L. Barstad, "Discrimination in Offshore and Marine DC Distribution Systems", in *Proc. of 2016 IEEE 17th Workshop on Control and Modeling for Power Electronics*, June 2016, Norway, pp. 1–7.

[4]   IEEE, *IEEE Std 1709 Recommended Practice for 1 kV to 35 kV Medium-Voltage DC Power Systems on Ship*, Nov 2010.

[5]   US Navy, *Naval Power Systems Technology Development Roadmap*, 2013.

[6]   HH Ferries Launches Battery-powered Ferry Duo, [online] https://world-maritimenews.com/archives/264404/hh-ferries-launches-battery-powered-ferry-duo/

[7]   L. Qi, J. Liang, and J. Lindtjorn, "Configuration Generation and Analysis of AC and DC Conceptual Designs of Shipboard Power Systems", in *Proc. of 2019 IEEE ESTS*, Aug 2019, Virginia, USA, pp 1–6.

[8]   J. Hansen, J. Lindtjorn, and K. Vanska, "Onboard DC Grid for Enhanced DC Operation in Ships", in *Proc. of 2011 Dynamic Positioning Conference*, Oct 11–12, 2011, Houston, TX, USA, pp. 1–8.

[9]   J. F. Hansen, J. O. Lindtjorn, U. Odegaard, and T. A. Myklebust, "Increased operational performance of OSVs by Onboard DC Grid", in *Proc. of 2011 Offshore Support Vessel*, August 2011, Singapore, pp. 1–8.

[10] L. Qi, A. Antoniazzi, and L. Raciti, "DC Distribution Fault Analysis, Protection Solutions, and Example Implementations", *IEEE Transaction on Industrial Applications*, Vol. 54, Issue: 4, 2018, pp. 3179–3186.

[11] X. Feng, L. Qi, and J. Pan, "A Novel Fault Location Method and Algorithm for DC Distribution Protection", *IEEE Transaction on Industrial Applications*, Vol. 53, Issue: 3, 2017, pp. 1834–1840.

[12] L. Qi, A. Antoniazzi, L. Raciti, and D. Leoni, "Design of Solid State Circuit Breaker Based Protection for DC Shipboard Power Systems", *IEEE Journal of Emerging and Selected Topics in Power Electronics, Special Issue on Emerging Electric Ship MVDC Power Technology*, 2017, pp. 260–268.

[13] R. D. Middlebrook, "Input Filter Considerations in Design and Application of Switching Regulators," in *Proc. of 1976 IEEE Industry Applications Society Annual Meeting*, Aug 1976, pp. 366–382.

[14] H. Liu, H. Guo, J. Liang, and L. Qi, "Impedance-based Stability Analysis of MVDC Systems Using Generator-Thyristor Units and DTC Motor Drives", *IEEE Journal of Emerging and Selected Topics in Power Electronics, Special Issue on Emerging Electric Ship MVDC Power Technology*, 2017, pp. 5–13.

[15] J. Liang, L. Qi, J. Lindtjorn, and F. Wendt, "Frequency Dependent DC Voltage Droop Control for Hybrid Energy Storage in DC Microgrids", in *Proc. of 2015 IEEE PES General Meeting*, Denver, CO, July 2015, pp. 1–5.

[16] S. Mashayekh, Z. Wang, L. Qi, J. Lindtjorn, and T. Myklebust, "Optimum Sizing of Energy Storage for an Electric Ferry Ship", in *Proc. of 2012 IEEE PES General Meeting*, San Diego, CA, July 2012, pp. 1–8.

Chapter 11

Conclusions

The purpose of this small chapter is to provide some concluding remarks from each of the main authors on their respective chapter in the book. The chapter authors had two questions to address for the future audience. First, what is it that the chapter authors would want the audience of this book to take away from their respective chapter? Hence, the reader will find a small executive summary of each chapter with the key takeaways and highlights. The next question to answer is what is forthcoming in the chapter area? Or, where do the authors see innovation happening to occur in the context of medium voltage DC architectures in the next 5–10 years? This question is also answered with the executive summary of each major chapter in this book.

11.1 MVDC architectures

Chapter 2 is dedicated to the unique challenges, and opportunities associated with the design of MVDC-interfacing power electronic converters (PECs) and their influences on MVDC architectural implementations. The power conversion architectures in multiple promising MVDC applications are elaborated in this chapter, including large vessel shipboard electrification, more-electric and all-electric aircraft, utility-scale solar power generation, wind power generation, and battery energy storage applications. MVDC electrical distribution system architectures are informed by the topological implementations of MVAC to MVDC PECs and isolated MVDC to low voltage power electronic transformers and the approach to short circuit fault mitigation. MVDC protective systems divide up into PEC- and Breaker-based approaches. PEC-based protection is more readily implemented through coordination between PECs and no load isolating switches, but have a lower level of recoverability and survivability when compared to Breaker-based. On the other hand, viability of Breaker-based protection is limited by technical challenges and perceived risks of solid state circuit breakers (SSCBs). Moreover, an MVDC corridor between points of large or utility scale renewable generation enhances the transfer capacity and can be used to increase overall distribution network power quality at points of usage. The MVDC corridor is especially attractive to municipal grids and geographical areas separated by water, where there is need to transfer energy from bulk renewable energy installations to end users over long distances.

To facilitate the further development of MVDC systems in various emerging applications, the following innovations and challenges are to be addressed in the next 5–10 years:

- **Electromagnetic compatibility (EMC):** As the voltage level increases and more fast-switching semiconductor modules such as SiC MOSFETs utilized in the MVDC systems, both common-mode and differential-mode EMI will become more severe than ever before. Thus, to meet various EMC standards such as DO 160 and ensure a stable system operation, dedicated EMI filters that exhibit satisfactory noise attenuation with low power losses and compact footprint will be of paramount importance.
- **Reliability:** Reliability requires more attention for the further development of MVDC systems, especially in safety-critical applications such as electric airplanes and shipboards. Any single device failures such as semiconductor faults may cause cascaded electrical faults or even accidents and disasters. Particularly, MVDC systems may have a much lower impedance than MVAC counterparts, which leads to a higher amplitude of the short-circuit current in a short duration. Online fault prognosis and diagnosis algorithms for the most vulnerable components such as semiconductor switches and capacitors can be developed to predict and identify any aging faults. Meanwhile, fault-tolerant MVDC PEC circuit topologies can be developed to tolerate an electric fault as soon as it is detected.
- **Dependability and Resiliency:** As more experience is gained from the integrated power and energy delivery networks of shipboard and aircraft electrification, it is expected that terrestrial power distribution will evolve from point-to-point MVDC energy transfer to meshed networks. The coordination between PEC implementations and MVDC protective devices during fault events will play a significant role in the assurance of resilience and dependable energy delivery as the MVDC-based power distribution systems evolve and become more prevalent in electrical power grids.
- **Multi-domain systematic optimization:** To achieve high energy efficiency, high power density, high reliability, and low cost, multi-domain systematic optimization based on artificial intelligence algorithms can be carried out for multi-objective designs. For a general MVDC system, the comprehensive design optimization can include the objectives and constraints in electrical, mechanical, dielectric insulation, and thermal domains.

11.2   DC architecture utilization

In Chapter 3, we saw that grid transition must precede sustainable energy transition toward greater electrification. Under such a scenario, distribution network operators (DNOs) must explore solutions to restructure the existing grid with the goal of capacity reinforcement, improved controllability, and efficient power redirection. DC-based technologies are proposed to realize the grid transition from purely AC to hybrid AC–DC networks to address the anticipated challenges posed by energy transition. Refurbishing the existing AC links to operate under DC conditions is shown to enhance the power transfer capacity by approximately 50% within the studied constraints at higher energy efficiency. Reconfigurability between such parallel operating AC and DC links can further increase the achievable capacity gains during $(n-1)$ contingencies, which relates to the capacity maintained with a single component failure in the system. Further, DC interlinks are introduced to weakly mesh the radial AC distribution networks for efficiently redirecting the power to minimize local demand mismatches, prevent branch overloads and increase availability of the grid.

In the future, therefore, it is suggested that DC technologies will play an important role in restructuring the medium voltage AC distribution grids for achieving higher flexibility, controllability, and inter-connectivity with enhanced capacity and efficiency. More research on reliability of such systems is relevant from the point of view of grid connected power electronics, modularity, reconfigurability, and greater meshing. The proposed concepts of this chapter can be extended to integrate renewable energy resources directly to the embedded DC links, making the system multi-terminal, and thus transition towards a universal DC grid.

11.3 Dual active bridge designs

As the key enabling technology for MVDC networks, the bidirectional isolated DC-to-DC converter, referred to as a dual active bridge (DAB) converter, has been extensively described in Chapter 4. Special focus has been put on the three-phase DAB converter, which features smaller-sized passive components and is thus more suitable in high-power applications. As one of the most attractive features of this topology, soft-switching operation with different modulation schemes is first introduced. Through comprehensive analysis and performance characterization, it is shown that operating the DAB converter with asymmetrical duty cycles plus a phase shift can provide many benefits in light-load conditions, such as a wider soft-switching range and minimized conduction losses. This allows for flexible and efficient operation of the DAB converter over a wide operating range. Thereafter, this chapter further addresses a key issue in the DAB converter, which is the saturation of the high frequency transformer. As the DAB converter is widely considered as the power electronic building block (PEBB) in future MVDC applications, highly robust operation against the transformer saturation under various conditions is crucial. The transformer saturation can be induced by either the transient DC-bias due to an unregulated dynamic behavior or steady-state DC-bias due to non-ideal switching characteristics. Advanced control techniques are presented to tackle this issue from two perspectives separately. The instantaneous flux and current control, developed in the $\alpha\beta$ reference frame, provides an intuitive and elegant approach to set the trajectories of both the transformer winding current and magnetizing flux linkage in transient conditions. On the other hand, closed-loop anti-saturation control with DC component estimation technique is also developed to compensate the steady-state

DC-bias in the transformer. To further explore the advanced functionality of the DC-to-DC converter in the DC grid protection, this chapter also includes dedicated content of the DC fault-ride through operation. This technology would have a high-potential impact on more proactive and electronic protection of DC grids, which even includes the possibility of breakerless protection. With the foundation set based upon the two-level, three-phase DAB designs, a small treatment of the three-level, three-phase DAB has been provided to conclude this chapter. The three-level DAB inherits almost all the advantages and advanced control techniques of the standard two-level topology, and features even higher control flexibility and higher power density. Considering both two-level and three-level, three-phase converters are nowadays standard components in MV drive applications, PEBB based on these converters with high efficiency, high robustness, compact design and fault limiting functions will be widely available at affordable prices in the next 5 years. Built on PEBBs, intelligent solid-state substations can be easily developed to enable flexible and interconnected MVDC networks with distributed power generation.

11.4 Multiport designs

The blending of MVDC and MVAC in distribution systems is inevitable. Just like in AC transmission systems where there are AC substations with transformers for interconnecting different AC voltage levels, there will arise a need for DC sub-stations that provide an analogous function. Looking forward, multiport DC power converters are an appealing solution to control power flows between multiple different MVDC systems at different or similar voltage levels. They can also play a much larger role than simply voltage level shifters courtesy of semiconductor technology, with the potential to provide several additional features such as flexible control of individual port power transfers and DC fault blocking. Chapter 5 explores some different classes of multiport DC power converters and discusses their operating characteristics and features along with their inherent advantages and tradeoffs. One exemplar topology with a high degree of modularity and scalability is chosen to demonstrate applications of multiport converters in MVDC systems using several case study simulations.

The first point to point MVDC links are just emerging now. As always with new technologies, it takes time and buildup of confidence and experience before wider deployment is possible. Consider, e.g., the first HVDC grid that has just been built in China following years of research by industry and academia around the world. Looking forward, the next critical step will be the first global demonstrator project of a multiport DC converter that connects together multiple different MVDC distribution links, which could foreseeably take place within the next 10 years. This would likely happen first for MVDC level (distribution level) versus HVDC (transmission level) due to the inherently lower voltage and power levels and thus lower capital investments involved with the former. But before this can take place, more research and studies need to take place as multiport DC converters for power distribution systems is still a relatively new concept.

11.5 Control and mode visualization of bidirectional converters

Chapter 6 is focused on galvanically isolated bidirectional DC/DC converters. The chapter begins by analytically determining the impedance stability constraints of a DAB interfacing an electric machine that is modeled as a constant power load. With the intention of stabilizing the naturally unstable plant, the authors perform a stability analysis of the regulated architecture with the chosen controller being the proportional-derivative (PD) controller. Results are presented showing the limitations of PD control for the architecture. To overcome these common limitations, the authors' layout a tutorial for future readers with emphasis on model reference controller (MRC) designs – a special class of control techniques found in modern control textbooks but often overlooked or applied to electrical systems. In this tutorial, the MRC is developed with assumptions and plant limitations emphasized for the technique to be applicable. The development includes compensator design and generalized parameter selection approaches to ensure a desired dynamic response.

With a chapter focused on bidirectional power converter designs, Chapter 6 also presents an average power flow analysis of a three-port converter containing a multi-winding transformer and controlled through phase-shift control. It is important to understand the purpose and operation of the stacked submodules within each arm at the medium voltage ports. The procedure for deriving average power flow is necessary because traditional switched methods are not viable due to the zero-average current and voltage values of the converter. The resulting power output plots as a function of phase angle for both of the ports provides a useful tool for these converter designs and the number of operating modes present based upon the color coordination provided in the text. Inductance is added to the medium voltage ports for improved performance during system faults, and ultimately, the design considerations for sizing the mutual inductors are modeled, analyzed, and shown how the ideal waveforms can deviate because of these added inductors into the power converter.

For future consideration, galvanically isolated bidirectional DC/DC converters have become one the most common areas of study for MVDC level power conversion. In particular, the solid-state transformer, which utilizes the bidirectional converters, has been extensively modeled and full-scale prototypes have been demonstrated in academia globally. Faults on MVDC systems are a paramount concern, with the lack of a DC circuit breaker at high voltages preventing a commercially available method for interrupting fault currents. Although solutions are proposed as seen in forthcoming chapters of this text. Nevertheless, DC systems have much higher controllability than AC systems due to the capabilities of inverters and converters. Future DC systems are expected to be highly dynamic, especially with the incorporation of renewable energy resources. Bidirectional DC converters are expected to play a key role in future reconfigurable power systems.

11.6 Magnetic design

As seen in Chapter 7, magnetic component design requires a multi-disciplinary approach. The physics of magnetic and electric fields, material properties, mechanical construction, and power converter design and control are all major elements that converge to have a significant impact on the performance of the device. In this chapter, the authors provided an introduction to these major topics for engineers to begin the design process with emphasis on medium frequency (MF) transformers for MVDC applications.

Component design by the magnetic equivalent circuit method, that relies on an analog between traditional electric circuit analysis and magnetization physics, provides an analytical pathway for device development. This magnetic circuit simplifies the various pathways for magnetic flux into discrete circuit branches. Thus, refinement of the magnetic model begins by including more non-ideal effects, e.g., geometric and construction, nonlinear materials, and stray and leakage flux pathways. Then, the behavior of the magnetic equivalent circuit is translated into traditional electric circuit elements: inductors. Additional circuit elements, resistors and capacitors, capture the contribution to power losses and electric field coupling of design choices. All of these effects can play an important role in the performance and size of transformers for MVDC applications, with advanced magnetic core materials such as amorphous and metal amorphous nanocrystalline (MANC) alloys requiring particular attention to core and transformer construction.

Verifying the analytical model of a design or developing the model for an existing component requires specialized techniques. Small signal models can capture many of the low power and high frequency parasitic but the true large signal model requires application relevant characterization. The large signal characterization provides clear insight into how the component will behave during the power circuit operation. Importantly, this characterization helps engineers understand the nonlinear behavior of the magnetic device due to the level of the magnetic field, which also involves a detailed coupling with the underlying magnetization processes within a given magnetic core material. Clearly, the ability to measure and model the realistic properties of non-linear magnetic core materials is foundational to the successful design of MVDC transformers, and magnetic components in general.

Magnetic component design is at the precipice of rapid growth. The development and maturation of wide bandgap (WBG) semiconductors has shifted the design burden back to the passive elements of a converter. Optimization of magnetic components ultimately requires refinement and fine detail development of the presented physics. Current research is focused on two pathways for optimization, (1) development of new materials for the previously inaccessible, WBG enabled, design space; (2) co-design of WBG converters and magnetic components that optimize at the system level, particularly given the major and in many cases dominant size and weight associated with the magnetics. Each of these thrusts will provide more design tools for component optimization. These two thrusts further highlight necessary synergies between material science and electrical engineering needed for advancement. Future work will depend on close partnerships in interdisciplinary teams that leverage each other's expertise to optimize the multiple applicable physics required for magnetic component design.

11.7 Architecture stability

Chapter 8 presents basic knowledge for MVDC system stability studies, including small-signal stability and large-signal stability, from three aspects: system modeling, analytical tools, and enhancement approaches. The small-signal stability is the ability of a system to maintain steady-state operating points under small disturbances, which can be analyzed based on linearized state-space models or impedance models. More specifically, the state-space models are generally implemented with eigenvalue analysis which is preferred for the global stability study. The impedance models are more often used for local stability analysis with five types of stability criteria introduced here, i.e., the Middlebrook criterion, the forbidden-region-based criteria, passivity-based criterion, Nyquist stability criterion, and generalized Bode criterion. These impedance-based stability analysis approaches are also good for systems of which the characteristics can only be obtained through modeling or measurement at the terminals. Based on the small-signal stability analysis approaches, it can be found that the main reason for instabilities in MVDC systems is due to insufficient system damping, e.g., the negative incremental impedance of constant power loads (CPLs). To stabilize such issues, both passive and active approaches can be implemented to provide enough damping by reshaping system impedances. The large-signal stability is the ability of a system to return to an acceptable equilibrium range under large disturbances, which can be analyzed through time-domain simulations and mathematical tools. Time-domain simulations can provide an intuitive exhibition of system transient performance, but they cannot provide a closed-form solution to estimate system stable region. Mathematical tools, e.g., the Lyapunov direct method, Takagi–Sugeno fuzzy-model-based method, and mixed potential method, can complement well the simulation tools to estimate the region of attraction of the system. Therefore, to avoid large-signal instability issues, the allowable disturbance magnitude or component values can be estimated by mathematical tools and verified through simulations. According to the large-signal stability analysis, it can be found that CPLs and faults are the leading causes of system instabilities. The CPL-induced large-signal stability issues can be removed by both hardware-based and control-based methods. And the fault-induced large-signal stability issues require proper protection strategies, including fault detection and fault interrupting schemes.

Further advances are needed and expected for the design, control, and operation of stable future MVDC systems. For small-signal stability, impedance measurement units that are cost-effective and easy-to-use with a wide measurement range are needed for MVDC equipment and systems. They should become more available with increased MVDC applications and advancements of new measurement techniques, including signal disturbance injection techniques that will be facilitated by new power electronics technologies such as high-frequency SiC technology. For large-signal stability, development of cost-effective, efficient, and reliable protection devices and strategies for MVDC systems is key and still in the early stages with many challenging and competing performance requirements. The performance requirements include voltage capability and the capability to interrupt DC fault current with no zero-crossing, and high and fast-rising magnitude. Effective fault detection methods to identify the type and location of DC faults through measurement or estimation of transient currents and voltages are needed for complete selective protections. Protection devices with fast response time and adequate power ratings should become a trend for the limitation of fault currents and isolation of fault areas, e.g., solid-state DC circuit breakers. For both the small and large signal stability, efficient modeling and simulation tools suitable for large, complex MVDC and/or hybrid AC/DC systems, will be needed and should become available in the next several years.

11.8　Solid state breakers

Chapter 9 reviewed circuit breaker technologies for MVDC system protection, including mechanical circuit breakers, solid state circuit breakers and hybrid circuit breakers. The fundamentals, benefits, and design considerations of the three MVDC circuit breaker technologies are explained and discussed. A further classification of each type of circuit breaker is provided with the circuit topology and brief description of operating principles. Advantages and disadvantages of different topologies are also discussed and compared.

　　Mechanical MVDC circuit breakers have low conduction losses, but the long fault tripping time causes high fault current level and limits the maximum operational voltage level. The high arcing energy causes wear or damage to the circuit breaker contacts during each short circuit fault current interruption, reducing its lifetime and requiring frequent maintenance service.

　　Solid state MVDC circuit breakers demonstrate ultra-fast current interruption speed and have low fault-stress on application systems. The zero arc energy exposure during current interruption enables less fire hazard and arc flash mitigation for people and equipment. However, the high conduction losses propose great challenges in the cooling system design, especially when the series connection of multiple power semiconductor devices is needed to reach the targeted voltage rating. WBG power semiconductor devices (SiC or GaN power devices) show superior device performances like low conduction resistances and high blocking voltage capability, making the solid state circuit breaker technology more feasible and promising.

　　By integrating the mechanical circuit breaker and solid state circuit breaker, hybrid MVDC circuit breakers combine both advantages of low conduction losses and fast fault current interruption speed, showing significant potential for MVDC applications. The current interruption time is highly determined by the mechanical switch contact separation speed, so the design and optimization of a mechanical switch with fast contact opening speed is one important research direction. The current commutation time between the mechanical switch and solid state switch is another key factor on current interruption time, and different techniques to achieve fast current commutation are proposed and need further investigation.

11.9 DC-based ships

As explored in Chapter 10, one of the most promising applications for DC is within marine vessel electric systems, including both commercial and naval vessels. This chapter covers various aspects, including design, operation, protection, and stability control of low voltage and medium voltage DC marine vessel electric systems. DC systems are especially beneficial for reliable and efficient integration of energy storage with reduced power electronics devices. Sizing and control of battery and super capacitor energy storage for DC marine vessel electric systems are also presented in Chapter 10. ABB's Onboard DC Grid^{TM} system has been deployed and in operation since 2013. Lessons learned from several years of field operation provide convincing evidence of the benefits of applying DC to shipboard power distribution systems.

Business cases for implementing low voltage DC on commercial and naval vessels are clear and has been verified by the field operation experience. The applications of MVDC have been researched for many years. Currently, technologies, mainly DC converters and medium voltage DC protection, are ready for deployment in medium voltage applications. However, high costs associated with MVDC equipment and devices has been the major issue for impeding commercial shipboard applications. Medium voltage DC for naval shipboard systems is appealing for its high-power handling capability satisfying military needs. In the next 5–10 years, new technologies to decrease the cost of the medium voltage converters and protection while maintaining and increasing their power density are critical to deploy and implement MVDC for marine vessel electric distribution systems.

Index